# Neurocytochemical Methods

# NATO ASI Series

## Advanced Science Institutes Series

*A series presenting the results of activities sponsored by the NATO Science Committee, which aims at the dissemination of advanced scientific and technological knowledge, with a view to strengthening links between scientific communities.*

The Series is published by an international board of publishers in conjunction with the NATO Scientific Affairs Division

| | |
|---|---|
| A Life Sciences<br>B Physics | Plenum Publishing Corporation<br>London and New York |
| C Mathematical and Physical Sciences<br>D Behavioural and Social Sciences<br>E Applied Sciences | Kluwer Academic Publishers<br>Dordrecht, Boston and London |
| F Computer and Systems Sciences<br>G Ecological Sciences<br>H Cell Biology<br>I Global Environmental Change | Springer-Verlag<br>Berlin Heidelberg New York<br>London Paris Tokyo Hong Kong<br>Barcelona Budapest |

## NATO-PCO DATABASE

The electronic index to the NATO ASI Series provides full bibliographical references (with keywords and/or abstracts) to more than 30000 contributions from international scientists published in all sections of the NATO ASI Series. Access to the NATO-PCO DATABASE is possible in two ways:

– via online FILE 128 (NATO-PCO DATABASE) hosted by ESRIN, Via Galileo Galilei, I-00044 Frascati, Italy.

– via **CD-ROM** "NATO-PCO DATABASE" with user-friendly retrieval software in English, French and German (© WTV GmbH and DATAWARE Technologies Inc. 1989).

The CD-ROM can be ordered through any member of the Board of Publishers or through NATO-PCO, Overijse, Belgium.

Series H: Cell Biology Vol. 58

# Neurocytochemical Methods

Edited by

André Calas
Daniel Eugène

Département de Cytologie
Institut des Neurosciences, CNRS
Université Pierre & Marie Curie
7 Quai Saint-Bernard
75252 Paris Cedex 05
France

Springer-Verlag
Berlin Heidelberg New York London Paris Tokyo
Hong Kong Barcelona Budapest
Published in cooperation with NATO Scientific Affairs Division

Proceedings of the NATO Advanced Research Workshop on Recent Advances in Neurocytochemical Methods held in Seillac, France, September 26–28, 1990

Additional material to this bood can be downloaded from http://extra.springer.com.

ISBN-13:978-3-642-84300-6 e-ISBN-13:978-3-642-84298-6
DOI: 10.1007/978-3-642-84298-6


Softcover reprint of the hardcover 1st edition 1991

Typesetting: Camera ready by authors

31/3140-543210 – Printed on acid-free-paper

**Seillac ARW on Recent Advances in Neurocytochemical Methods**

Pr J. Taxi (right) dedicatee of the ARW together with Pr R. Couteaux (left)

# PREFACE

The great strides made in the field of morphological methods during the past decades have perhaps found their most spectacular expression in the functional exploration of the nervous system. In comparison with other tissues, nerve tissue displays three kinds of specificity : structural, because of the unique organization of the neuronal networks ; chemical as shown, for example, by the informative molecules exchanged between the nerve cells, and of course functional, thanks to the particular metabolic and electrophysiological characteristics of the neurons. Although for a long time the structural properties of the nervous system were generally considered to constitute the only field to which morphological techniques could be applied, we are to-day justified in believing that they can also explore the nerve tissue through its specific chemical and functional aspects, thanks to the development of immunocytochemistry and *in situ* hybridization, to the elaboration of the deoxyglucose method, to the use of voltage or ion sensitive dyes, and to the progress made in the application of *in vivo* techniques like PET. These methods have evolved so fast, the technical and fundamental problems they raise are so numerous and stimulating, and the importance of the complementary data they provide is so obvious, that we thought it was a good time to organize a new meeting between distinguished specialists in the neurocytochemical field. In doing so, our aim was not only to define the state-of-the-art as regards the morphofunctional techniques applied to the nervous system, but also to promote useful discussions and try to formulate proposals for future research. As all these goals correspond to those of the NATO Advanced Research Workshops, NATO accepted our application to organize such a meeting, which was held in Seillac from September 26th to 28th 1990.

To achieve our goals, the meeting was planned to include long periods of general discussions, whose interest justifies their summary in this book by the moderator of the session, who, as in the late NRP Bulletin, gives his personal account of the topics discussed. Another original aspect of the meeting was the evening poster session, which allowed the explanation and debate of additional and/or recent observations of particular importance.

Here again, we thought it of great interest, in the Proceedings of the meeting, not only to summarize this evening session but to give some of the younger participants an opportunity of presenting their data in the form of a short paper.

The last original aspect was the title of the Workshop. If we extend the concept of cytochemistry to the identification of all cellular and tissular constituents from the anatomical to the molecular level, and to the *in situ* study of their properties, of course Neurocytochemistry naturally seemed to us the term that defined the field dealt with at the meeting most appropriately.

One of the concluding remarks made at the gathering concerned the need for cooperation between those using the morphological approaches to the study of the nervous system, and those using other approaches such as neurochemistry, neuropharmacology or molecular genetics, since all their respective scales of specificity, sensitivity and resolution are now closely linked. Thanks to these links, there is reason to hope that the combined use of these approaches will reveal as yet unsuspected facets of the organization and function of the brain. In that case, the problem will increasingly be not so much to provide answers, but to formulate the right physiological questions.

It is in this spirit that throughout a scientific career devoted to the autonomic nervous system, our teacher and predecessor, Professor J. Taxi, has always refused to separate structure and function, and it is to him that most rightfully, the present NATO Workshop was dedicated.

A. CALAS D. EUGENE

**ORGANIZING COMMITTEE**


CALAS, A. — Département de Cytologie, Institut des Neurosciences, CNRS, Université Pierre & Marie Curie, 7 Quai Saint-Bernard, F-75252 Paris Cedex 05, France.

DE MEY, J. — EMBL, Department of Cell Biology, Postfach 10 2209, D-6900 Heidelberg, Germany.

DROZ, B. — Institut d'Histologie et d'Embryologie, Rue du Bugnon 9, CH-1005 Lausanne, Switzerland.

STORM-MATHISEN, J. — Anatomical Institute, University of Oslo, Karl Johans Gate 47, N-0162 Oslo 1, Norway.


**MAIN CONTRIBUTORS**


BARON, J.C. — INSERM U. 320 and CYCERON, B.P. 5027, F-14021 Caen Cedex, France.

BOSLER, O. — Laboratoire de Neurobiologie, CNRS, B.P. 71, F-13402 Marseille Cedex 9, France.

COURAUD, J.Y. — Section de Pharmacologie et d'Immunologie, DRIPP, CEN/Saclay, F-91191 Gif-sur-Yvette Cedex, France.

COUTEAUX, R. — Département de Cytologie, Institut des Neurosciences, CNRS, Université Pierre & Marie Curie, 7 Quai Saint-Bernard, F- 75252 Paris Cedex 05, France.

FONNUM, F. — Norwegian Defence Research Establishment, Division for Environmental Toxicology, P.O. Box 25, N-2007 Kjeller, Norway.

MATTHEWS, M.R. — Department of Human Anatomy, University of Oxford, South Parks Road, GB-Oxford, OX1 3QX, England.

Mc MAHAN, U.J. — Department of Neurobiology, Stanford University School of Medicine, Stanford, California, 94305, U.S.A.

MIRSKY, R. — Department of Anatomy and Developmental Biology, University College London, Gower Street, GB-London WC1E 6BT, England.

| | |
|---|---|
| MOREL, N. | Département de Neurochimie,<br>Laboratoire de Neurobiologie Cellulaire et Moléculaire, CNRS,<br>F-91198 Gif-sur-Yvette Cedex, France. |
| MULLER, W. | Roche Institute of Molecular Biology,<br>Roche Research Center, 340 Kingsland Street,<br>Nutley, New Jersey 07110, USA. |
| OTTERSEN, O.P. | Department of Anatomy, Institute of Basic Medical Sciences,<br>University of Oslo, P.O. Box 1105 Blindern,<br>N-0317 Oslo 3, Norway. |
| SEQUIER, J.M. | Merck Sharp & Dohme Research Laboratories,<br>Neuroscience Research Centre,Terlings Park, Eastwick Road,<br>GB-Harlow, CM20 2QR, England. |
| SOMOGYI, P. | Medical Research Council,<br>Anatomical Neuropharmacology Unit,<br>University of Oxford, South Parks Road.,<br>GB-Oxford OX1 3QT, England. |
| STEINBUSCH, H.W.M. | Department of Pharmacology,<br>Faculty of Medicine, Vrije Universiteit,<br>NL-Amsterdam, The Netherlands. |
| STREIT, P. | Brain Research Institute,<br>University of Zürich, August-Forel-Strasse 1,<br>CH-8029 Zürich, Switzerland. |
| TAXI, J. | Département de Cytologie, Institut des Neurosciences,<br>CNRS, Université Pierre & Marie Curie, 7 Quai Saint-Bernard,<br>F-75252 Paris Cedex 05, France. |
| UHL, G.R. | Laboratory of Molecular Neurobiology , ARC/NIDA,<br>and Departments of Neurology and Neuroscience,<br>Johns Hopkins School of Medicine, P.O. Box 5180,<br>Baltimore, Maryland 21224, USA. |
| VANDESANDE, F. | Department of Biology,<br>Catholic University of Louvain, Naamsestraat 59,<br>B-3000 Louvain, Belgium. |
| WALLEN, P. | Nobel Institute for Neurophysiology,<br>Karolinska Institutet, P.O. Box 60400,<br>S-104 01 Stockholm, Sweden. |

| | |
|---|---|
| WEISS, D.G. | Institut für Zoologie,<br>Technische Universität München, Lichtenbergstr. 4,<br>D-8046 Garching, Germany. |
| ZOLI, M. | Institute of Human Physiology,<br>University of Modena, Via Campi 287<br>I-41100 Modena, Italy. |

## CONTENTS

## SHORT PAPERS

# DEVELOPMENT and APPLICATION of ANTIBODIES to PRIMARY (DA, L-DOPA) and SECONDARY (cGMP) MESSENGERS: A technical report

Harry W.M. STEINBUSCH, S. Peter Van VLIET, John G.J.M. BOL and Jan De VENTE

Dept. Pharmacology, Faculty of Medicine, Free University,
1081 BT Amsterdam, The Netherlands

In the last decade antibodies have been raised against a large number of amine- and amino-acid neurotransmitters and candidates. These antibodies have been used in tracing neuronal connections in the central nervous system and locating the cell bodies and varicose fibers synthesizing these molecules. There are a number of specific difficulties inherent to the use of antibodies against these small, water soluble, molecules. To obtain these antibodies, the small haptens have to be coupled to a larger protein molecule to acquire an immunogenic conjugate. Two fixatives, formaldehyde and glutaraldehyde, have been used in most cases, and highly specific antibodies have been raised against conjugates prepared with both fixatives. In visualizing a molecule cross-linked to tissue constituents the fixative applied is probably the most important factor in combination with the antibody used. Ideally, the procedure for cross-linking small molecules to tissue constituents should be strictly identical with the procedure to prepare the immunogen against which the antibody is raised. The fixative might alter the structure of the hapten and the antibody against this molecule might not recognize the free, unmodified hapten in solution. Furthermore, in the cases of formaldehyde and glutaraldehyde, the fixatives are part of the antigenic determinant, together with an unknown part of the protein molecule to which the hapten is coupled. One more problem is the lack of knowledge about the amount of hapten actually retained by the tissue after fixation. The specificity of an antibody raised against an antigen is the most important factor in determining the usefulness in all its applications. A commonly applied testing criterion is the sensitivity and specificity of a particular antibody in a radioimmunoassay. However, in immunocytochemistry the situation is more complex because one has to deal with an antigen that might be modified in a decisive way by some chemical agent (the fixative) in a surrounding which is not defined (the cellular matrix). Thus specificity testing by radioimmunoassay is of no value for immunocytochemistry, even disregarding the question of concentrations normally used in a radioimmunoassay compared to local tissue concentrations (Schipper and Tilders, 1983; Steinbusch et al, 1986 ; Berkenbosch et al, 1986; Steinbusch and Tilders, 1987). There are a few routinely performed control experiments for immunocytochemistry. First, there is the use of a control serum, i.e. as non-immune or a pre-immune serum which should not give the desired staining reaction. Second, an absorption or immunoinhibition test, both liquid

NATO ASI Series, Vol. H 58
Neurocytochemical Methods
Edited by A. Calas and D. Eugène

and/or solid phase, may be performed. Third, the specificity of the antiserum should be established by using a model system mimicking the procedures and circumstances for tissue processing as closely as possible. One other test which one might employ, although this is not always possible, is the application of a particular antiserum in a biological model system. This has been done in many instances, e.g. for biogenic amines and cGMP.

Specificity of the obtained antibody is often determined at concentrations which are not realistic in comparison to tissue concentrations. Recently, Schipper and Tilders (1983) have developed a model system for establishing the relationship between immunostaining intensity and antigen concentrations. The models comprise of a gelatin matrix to which various concentrations of water soluble substances (monoamines or second messengers) can be added. Fixation, embedding and staining conditions of these models are equivalent to those used for biological preparations. Immunostaining in sections of these models is distributed consistent throughout the gelatin matrix, which facilitates quantitative estimations of the staining intensity. A linear relationship can be found between the concentration of the monoamines or cyclic nucleotides added to the gels and the immunostaining intensity after incubation with the antisera (Schipper and Tilders 1983, De Vente et al. 1987). Such linear relationship between concentration and immunostaining intensity can also be found using neuropeptides as antigens (Berkenbosch et al. 1986). Thus, within certain concentration limits, immunofluorescence intensity can be used for obtaining quantitative information on local concentrations of antigens. Therefore the model system used to study antibody specificity should mimic the situation of the tissue fixation. In this report we want to present data concerning the production and application of antibodies against primary and secondary messengers. As primary messengers are noted all neurotransmitters involved in synaptic or non-synaptic regulating systems. Secondary messengers as cyclic GMP or cyclic AMP are regarded synaptic modulating systems, which are linked to processes involved after a cyclase stimulation. The major monoaminergic systems are the dopaminergic, noradrenergic and the serotonergic system. Antibodies against monoamines have been raised either against glutaraldehyde or formaldehyde conjugates. However, the recent observations of purely aromatic DOPA decarboxylase immunoreactive cells have resulted in some interest in the possibility of a new aminergic system which uses L-DOPA as its major component.

L-Dopa is the precursor for the synthesis of dopamine, noradrenaline and adrenaline, all primary messengers. Accepting the presence of strictly aromatic aminoacid decarboxylase immunoreactive($_i$) cells it might be possible that there are purely L-DOPA$_i$ cells present within the CNS. To elucidate this matter an antibody was raised in rabbits against a 3,4-dihydroxyphenylalanine (L-DOPA)-glutardialdehyde-thyroglobulin complex. The specificity of this antibody was tested on nitrocellulose strips. In addition we compared dopamine antibodies raised against formaldehyde and glutaraldehyde conjugated DA. The efficiency of various fixatives to couple dopamine and the specificity of the antibodies obtained was studied in two model systems. Both types of antibodies were applied to brain tissue sections and the results were compared with those obtained in the model systems.

Monoamines and neuropeptides can be considered as primary messengers, whereas cyclic nucleotides and phosphoinositoles are examples of second messengers. In the CNS many of the receptors acted upon by neurotransmitters are localized on neurons and mediate a variety of processes in neuronal communications. These receptors are linked to guanylate- or adenylate cyclase systems, which by themselves will increase/decrease levels of resp. cGMP or cAMP. The prime interest of our immunohistochemical localization of cGMP is not only their localization, but to provide evidence for the neuronal activity at the postsynaptic side. In other words to proceed one step further from static neuroimmunocytochemistry to a more dynamic approach. To answer questions, like the effects of different transmitters on their postsysnaptic receptors or in particular the post-synaptic effect of co-localized neurons, antibodies to second messengers could fulfill an important role in this matter. Biochemical determination of cGMP revealed its presence in the mammalian brain. The exact localization together with the differences between cellular concentrations in various cell types and the subcellular distribution of cGMP is, with a few exceptions, not known. We studied the localization of cGMP in the CNS using immunocytochemistry. Using brain tissue slices incubated in vitro prior to fixation with formaldehyde, we manipulated cGMP levels and were able to visualize cGMP in specific cellular structures. Although the physiological function of cGMP in visual transduction has been unequivocally demonstrated (Fesenko et al. 1985), and the role of CGMP in relaxation of smooth muscle is presently rapidly clarified (Ignarro 1989; Matsunaga and Furchgott 1989; Murad 1986; Palmer et al. 1988), the role of cGMP in the central nervous system remains elusive. Difficulties in assigning a role for cGMP in the CNS partly relate to the physically and biochemically different forms of guanylate cyclase (GNC) which are present in the CNS (Waldman and Murad, 1987). Thus a detailed knowledge about the localization of the soluble and particulate GNC seems a prerequisite to evaluate biochemical and/or physiological studies on the role of cGMP. The particulate GNC can be activated by atrial natriuretic factor (ANF) (Hamet et al, 1984), whereas the soluble GNC is activated by sodium nitroprusside (SNP) (Waldman and Murad, 1987). In recent years we exploited an immunocytochemical procedure to visualize formaldehyde-fixed cGMP in a variety of tissues, e.g. the brain, the aortic smooth muscle, the kidney and the retina of the rat (Berkelmans et al. 1989; De Vente et al. 1987, 1988, 1989a,c; Steinbusch and De Vente, 1987; Steinbusch et al., 1988). Thus we found ANF-responding, cGMP-producing cells in e.g. the hippocampus, which were almost exclusively astrocytes. At basal levels a weak cGMP staining was associated with blood vessels and cell bodies in the pyramidal layer of the hippocampus. In the presence of isobutyl-methylxanthine (IBMX) cGMP immunostaining increased in blood vessels and cell bodies, whereas an occasional nerve fiber was observed. Sodium nitroprusside alone did not appreciable increase cGMP, but in combination with IBMX or theophylline an intense staining of the blood vessel walls and in addition to relative few astrocytes a dense neuronal network throughout the hippocampus was observed. In the developing cerebellum we found cGMP-immunoreactivity in astrocytes and Bergman fibers after stimulation with NMDA, and ANF. Sodium nitroprusside also stimulated cGMP content in Bergman fibers and astrocytes, but in addition we found cGMP in mossy fibers. Using quantitative immunofluorescence of cGMP as a post- or presynaptic marker we have the possibility to measure effects of monoamines or neuropeptides in individual neurons. In conclusion, it is worthwhile to mention that immunocytochemistry of second messengers can provide information about the dynamic activity of a cell, whereas antibodies against primary messengers can be used as a tool to study neuronal connectivity.

## METHODOLOGICAL CONSIDERATIONS

It is our general strategy that if we want to develop a new antibody to study first the effects of a fixation procedure on the retention of a compound under investigation. After finding the optimal fixation procedure, which means the fixative giving the highest retention, a conjugate is made following the identical procedure. It is always our aim that the production of a conjugate and the tissue fixation should be almost identical. Therefore we will follow in this subchapter this procedure divided between the primary and the secondary messengers.

**Fixation efficiency of amines** Gelatin models were prepared as a solution of 5 % gelatin, 0.1 M phosphate buffer (pH 8.0), 1 M sucrose and 0.3 % $Na_2S_2O_5$ in distilled water, containing the amines in the indicated concentrations with the incorporation of a tracer amount of the $^3$H-labeled amine at each concentration. The gels were frozen using $CO_2$ gas and entirely cut up into 20 μm sections by use of a cryostat. The sections were collected in a test tube and subjected to a washing and incubation procedure similar to the immunocytochemical procedure. After a final wash the sections were dissolved in 1 ml tissue solubilizer and the radioactivity was determined. For each fixed gel an unfixed control containing an identical concentration of the amines was treated in the same way. Remaining radioactivity was expressed as percentage of the original applied activity. Nitrocellulose models were prepared containing the amines in the indicated concentrations with the incorporation of a tracer amount of the $^3$H-labeled amine at each concentration. The strips were subjected to a washing and incubations procedure similar to the immunocytochemical procedure. After a final wash the spots were cut out and radioactivity was determined. For each concentration an unfixed control was included. Remaining radioactivity was expressed as percentage of the original applied activity.

**Preparation of dopamine conjugates** Formaldehyde-conjugation. The reaction mixture consisted of the following: 2 ml 25 mg/ml bovine thyroglobulin, 1 ml 100 mM DA-HCl, 6 ml 0.1 M acetate (pH 4.0) and 1 ml 40% freshly depolymerized paraformaldehyde (PFA). The reaction mixture turned yellow upon addition of the formaldehyde. The reaction was terminated by repeated ultrafiltration. The conjugate was freeze-dried and stored at -20°C. Spectrophotometrically a molar ratio was found of 40 molecules of DA on one thyroglobulin molecule. Glutaraldehyde-conjugation. The preparation of the DA-glutaraldehyde (GA) protein conjugate was based upon the method described by Geffard et al (1984) and Mons and Geffard (1987) with slight modifications in the concentrations of the reactants. The reaction mixture consisted of the following: 6 ml 5 mg/ml DA-HCL (containing a trace amount of $^3$H-DA) in distilled water containing 3 mg/ml sodium metabisulfite ($Na_2S_2O_5$), 3 ml 30 mg/ml bovine thyroglobulin in 3 M acetate (pH 8.0), and the reaction was started by the addition of 3 ml 1.5 % glutaraldehyde. The reaction was stopped after 4 min by the addition of 3 ml 30 mM $NaBH_4$. The pale greenish-yellow conjugate was dialyzed against frequent changes of distilled water at 4°C in the dark. Thereafter the conjugate was freeze-dried and stored at -20°C. For the conjugate a molar ratio was found of 80 - 100 DA molecules on one thyroglobulin molecule.

**Immunization procedure** Polyclonal antibodies were raised in adult male New Zealand rabbits (TNO, Rijswijk, The Netherlands) and in young adult goats. Rabbits were immunized with 1 mg conjugate suspended in 0.5 ml saline and 0.5 ml complete Freunds adjuvant. Before immunization preimmune serum of each rabbit was collected and stored. The conjugate suspensions were injected intradermally at 8 sites at the back of the animals. Booster injections of 1 mg conjugate suspended in 0.5 ml saline and 0.5 ml incomplete Freund adjuvant were given intramuscularly at three weeks intervals. Ten days after the booster injections blood samples were taken and the serum tested. The blood was left at room temperature without heparin or citrate for 30 min and left at 4$^0$C over the night. Hereafter the blood was centrifugated for 10 min at 2000 rpm. The serum was decantated, 3 mg GA-treated thyroglobulin per ml was added, incubated for 30 min at 37$^0$C and left over the night at 4$^0$C. The next morning the serum was centrifugated for 10 min at 3000 rpm and the antiserum obtained was collected and frozen at -20$^0$C. Goats were immunized with 7 mg DA-glutaraldehyde-thyroglobulin conjugate suspended in 1 ml saline and 1 ml complete Freunds adjuvant. The protocol was further the same as described above for the rabbits. Ten days after the 5th booster injection the rabbits were bled.

**1. Serumspecificity** of antisera was established using immunoinhibition test. Antisera were diluted in TBS containing 0.5% Triton X-100 (TBST) + 1% $Na_2S_2O_5$ (pH 7.4). Subsequently antisera were incubated with serial dilutions of 1mg/ml, 0.1mg/ml and 0.01mg/ml conjugated L-DOPA-GA-thyroglobulin, DA-PFA-thyroglobulin or DA-GA-thyroglobulin for 1 hr at 37°C, and hereafter over the night at 4°C. Precipitates were removed by centrifuging for 10 min at 3000 rpm at 4°C.

**2. Gelatin model system incubation for primary messengers** This procedure has been described before and was slightly modified from Schipper and Tilders (1983). In brief, a solution of 10% (wt./vol.) gelatin in distilled water was prepared. Solutions of DA, noradrenalin, serotonin, histamine and adrenalin were made up (0.2 μM - 10 mM) in 0.2 M phosphate buffer, pH 8.0, containing 0.6% sodium metabisulfite and 2 M sucrose and mixed with equal volumes of the gelatin solution. Aliquots of these solution were pipetted into Tygon tubing. After gelation, the gels were fixed for 16 h at room temperature in a solution of 4% (wt./vol.) freshly depolymerized PFA and 0.5% GA in 0.1 M phosphate buffer, pH 8.0, containing 1 M sucrose. This system cannot be used in combination with a 5% glutaraldehyde fixation, since the glutaraldehyde will fix the gelatin so strongly that the gels cannot be removed from the tube. The combined PFA-GA fixed gels were removed from the tubing, embedded in OCT compound and frozen in isopentane prechilled with liquid $N_2$. 10 μm thick cryostat sections were cut and mounted on slides coated with chromalum gelatin. Slides were stored at -20°C. Before application of the primary antisera, the sections were washed with 1% $NaBH_4$ and 1% sodium metabisulfite in Tris-Buffered-Saline (TBS) for 30 min.

**3. Spot-blot model system for primary messengers** To test the specificity of the newly raised antibodies to the primary messengers a model was used applying nitrocellulose sheets which analogues the model described by Larsson (1982). Dilutions (10mM, 1mM, 0.1mM en 0.01mM) of L-Phenylalanine, L-Tyrosine, L-Dihydroxyphenylalanine (L-DOPA), 3-Hydroxy-tyramine-hydrochloride (Dopamine), L-Noradrenaline-hydrogentartrate (NA), Adrenaline-hydrogentartrate (A) and Serotonin-creatininesulfate (5-HT) resp. were made in a solution of 1% Bovine Serum Albumin (BSA) in 50mM Tris buffer + 0.5% Natriumdisulfite ($Na_2S_2O_5$) (pH 7.0). For testing the antisera against DA and L-DOPA, glutaraldehyde fixation of the strips was either by immersion in 5% glutaraldehyde in 0.1 M phosphate buffer (pH 7.4) for 5 min at room temperature, or by vapor fixation heating 25% glutaraldehyde in the presence of the nitrocellulose strips for 1 hour at 80°C. Hereafter, the strips were washed for 3 times 10 min with a 1% BSA-solution in 50mM TBS + 0.5% $Na_2S_2O_5$ (pH 7.4), to prevent non-specific binding of the antibodies. Therefore, glutaraldehyde fixed models were incubated for 30 min with 1% $NaBH_4$ and 1% sodium metabisulfite in TBS. Subsequently the model strips were soaked for 1 hour in a solution of 1% BSA to block further protein binding to the nitrocellulose. After 3 times 5 min washing in TBS + 0.5% $Na_2S_2O_5$ the glutaraldehyde conjugates were reduced by treating the strips with 10mM sodiumborohydride in TBS + 0.5% $Na_2S_2O_5$ (pH 7.4) for 30 min. The strips were incubated overnight at 4°C or for 1 hour at 37°C with the respective antibodies diluted in TBS, containing 1% sodium metabisulfite. This was followed by a 3 times 5 min wash with TBS, TBS-T, and TBS preceeded by a 30 min incubation with the secondary antibody. Hereafter the procedure was continued with a rinse with TBS, Tris-HCl and Tris-HCl for 5 min and incubation with a peroxidase anti-peroxidase (PAP)-labeled antibody for 30 min as described previously. After washing the immunoreactivity was visualized by means of a diamidino benzidine (DAB) staining.

**Tissue preparation for DA/L-DOPA-immunocytochemistry** Male Wistar rats were anesthetized with nembutal (40 mg/kg, ip). Formaldehyde fixation They were preperfused through the ascending aorta for one minute (70 mm Hg) with $Ca^{2+}$-free Tyrode's buffer at room temperature, which was aerated for 15 min with 95% $O_2$/5% $CO_2$. The formaldehyde-based fixative consisted of ice cold 4% freshly depolymerized paraformaldehyde, 0.05% glutaraldehyde and 0.2% picric acid in 0.1 M phosphate (pH 7.4). During 20 - 30 min 500 ml of this fixative was perfused through the animal. Thereafter brains were dissected out and fixed for 2 hours at 4°C in the same fixative without glutaraldehyde. Glutaraldehyde fixation They were preperfused through the ascending aorta for 30 seconds (120 mm Hg) with $Ca^{2+}$- and glucose-free Krebs-Ringer buffer, containing 1 % $Na_2S_2O_5$ at room temperature, which was aerated for 15 min with 95% $O_2$/5% $CO_2$. Hereafter, the animals were perfused under high pressure (120 mm Hg) with 500 ml freshly made, ice cold, 5% glutaraldehyde and 1% sodium metabisulfite in 0.05 M cacodylate buffer (pH 7.1) (perfusion time 5 minutes). Brains were removed and postfixed in the same fixative for 30 min at 4°C.

**Tissue sectioning for glutaraldehyde fixed tissue** For Vibratome sectioning (50 μm) the tissue can be cut immediately in Tris-buffered saline (TBS) including 1% sodiummetabisulfite (TBS-S) at 4°C. For cryostat sectioning (10 μm) the brains were transferred for at least one night to TBS-S containing 5% sucrose and 1% $Na_2S_2O_5$ or for freeze microtome sections (35 μm) to TBS-S containing 20% sucrose and 1% $Na_2S_2O_5$.

**Tissue incubation for glutaraldehyde fixed tissue** Vibratome sections were incubated with DA-antisera at a dilution of 1:2000 or with L-DOPA-antiserum, diluted 1:1500, at 4°C. Dilutions were made in TBS-T and 1% $Na_2S_2O_5$ (pH 7.4). Incubations were routinely free floating for two nights at 4°C. In the case of glutaraldehyde fixed material, the sections were pretreated with 10 mM $NaBH_4$ and 1% sodium metabisulfite in TBS for 30 min, to reduce double bonds introduced by the glutaraldehyde fixation and all dilutions and incubations were done in buffer containing 1% sodium metabisulfite till the washing steps after the first antibody incubations. The stainings procedure consisted further of an incubation with a secondary antibody, diluted Goat-Anti-Rabbit IgG or Donkey-Anti-Goat IgG 1:60 in TBST for 2 hrs, washing again in TBS, TBS and TBST, for 15 min resp. at room temperature. Hereafter the sections were incubated with rabbit or goat peroxidase-antiperoxidase complex diluted 1:600 in TBST, and finally washed in TBS, Tris-HCl and Tris-HCl during 15 min resp. Staining was accomplished by treating the sections with a solution of 0.05% DAB and 0.01% $H_2O_2$ in Tris buffer. The sections were finally collected on chrome-alum gelatin coated slides. Cryostat sections were cut at -20°C with a thickness of 6-10 µm and also collected on chrome-alum gelatin coated slides. The sections were washed for 30 min in TBS + 1% $Na_2S_2O_5$ (pH 7.4), and subsequently washed for 30 min in 0.75% $NaBH_4$ solution in TBS + 1% $Na_2S_2O_5$. After 2 rinses during 10 min in TBS + 1% $Na_2S_2O_5$ and 1 rinse of 10 min in TBST + 1% $Na_2S_2O_5$ (pH 7.4), all at room temperature, the sections were incubated for 48-72 hrs at 4°C with rabbit DA- (diluted 1:1000), goat DA- (diluted 1:1000) or rabbit L-DOPA-antiserum (diluted 1:200). Some sections were incubated with a cocktail of goat DA-antiserum (1:1000) and rabbit L-DOPA-antiserum (1:200). All antisera were diluted in TBST + 1% $Na_2S_2O_5$. After incubation all sections were washed at room temperature 2 times 10 min in TBS, and 1 time for 10 min in TBST. Subsequently the sections were incubated for 60 min in the dark with fluorescein isothiocyanate (FITC)-conjugated Horse-Anti-Rabbit IgG (1:20 Centr. Lab. Bloedtr. Dienst, Amsterdam). For double staining experiments a cocktail was used of FITC-conjugated Donkey-Anti-Goat IgG (1:20 Nordic) and tetramethyl rhodamine isothiocyanate isomer R (TRITC)-conjugated Swine-Anti-Rabbit IgG (1:20 Nordic, The Netherlands), all diluted in TBST. After rinsing in TBS the sections were collected in a mixture of TBS and glycerol (1:2) containing 0.1% phenylenediamine.

**Preparation of conjugates against secondary messengers** The cGMP-formaldehyde-thyroglobulin conjugate has been described in detail elsewhere (De Vente et al. 1987).

**Serum specificity** of antisera was also established using an immunoinhibition test. cGMP antisera and preimmune sera were appropriately diluted with TBS-T in the presence of free cGMP or the cGMP-formaldehyde-thyroglobulin conjugate. The mixtures were incubated for 1 h at 37°C, followed by 18 h incubation at 4°C. Possible precipitates were removed by centrifugation at 20,000 g for 15 min at 4°C. The preincubated antisera were applied to model sections, cryostat sections, and the 100 μm thick aorta ring preparations.

**Gelatin model system incubation for cGMP** This procedure has been described before and was slightly modified from Schipper and Tilders (1983). In brief, a solution of 10% (wt./vol.) gelatin in distilled water was prepared. Solutions of cGMP or other nucleotides and nucleosides were made up (0.2 μM - 10 mM) in 0.2 M phosphate buffer, pH 7.4, and 2 M sucrose and mixed with equal volumes of the gelatin solution. Aliquots of these solution were pipetted into Tygon tubing. After gelation, the gels were fixed for 16 h at room temperature in a solution of 4% (wt./vol.) freshly depolymerized formaldehyde and 0.5% glutaraldehyde in 0.1 M phosphate buffer, pH 8.0, containing 1 M sucrose. The gels were removed from the tubing, embedded in OCT compound and frozen in isopentane prechilled with liquid $N_2$. 10 μm thick cryostat sections were cut and mounted on slides coated with chromalum gelatin. Slides were stored at -20°C. Before application of the primary antisera, the sections were washed with 1% $NaBH_4$ and 1% sodium metabisulfite in TBS for 30 min. Microfluorimetric measurements The intensity of the emitted fluorescence light was measured by use of an automated microfluorimeter (MPV II, Leitz), as described in detail previously. For orientation in the preparation we used the transmitted excitation pathway. The intensity of the excitation light was reduced to an acceptable low level to diminish photodecomposition of the fluorophore. The intensity of the cGMP-immunofluorescence was quantitatively assessed in the model sections, using a measuring spot of 113 $\mu m^2$.

**Spot-blot model system for cGMP** To test the specificity of the antibodies against the secondary messengers solutions containing 2 mg/ml of protein in distilled water of the following proteins were prepared: human serum albumin, bovine serum albumin, hemocyanin, gelatin, trypsin and pepsin. These solutions were diluted 1:1 with each concentration of a serial dilution of cGMP in distilled water, resulting in a solution containing 1 mg/ml protein which was 1 μM - 10 mM in cGMP. Aliquots of 3 μl were spotted onto Zetaprobe or nitrocellulose sheets and air dried. For testing the cGMP-antisera the strips were incubated for 2 h in a solution of 4% freshly depolymerized formaldehyde in 0.1 M phosphate buffer, pH 7.4. Hereafter, the strips were washed for 3 times 10 min with a 1% BSA-solution in 50mM Tris-Buffered-Saline (TBS) (pH 7.4). Subsequently the model strips were soaked for 1 hour in a solution of 1% bovine serum albumin (BSA) to block further protein binding to the nitrocellulose. After washing 3 times 5 min with TBS the strips were incubated overnight at 4°C or for 1 hour at 37°C with the respective antibodies diluted in TBS. All strips were washed and incubated in TBS, TBS-T, and TBS followed by a 30 min incubation with the secondary antibody, followed by the same washing procedure and incubation with a PAP-labeled antibody for 30 min as described previously. After washing the immunoreactivity was visualized by means of a DAB staining.

**Tissue treatment and incubation for cGMP-immunofluoresence** Male Wistar rats (180-220 g) were decapitated and the brains were removed immediately and transferred into ice cold incubation buffer of the following composition (mM): NaCl 121.1; KCl 1.87; $KH_2PO_4$ 1.17; $MgSO_4$ 1.15; $NaHCO_3$ 24.9; $CaCl_2$ 1.2; glucose 11.0; aerated with 5% $CO_2$/95% $O_2$, pH 7.4. containing 1 mM IBMX unless stated otherwise. Brain tissue was placed on a filter paper on a precooled (4°C) stainless steel table, and transversal 300 μm slices were cut with a McIlwain tissue chopper. Slices were separated from each other under a microscope while being submerged in ice-cold Krebs-Ringer bicarbonate buffer (KRB) and IBMX, pH 7.4, aerated with 5% $CO_2$/95% $O_2$. Slices were partitioned in a multiwell tissue culture chamber and were incubated for 40 min in KRB + IBMX at 37°C under an atmosphere of 5% $CO_2$/95% $O_2$. Incubation time before adding drugs was 30 min except for the phosphodiesterase inhibitor IBMX (1 mM) which was present from the start of the incubation. Incubations in the presence of drugs (including controls) were terminated after 10 min by adding the ice cold fixative solution (final concentration 4% formaldehyde, 5% sucrose, and 0.1 M phosphate, pH 7.4) for 2 hrs, followed by 1.5 hr wash in cold Tris-buffered saline (TBS), containing 5% sucrose (pH 7.4). Immediately thereafter 10 μm thick sections were cut on a cryostat at -20°C, thawed onto chromalum gelatin-coated slides and processed immediately for immunohistochemistry. The slides were allowed to reach room temperature and dry. They were rehydrated and washed for 3x5 min in TBS. Sections were incubated overnight in the cold with cGMP-antiserum, diluted 1:300 in TBS-T. Thereafter they were washed 3x5 min in TBS and incubated for 1 hr at room temperature with fluorescein isothiocyanate (FITC)-conjugated sheep anti-rabbit immunoglobulins diluted 1:20 with TBS-T. Finally the slides were washed again as described above and mounted in TBS-glycerin (1:3 vol./vol.). It was demonstrated before that by following this protocol, at least 20% of the initial soluble cGMP pool is retained on a protein matrix (De Vente et al. (1987).

For sequential staining the cGMP-immunostaining was removed according to Tramu et al (1978). Solutions of 2.5% $KMnO_4$, 5% $H_2SO_4$ and distilled water were mixed in various combinations. The optimal combination was 1:1:50 in the above mentioned order. Sections were dipped in this solution for 1 min, followed by a 1 min rinse in 0.5% $Na_2S_2O_5$ and therafter washed for 3 min in running tap water. Glial fibrillary acidic protein (GFAP) was visualized using a GFAP antiserum, in the usual way (De Vente et al. 1989d).

**RESULTS**

The apparent **affinity** of an antiserum for an antigen fixed in the model generally differs clearly from the one that is available in free solution (Schipper and Tilders, 1983). Immunostaining of water-soluble substances in the gel demands fixation and is dependent on the retention of these substances in the gel (see Table I). Since the retention of the various monoamines is dependent on the fixative used, it is not possible to use directly the quantitative data obtained from the MPV-II system. The gelatin model gives in this respect an impression upon the specificity of an antiserum. Changes in immunostaining intensity may reflect alterations in concentration of antigens. In addition, using the gelatin model the apparent affinity of an antiserum for fixed antigens can be accessed. Another important factor, which contributes to differences in apparent affinities for fixed and free antigens is a lower accessibility of the antibodies to the fixed antigens. To increase penetration of antibodies in biological preparations, detergents like Triton X-100 are frequently used. Changing the concentration of the gelatin concentration also exerts dramatic effects on the amount of the immunostaining intensity: lower concentrations of gelatin increase, whereas higher concentrations decrease or even prevent the immunostaining. It is also worth noting in this respect, that changing of the section thickness has no effect on the intensity of the immunostaining signal, showing that penetration of antibodies in the gel may exist over a range of less than 10 μm for gelatine gels fixed with 4% paraformaldehyde (= standard section thickness). Currently, there is no method available which gives a better impression of the affinity of an antiserum.

Table I. PFA: paraformaldehyde, GA: glutaraldehyde, ECDI: ethyl carbodiimide

| | | Retention % | | |
|---|---|---|---|---|
| | | 5-HT | DA | NA |
| Fixative | | | | |
| 4% PFA | (pH 6.0) | 56 | | 25 |
| 4% PFA | (pH 7.4) | 45 | 6 | 15 |
| 4% PFA + 0.1% GA | (pH 7.4) | 59 | | 26 |
| 4% PFA + 0.25% GA | (pH 7.4) | 57 | 25 | 36 |
| 4% PFA + 0.05% GA | (pH 7.4) | | 10 | |
| 4% PFA + 0.5% GA | (pH 6.0) | 59 | 29 | 39 |
| 4% PFA + 0.5% GA | (pH 7.4) | 57 | 29 | 35 |
| 4% PFA + 5% GA | (pH 7.4) | 25 | 29 | 33 |
| 0.5% GA | (pH 7.4) | 40 | 29 | 19 |
| 5% GA | (pH 7.4) | 37 | 45 | 32 |
| 0.5% ECDI | (pH 7.4) | 37 | | 28 |

<u>Specificity of the antibody against DA and L-DOPA</u> The specificity tests on the nitrocellulose strips indicated that the goat antiserum against conjugated DA shows a slight cross-reactivity towards L-DOPA. At relative low dilutions of the goat DA-antiserum, 1:1000, there is a strong immunoreactivity on the DA spots. In contrary there is only a weak immunoreactivity on the L-DOPA-, NA- and the A-spots. If the DA-antiserum is further diluted, 1:9000, only a positive staining is noted in the DA spot (Figure 1). The immunoinhibition tests for the DA and the L-DOPA antisera also revealed the specificity of the antisera.

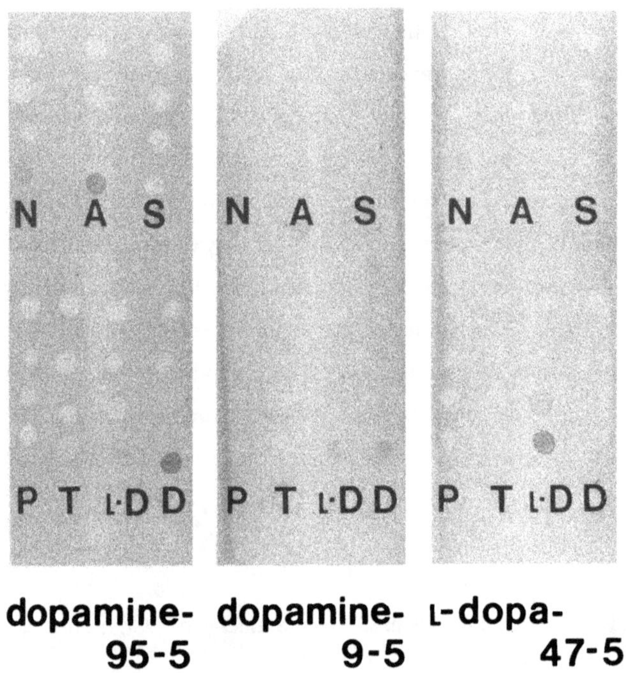


Figure 1 : Spot-blots using rabbit DA antiserum (9-5), goat DA antiserum (95-5) and rabbit - DOPA antiserum (47-5). Phenylalanine (P), tyrosine (T), L-Dopa (L-D), dopamine (D), noradrenaline (N), adrenaline (A) and serotonin (S) were used at the following concentrations: 10mM 1mM 0.1mM 0.01mM.

**Distribution of L-DOPA- and DA-immunoreactive neurons in the rat brain**

We observed high intensity of L-DOPA-immunoreactivity in already previously known dopaminergic- and noradrenergic cell bodies of e.g. the Zona Incerta, Substantia Nigra and Locus Coeruleus (Figs 2,3). In addition, we demonstrated in the nucleus retroventricularis the appearance of a larger number of L-DOPA-immunoreactive ($_i$) cells comparing to the number of $DA_i$ cells. A general difference between L-DOPA- and $DA_i$ neurons is that in L-$DOPA_i$ neurons the dendrites and cell bodies are more intensely stained than the varicose fibers, while

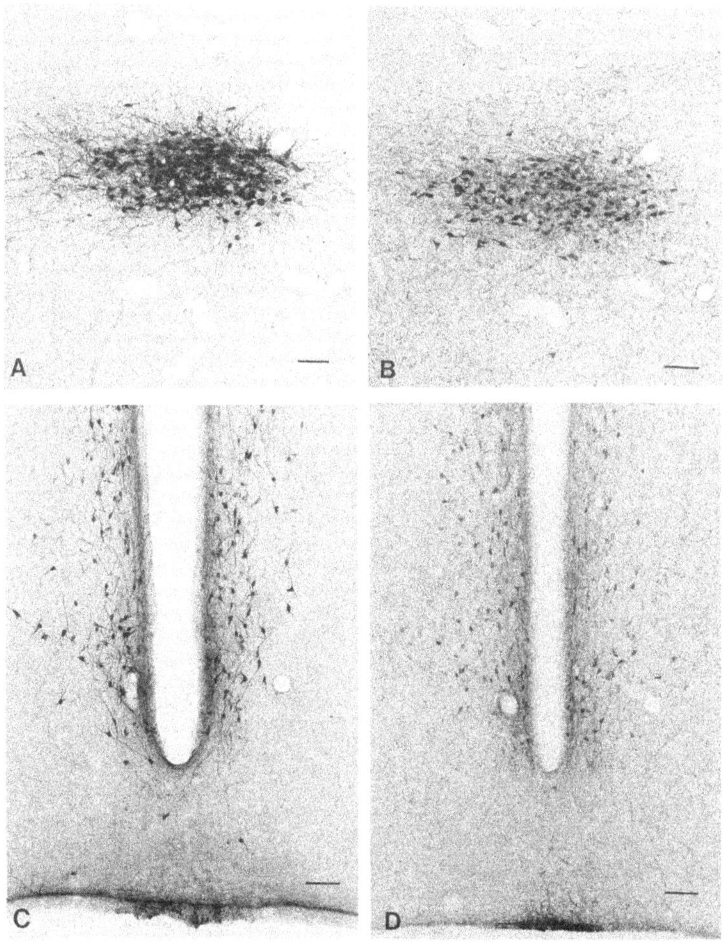

Figure 2. Photomicrographs of transverse 50 μm Vibratome sections through the region of the zona incerta (A,B) and the rostral part of the periventricular hypothalamic area (C,D) after incubation with an antiserum to dopamine (B,D) or L-DOPA (A,C). Note the difference between the DA and the L-DOPA-immunocytochemistry in the distinct labeling of the cell bodies by the L-DOPA antiserum and the detailed staining of the varicose fibers with the DA-antiserum. Bars indicate 50 μm.

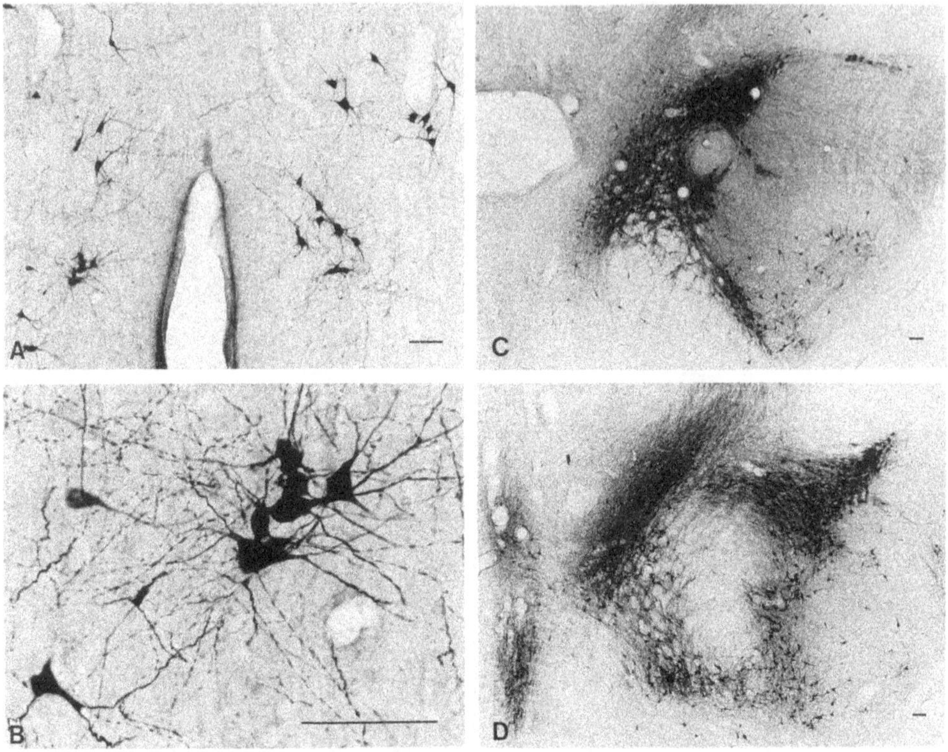


Figure 3. Photomicrographs of transverse (A,B) and horizontal (C,D) 50 μm Vibratome sections through the dorsomedian hypothalamic region (A,B) and the substantia nigra/ ventral tegmental area (C,D) after incubation with an antiserum to L-DOPA (A,B) or dopamine (C,D). A,B: L-DOPA-positive neurons are demonstrated in the A11 cell group. C,D: A large number of $DA_i$ cell bodies are localized in the substantia nigra, pars compacta and the area tegmentalis of Tsai. Bars indicate 50 μm.

in DA neurons the varicose fibers and dendrites are relatively more intense immunoreactive in contrary to the cell bodies. Alternating staining of sections through the nucleus retroventricularis with L-DOPA- and DA-antiserum shows that in the rostral part relatively more L-DOPA- than DA neurons are immunoreactive, while in the caudal part the opposite is apparent. In addition some double staining experiments were performed in which we applied a mixture of rabbit anti-L-DOPA and goat anti-DA antiserum. With this combination we observed in the Zona Incerta an approximately 95% overlap of L-DOPA$_i$ and DA$_i$ cells (Fig. 2). In conclusion these results suggests the existence of a separate L-DOPA$_i$ neuronal system. This system is strictly localized in the hypothalamus and further investigations are necessary to show functional implications.

**Distribution of L-DOPA- and DA-immunoreactive varicose fibers in the rat brain** Vibratome sections incubated with the L-DOPA- as well as the DA-antisera revealed smooth as well varicose fibers. Cryostat sections treated for FITC or TRITC immunofluorescence showed cell bodies and few fibers. The DA$_i$ cell bodies are less intensely stained than the L-DOPA$_i$ cell bodies, this is in contrary to the fiber systems. DA$_i$ fibers are much easier to recognize and more abundant than L-DOPA$_i$ fibers. The latter ones are more confined to area immediately surrounding the cell body areas (Figs 2,3).

**Basal guanylate cyclase activity** As already mentioned briefly above, immunostaining of basal cGMP levels in the rat brain was generally of low intensity, with a few exceptions (De Vente et al. 1987): the islands of Cajal, the cerebellum, and the olfactory bulb. For the latter area. The basal cGMP immunostaining in the olfactory bulb in 100 μm thick slices incubated in vitro. Intense cGMP-immunostaining was found in cell bodies of the inner plexiform layer. These cell bodies had several processes with intense cGMP-immunostaining; in the immediate surroundings of these cells an extracellular pool of cGMP was observed. The addition of 1 mM IBMX had a stimulating effect of cGMP levels in blood vessels walls and in thin, varicose fibers of the external plexiform layer of the olfactory bulb. Incubation of slices in the presence of 10 μM SNP in combination with 1 mM IBMX enhanced cGMP-immunostaining still further in blood vessel walls, and thin, varicose fibers. Using the approach of perfusion fixation of the whole animal, cGMP-immunoreactivity was found preserved in cellular structures of the olfactory bulb. An overview of the cGMP-immunostaining in the olfactory bulb has been presented and demonstrate the absence of cGMP-immunostaining in the cortical area. Thin varicose fibers, which could be traced for over 100 μm, were observed in the internal plexiform layer. Cells with long processes having varicosities were observed in the granule cell layer; these cells were found throughout the granular layer of the olfactory bulb. Control experiments using the preimmune serum showed some non-specific staining. Preabsorbtion of the cGMP-antiserum with cGMP-formaldehyde-thyroglobulin conjugate abolished the staining to the same non-specific staining as observed with the preimmune serum. We also showed that after perfusion fixation cGMP-immunoreactivity in the cerebellum is found in the same structures as seen after in vitro incubation. Thus perfusion fixation might yield some information on the localization of cGMP in the basal situation. Nevertheless, it is known that anesthetics have a profound influence on the levels of cGMP in the CNS. We also experienced not so much contradicting as well as varying results using the perfusion fixation and these reasons made us decide to abandon this procedure and use only the intro incubation set up.

**General stimulation of guanylate cyclase by sodium nitroprusside** The effect of stimulation of the guanylate cyclase by 10 μM SNP is shown in the MPO, the hippocampus, the lateral septal area, and in the caudate putamen. SNP stimulation was apparent in an abundant amount of thin fibers containing numerous varicosities. $cGMP_i$ cell bodies in the hippocampus were astrocytes, with a few exceptions. Not all astrocytes in the hippocampus were immunoreactive for cGMP after SNP stimulation. Under these same conditions we found in the hypothalamic paraventricular nucleus, and the medial preoptic area (MPO) stimulation cGMP production by SNP in cell bodies which do not stain for GFAP. No sequential staining was done for the septal area and the caudate putamen. Thus the soluble GNC is found primarily in varicose fibers in the rat brain. This observation confirms the hypothesis put forward by Ferrendelli more than 20 years ago (Ferrendelli et al., 1970). In addition the soluble GNC is found scattered through the brain in subpopulations of astrocytes and in few neuronal cell bodies, which could not be further identified (Fig.4).

**Receptor specific stimulation of guanylate cyclase by atrial natriuretic factor** ANF stimulated cGMP production reached a maximum in 5 to 10 minutes; SNP stimulated cGMP production was maximal after 10 minutes. We have shown numerous examples of the stimulation of GNC by ANF in the ANF-responding, cGMP-producing cells in the rat brain in previous publications. We present one example of ANF-responding cells in the MPO and ANF-responding, cGMP producing cells in the olfactory bulb. Sequential staining of the same sections with an antiserum against GFAP demonstrates that the responding, cGMP-producing cells are astrocytes. This was also documented for the hippocampal area, the lateral septum, and the medial amygdala (De Vente et al. 1988, 1989a, 1990) and the olfactory bulb.

**Localization of cGMP in the adult rat brain** In our initial studies perfusion fixated rats were used (de Vente et al. 1987) and applied the cGMP antisera to 50 μm vibratome slices. This

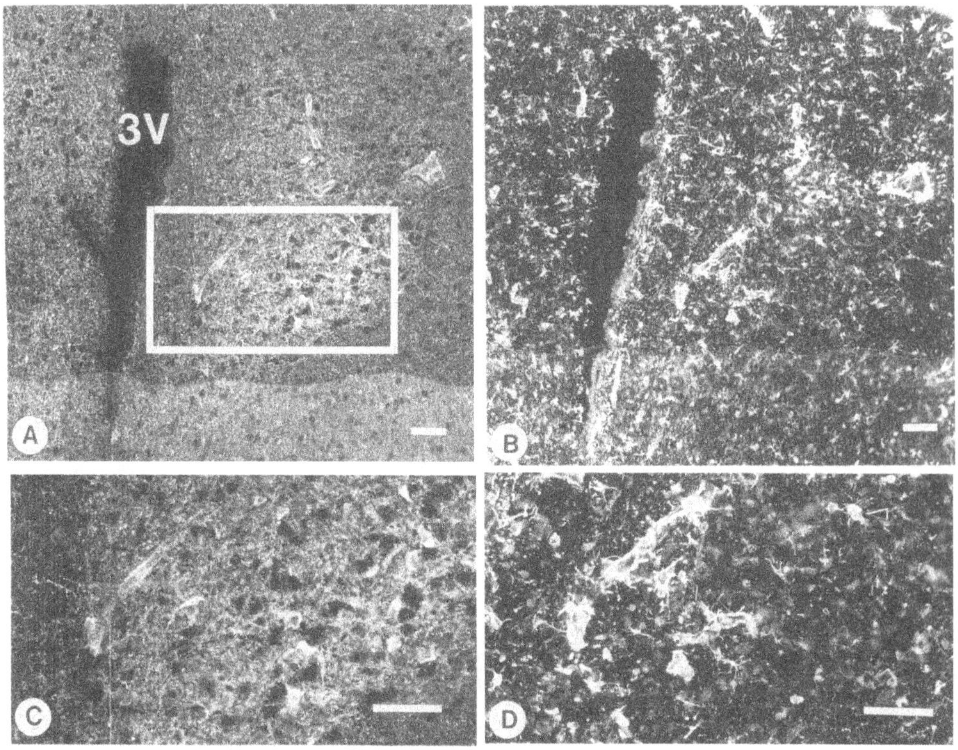


Figure 4: cGMP- (A,C) and GFAP- (B,D) immunostaining in the area of the nucleus paraventricularis incubated in vitro in the presence of 1 mM IBMX and 10 µM SNP; B: same area as shown in A), after removal of cGMP-immunostaining and subsequent incubation with GFAP-antiserum. There is no overlap between cGMP and $GFAP_i$ cells. D: Arrows indicate astrocytes which are cGMP-negative and GFAP-positive. Bars 100 µm.

method gives often varying results and, when compared to the slice method outline below, has little versatility. Nevertheless, the results obtained with both methods are in agreement with each other, as demonstrated for the cerebellum (De Vente et al. 1989b) and in the olfactory bulb. In one study (Berkelmans et al. 1989) the stimulation of the soluble guanylate cyclase by SNP was shown after whole body perfusion fixation with formaldehyde: the effect of SNP was demonstrated in aorta, kidney, retina and those brain tissues which are outside the blood brain barrier. From these studies we concluded that either in the basal situation cGMP levels are too low to be detected, with the exception of the cerebellum and the olfactory bulb, or the fixation of cGMP is too slow the conjugate enough nucleotide to the protein matrix to be detected.

Localized effects on cGMP metabolism are best studied using in vitro methods. The cerebellum presents an exception in that an extensive area of cGMP-immunostaining of high intensity is found in the basal situation in the molecular, the Purkinje cell layer, and the granular layer. Sequential staining for GFAP revealed that cGMP is seen in Bergmann glia and astrocytes in the granular layer. Occasionally, cGMP-immunoreactivity is seen in thin varicose fibers which are not GFAP positive (De Vente et al. 1989b,1990b). No cGMP-immunoreactivity was observed in Purkinje cells. Based on in vivo experiments, Purkinje cells were proposed to be the major sites of cGMP production in the cerebellum (Biggio and Guidotti 1976; Mao et al. 1975). This hypothesis was supported by: a. microdissection studies showed that cGMP levels in the cerebellum were highest in the Purkinje cell layer (McCandless et al. 1979 a,b; Rubin and Ferrendelli 1977); b. immunocytochemical studies showed the abundant presence of the G-substrate protein in the cerebellar Purkinje cell (DeCamilli et al. 1984; Levitt et al. 1984); c. (immuno)histochemical studies showed that guanylate cyclase was present in all cell types in the cerebellum (Ariano et al. 1982; Poeggel and Luppa, 1988; Zwiller et al. 1981). However, immunocytochemical data pointed also to glial cells as structures involved in generating cGMP (Chan-Palay and Palay 1979; Cumming et al. 1977, 1979). Biochemical data, using cell fractionation techniques in combination with specific lesioning of cerebellar structures (Bunn et al. 1986; Garthwaite and Garthwaite 1987), provided evidence which strongly suggested to glial cells to be the site of origin of the large cGMP pool in the cerebellum. This hypothesis was confirmed by our experiments using the cGMP-antibody.

Stimulation of cGMP production by ANF or SNP in the adult cerebellum is not clearly observed in the glial structures: the basal staining is too irregular seen throughout the entire cerebellum. However, SNP stimulates cGMP in thin varicose fibers and also in cell bodies which are not GFAP positive (De Vente et al. 1989b,1990b). We demonstrated that in the molecular layer areas with intense cGMP-immunostaining are immediately adjacent to areas where the cGMP-

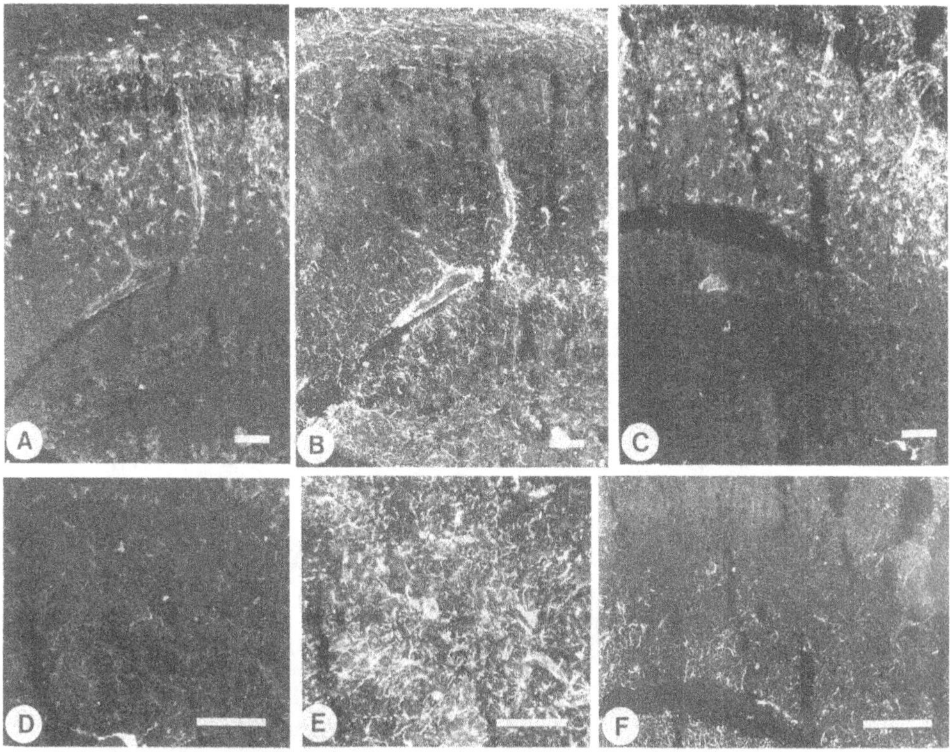


Figure 5: cGMP-immunostaining in the immature (day 4 postnatal) hippocampus of the rat after in vitro incubation in the presence of 1 mM IBMX: A,D: basal condition; B,E: GFAP-immunostaining of the same section as shown in A,D, after removal of the cGMP-immunostaining using the Tramu method; C,F: addition of 10 μM SNP; D,E,F are higher magnifications. Bars 100 μm.

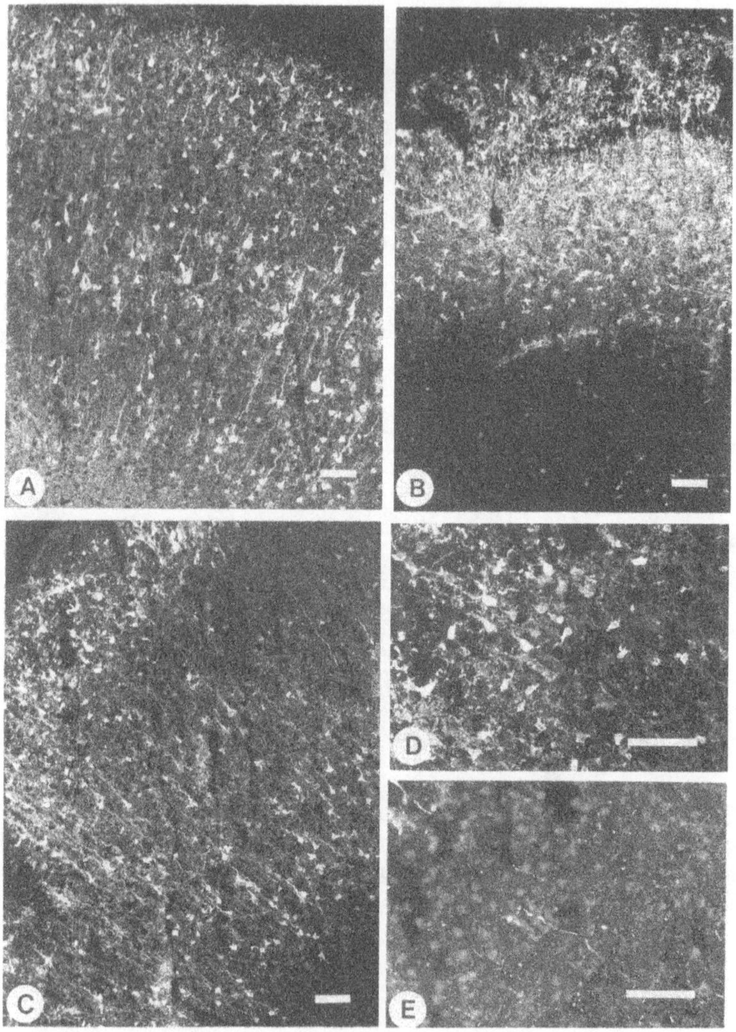


Figure 6: cGMP-immunostaining in the adult cerebral cortex (A,C,D) or hippocampus (B) incubated in vitro in the presence of 0.5 mM NMDA (A,B) or sodium nitroprusside (C,D). E: GFAP-immunostaining of the same section of D. Bars 100 μm.

immunostaining is absent (De Vente et al. 1989b) and there is no relationship observed between areas where the cGMP-immunostaining is found and known projection areas.

**Stimulation of GNC in the immature rat brain** We performed many experiments under varying conditions in an attempt to demonstrate the stimulation of cGMP production in the adult rat brain by excitatory amino acids and analogs, like glutamate, NMDA and kainic acid. These experiments were all unsuccessful. However, in the immature rat brain we found a massive stimulation of cGMP production by NMDA and kainic acid in several brain regions. This we demonstrate in the parietal cortex of the 4 day old rat (Fig. 5). In the basal situation we observed scattered cGMP-positive cell bodies; stimulation of cGMP production by 10 μM SNP or 0.5 mM NMDA. Cells showing cGMP-immunoreactivity after NMDA or SNP look very much alike, but this issue is still subject of investigations. Although GFAP-positive radial fibers are present at day 4 postnatal it cannot be stated with certainty whether these fibers are $cGMP_i$ or not, due to tissue shrinkage and distortion during the Tramu treatment (Tramu et al. 1978). There is a gradual decline in the cGMP-response of cells reactive to NMDA, at day 21 only few cells are responsive and in the 6 week old rat no effect of NMDA on cGMP-immunostaining is observed. The effect of SNP is still apparent, but is seen in thin varicose fibers with only few cell bodies being $cGMP_i$.

In the immature cerebellum (day 10 postnatal) we also found a highly reactive cGMP-generating system. In the basal situation small patches of very intense cGMP-immunofluorescence were observed in the inner granular layer. Stimulation of guanylate cyclase with ANF or sodium nitroprusside, resulted in the appearance of intense cGMP-immunostaining in the molecular layer in Bergman fibers, in the Purkinje cell layer in Bergman cell bodies, in the inner granular layer and white matter in astrocytes, whereas no cGMP-immunostaining was observed in the external granular layer. Kainic acid stimulated cGMP in Bergman glia and astrocytes of the inner granular layer; no effect of kainic acid was observed in the white matter or deep nuclei. Stimulation of cGMP production by NMDA resulted in a similar staining pattern as found after kainic acid stimulation (De Vente et al. 1990b). In the immature hippocampus (day 10 postnatal) we find again a highly reactive cGMP generating system. ANF, NMDA and SNP stimulate cGMP in cell bodies with numerous interwoven processes. These cells have the appearance of astrocytes; no cGMP-immunoreactive pyramidal cells were observed in the hippocampus (Fig. 6).

## DISCUSSION

**Localization of primary messengers** Staining of Vibratome sections with L-DOPA or DA-antisera reveals cell bodies, situated in the well defined dopaminergic cell areas, A7 untill A15. This is a typical signal since all dopaminergic will have L-DOPA as one of its precursors. However, recent studies using antibodies to the catecholaminergic enzymes have shown that there are tyrosine-hydroxylase (TH)-immunopositive cells, which are also aromatic dopa decarboxylase (ADDC)-immunopositive, in contrary there are also TH-positive cells which are AADC-immunonegative. Finally, a last group of neurons has been described which are only AADC-positive and TH-positive. Therefore it will be important to combine our DA- and L-DOPA data with results obtained with antisera against the catecholaminergic enzymes. Our observations indicate that there are two types of neurons. Neurons which are both L-DOPA- as well as DA-positive and neurons which are L-DOPA-positive and DA-negative. These cell bodies are intermingled between the dopaminergic cell areas. The L-DOPA$_i$- and DA$_i$ cells do not differ in size and shape. The L-DOPA$_i$ (varicose) fibers seem thicker and contain less varicosities, and moreover there are less L-DOPA$_i$ fibers than DA$_i$ fibers. L-DOPA$_i$ fibers are confined to cell bodies areas, and are not distributed throughout the CNS a seen with the DA-antiserum. This can be explained by the fact that L-DOPA in neurons in principal is situated within the cell soma, from which DA through via the fibers will be transported to the synapses. One other possibility for the limited fiber staining for L-DOPA could be low concentrations of L-DOPA within the thin varicose fibers or the fixation technique used is not appropriate for the retention of L-DOPA. This limited concentration is too low to be detected by our immunocytochemical procedure. It also might be possible that some cells have a different turnover to convert the different intermediate products in the catecholamiergic pathway. This could imply that a certain neuron could contain much more L-DOPA than DA, while in another neuron the L-DOPA will be rapidly converted to DA. This will result in a relative strong DA staining in contrary to a relative weak L-DOPA staining. In addition the speed of release of a neurotransmitter or the activity of the neuron can influence the concentration of the monoamine involved in this particular cell.

**Colocalization of primary messengers** Several studies have already suggested the presence of purely L-DOPA-neurons. Jaeger et al. (1983, 1984), Skagerberg et al. (1988) and Meister et al. (1988) showed the presence of dopaminergic cell bodies which were TH-positive and AADC-negative. Mons et al. (1988) presented the existence of L-DOPA$_i$ neurons. We are able to confirm their results using a combined L-DOPA and DA-immunocytochemical combination.

**Methodological aspects of immunocytochemistry for secondary messengers** The first immunocytochemical procedure to visualize cyclic nucleotides was introduced by Wedner et al (1972) using antibodies developed by Steiner et al (1972). These antibodies to cAMP and cGMP were raised against succinylated cyclic nucleotides coupled to carrier protein by the carbodiimide reactions. These antibodies proved to be effective in a radioimmunoassay for cAMP and cGMP. Immunocytochemical demonstration of the cyclic nucleotides required the use of unfixed material, although it was shown that for cGMP-immunocytochemistry a formaldehyde fixation might even improve the results (Ariano et al 1982; Chan-Palay and Palay 1979). However, it was demonstrated that using unfixed tissue material the nucleotides could not be retained by the tissue (Cumming et al. 1977; Steiner et al. 1972). Thus there exists a discrepancy between loss of cyclic nucleotides from the tissue and the specific demonstration of these compounds in the tissue using Steiner antibodies. Steiner et al. (1972) assumed that their antibodies recognized the cyclic nucleotides bound to their respective protein kinase. Although this hypothesis seems probable it was never rigorously tested. To demonstrate small water, soluble molecules by means of immunocytochemistry, it is obligatory to fix these molecules to a protein matrix. However, this necessitates the raising of antibodies against these molecules conjugated to an immunogenic protein in a similar way as used for tissue fixation (Berkenbosch et al. 1986; De Vente et al. 1987; Schipper and Tilders 1983; Steinbusch and Tilders, 1987) in order to forestall difficulties relating to changes brought about in the target molecules by the fixative. In our hand formaldehyde was the most suitable reagent to fix cGMP to the protein matrix (De Vente et al. 1987) and antibodies were raised in rabbits against a cGMP-formaldehyde-thyroglobulin complex. It is possible to fix cGMP to a protein matrix using a vapour fixation technique (De Vente et al. 1989c). In the two models systems, the nitrocellulose strip (Larsson 1982) and the gelatin model (Schipper and Tilders 1983) we demonstrated the specificity of our antibodies towards other formaldehyde-fixed nucleotides and nucleosides (De Vente et al. 1987,1989c) and we also showed the Steiner antibodies were much less efficient in binding to formaldehyde-fixed cGMP (De Vente et al. 1989c). Glutaraldehyde is a preferred fixative for the demonstration of monoamines and amino acids and gives a good tissue fixation (Geffard et al. 1984; Steinbusch et al. 1986). Although a glutaraldehyde-cGMP conjugate was reported and used for raising monoclonal antibodies against cGMP, we found it an unsuitable reagent. Glutaraldehyde reacts slower and less efficient than formaldehyde in coupling cGMP to the protein (De Vente et al. 1987). A combination of formaldehyde and glutaraldehyde would yield a potential interesting fixative in view of the possibilities of sequential staining for cGMP and neurotransmitters.

Interpretation of the available data on cGMP in the central nervous system has been hampered by the lack of detailed knowledge about the precise cellular localization of the sites of synthesis and degradation of cGMP. There exist at least two different enzymes (Waldman and Murad, 1988), the particulate bound guanylate cyclase and the soluble guanylate cyclase. These enzymes are different in respect to their physical characteristics and different in regard to their localization. In addition, cGMP levels are generally 5 to 50 times as low as cAMP levels (Ferrendelli et al. 1970). Until the discovery of ANF (DeBold et al. 1981; Hamet et al. 1984) there was no specific tool to stimulate the particulate guanylate cyclase (Flynn and Davies, 1985; Inagami, 1989;

Leitman and Murad, 1987). Although there have been reports which demonstrate the stimulation of the particulate enzyme by sodium nitroprusside (Waldman and Murad 1987), our data shows that this is not a general mechanism in the central nervous system of the rat. Our finding that ANF and sodium nitroprusside stimulate cGMP production in the same cellular structures in the cerebellum is best explained by assuming that both enzymes are present in these cells at this age. There has been reports which described the development of guanylate cyclase activity in granule cells after 4 days in culture (Novelli and Henneberry 1987). A role for glial cells in regulating ionic composition of the extraneuronal compartment is a long-standing hypothesis (Orkand 1977; Somjen 1988). Recent evidence points to as function of cGMP in the control of ion channel gating (Fesenko et al. 1985; Nakamura and Gold 1987; Schulz et al. 1986) or control of intracellular $Ca^{2+}$ homeostasis as shown for smooth muscle relaxation (Furukawa et al. 1988; Karaki et al. 1988; Twort and Van Breemen 1988). cGMP might function in a similar fashion in the central nervous system in controling the ionic environment and $Ca^{2+}$ fluxes (Ogura et al. 1986; Paupardin-Tritsch et al. 1986).

Acknowledgements. This study was supported by grant no. 900-534-055 from The Netherlands Foundation of Medical Research (MEDIGON). We thank Henk Nordsiek for processing the photomicrographs. The gift of the antiserum against GFAP by Dr. F.C.S. Ramaekers, Department of Pathology, Catholic University, Nijmegen, The Netherlands, is gratefully acknowledged.

## REFERENCES

Ariano MA, Lewicki JA, Brandwein HJ and Murad F (1982) Immunohistochemical localization of guanylate cyclase within neurons of rat brain. Proc Natl Acad Sci USA 79: 1316-1320

Berkelmans HS, Schipper J, Hudson L, Steinbusch HWM and De Vente J (1989) cGMP immunocytochemistry in aorta, kidney, retina, and brain tissues of the rat after perfusion with nitroprusside. Histochemistry 93: 143-148

Berkenbosch F, De Vente J, Schipper J and Steinbusch HWM (1986) Quantitative immunocytochemistry of monoamines, neuropeptides and second messengers in non-biological and biological models. In: HWM Steinbusch (Ed) Monoaminergic Neurons at Light and Electron microscopical Levels. IBRO-Handbook Series: Methods in the Neurosciences, Vol. 10, Wiley and Sons, Chicester, UK

Biggio G and Guidotti A (1976) Climbing fiber activation and 3',5'-cyclic guanosine monophosphate (cGMP) content in cortex and deep nuclei of cerebellum. Brain Res 107: 365-373

Bunn SJ, Garthwaite J and Wilkin GP (1986) Guanylate cyclase activities in enriched preparations of neurones, astroglia and a synaptic complex isolated from rat cerebellum. Neurochem Int 8: 179-185

Chan-Palay V and Palay SL (1979) Immunocytochemical localization of cyclic GMP: light and electron microscopic evidence for involvement of neuroglia. Proc Natl Acad Sci USA 76: 1485-1488

Cumming R, Arbuthnott G and Steiner AL (1979) Characterization of immunofluorescent cGMP-positive fibres in the central nervous system. J Cycl Nucl Res 5: 463-467

Cumming R, Eccleston D and Steiner AL (1977) Immunohistochemical localization of cyclic GMP in rat cerebellum. J Cycl Nucl Res 3: 275-279

De Vente J, Schipper J and Steinbusch HWM (1989c) Formaldehyde fixation of cGMP in distinct cellular pools and their recognition by different cGMP-antisera: an immunocytochemical study into the problem of serum specificity. Histochemistry 91: 401-412

De Vente J, Manshanden CG, Sikking RA, Ramaekers FCS and Steinbusch HWM (1990) A functional parameter to study heterogeneity of glial cells in rat brain slices: cGMP production in ANF-responsive cells. Glia 3: 43-54

De Vente J, Bol JGJM and Steinbusch HWM (1989b) Localization of cGMP in the cerebellum of the adult rat: an immunohistochemical study. Brain Res 504: 332-337

De Vente J, Bol JGJM and Steinbusch HWM (1989a) cGMP-producing, atrial natriuretic factor-responding cells in the rat brain. An immunocytochemical study. Europ J Neurosci 1: 436-460

De Vente J, Steinbusch HWM and Schipper J (1987) A new approach to immunocytochemistry of 3'5'-cyclic guanosine monophosphate: preparation, specificity and initial application of a new antiserum against formaldehyde-fixed 3'5'-cyclic guanosine monophosphate. Neuroscience 22: 361-373

De Vente J, Wietsma JJ, Bol JGJM, Schipper J, Malhotra SK and Steinbusch HWM (1989d) Potassium increases cGMP in glial cells in rat hippocampus slices: a combined biochemical and immunocytochemical study. Neurochem Res Comm 4: 25-31

De Vente J, Bol JGJM, Hudson L, Schipper J and Steinbusch HWM (1988) Atrial natriuretic factor-responding and cyclic GMP producing cells in the rat hippocampus: a combined micropharmacological and immunocytochemical approach. Brain Res 446: 387-395

DeBold AJ, Borenstein HB, Veresse AT and Sonnenberg H (1981) A rapid and potent natriuretic response to intravenous injection of atrial myocardial extract in rats. Life Sci 28: 89-94

DeCamilli P, Miller PE, Levitt P, Walter U and Greengard P (1984) Anatomy of cerebellar Purkinje cells in the rat determined by a specific immunohistochemical marker. Neuroscience 11: 761-817

Ferrendelli JA, Steiner AL, McDougal DB, Kipnis DM (1978) Biochem Biophys Res Commun 41: 1061-1067

Fesenko EE, Kolesnikov and Lubarsky AL (1985) Induction by cGMP of cationic conductance in plasma membrane of retinal rod outer segment. Nature 313: 310-313

Flynn SG and Davies PO (1985) The biochemistry and molecular biology of atrial natriuretic factor. Biochem J 232: 313-321

Furukawa KI, Tawada Y and Shigekawa M (1988) Regulation of the plasma membrane $Ca^{2+}$ pump by cyclic nucleotides in cultured vascular smooth muscle cells. J Biol Chem 263: 8058-8065

Garthwaite J and Garthwaite G (1987) Cellular origins of cyclic GMP response to excitatory amino acid receptor agonists in rat cerebellum in vitro. J Neurochem 48: 29-39

Geffard M, Kah O, Onteniente R, Seguella P, Le Moal M and Delaage M (1984) Antibodies to dopamine: Radioimmunological study of specificity in relation to immunocytochemistry. J Neurochem 42: 1595-1599

Hamet P, Tremblay J and Pang SC (1984) Effect of native and synthetic atrial natriuretic factor on cyclic GMP. Biochem Biophys Res Comm 123: 515-527

Ignarro LJ (1989) Endothelium-derived nitric oxide: actions and properties. FASEB J 3: 31-36

Inagami T (1989) Atrial natriuretic factor. J Biol Chem 264: 3043-3046

Jaeger CB, Ruggiero DA, Albert VR, Park DH, Joh TH and Reis DJ (1984) Aromatic L-amino acid decarboxylase in the rat brain: immunocytochemical localization in neurons of the brain stem. Neuroscience 11: 691-713

Jaeger CB, Teitelman G, Joh TH, Albert VR, Park DH and Reis DJ (1983) Some neurons of the rat central nervous system contain aromatic-L-amino acid decarboxylase but not monoamines. Science 219: 1233-1235

Karaki H, Sato K, Ozaki H and Murakami K (1988) Effects of sodium nitroprusside on cytosolic calcium level in vascular smooth muscle. Europ J Pharmacol 156: 259-266

Larsson L-I (1982) Antigen-defined immunocytochemistry. In: Modern Methods in Pharmacology, Alan R. Liss, New York, p 1

Leitman DC and Murad F (1987) Atrial natriuretic factor receptor heterogeneity and stimulation of particulate guanylate cyclase and cyclic GMP accumulation. Endocrinal Metab Clin North Am 16: 79-105

Levitt P, Rakic P, DeCamilli P and Greengard P (1984) Emergence of cyclic guanosine 3',5'-monophosphate dependent protein kinase immunoreactivity in developing rhesus monkey cerebellum: correlative immunocytochemical and electron microscopic analysis. J Neurosci 4: 2552-2564

Mao CC, Guidotti A and Landis S (1975) Cyclic GMP reduction of cerebellar concentrations in 'nervous' mutant mice. Brain Res 90: 335-339

Matsunaga K and Furchgott RF (1989) Interactions of light and sodium nitrite in producing relaxation of rabbit aorta. J Pharmacol Exp Ther 248: 687-695

McCandless DW, Feussner GK, Lust WD and Passonneau JV (1979a) Metabolite levels in brain following experimental seizures: the effects of isoniazid and sodium valproate in cerebellar and cerebral cortical layers. J Neurochem 32: 755-760

McCandless DW, Feussner GK, Lust WD and Passonneau JV (1979b) Metabolite levels in brain following experimental seizures: the effects of maximal electroshock and phenytoin in cerebellar layers. J Neurochem 32: 743-753

Meister B, Hokfelt T, Steinbusch HWM, Skagerberg G, Lindvall O, Geffard M, Joh TH, Cuello AC and Goldstein M (1988) Do tyrosine hydroxylase-immunoreactive neurons in the ventrolateral arcuate nucleus produce dopamine or only L-DOPA. J Chem Neuroanat 1: 59-64

Mons N, Danel N and Geffard M (1988) Visualization of L-dihydroxyphenylalanine in rat brain by using specific antibodies. Brain Res 451: 403-407

Mons N and Geffard M (1987) Specific antisera against the catecholamines: L-3,4-dihydroxyphenylalanine, dopamine, noradrenaline, and octopamine tested by an enzyme-linked immunosorbent assay. J Neurochem 48: 1826-1833

Murad F (1986) Cyclic guanosine monophosphate as a mediator of vasodilatation. J Clin Invest 78: 1-5

Nakamura T and Gold GH (1987) A cyclic nucleotide-gated conductance in olfactory receptor glia. Nature 325: 442-444

Novelli A and Henneberry RC (1987) cGMP synthesis in cultured cerebellar neurons is stimulated by glutamate via a $Ca^{2+}$-mediated, differentiation-dependent mechanism. Brain Res 34: 307-310

Ogura A, Ozaki K, Kudo Y and Amano T (1986) Cytosolic calcium elevation and cGMP production induced by serotonin in a clonal cell of glial origin. J Neurosci 6: 2489-2494

Orkand RK (1977) Glial cells. In: Handbook of physiology - The nervous system I. ER Kandel (ed) American Physiological Society, Bethesda, Md, p. 855-875

Palmer RMJ, Ashton DS and Moncada S (1988) Vascular endothelial cells synthesize nitric oxide from L-arginine. Nature 333: 664-666

Paupardin-Tritsch D, Hammond C and Gerschenfeld HM (1986) Serotonin and cyclic GMP both induce an increase in the calcium current in the same identified molluscan neurons. J Neurosci 6: 2715-2723

Poeggel G and Luppa H (1988) Histochemistry of nucleotidyl cyclases and cylic nucleotide phosphodiesterases Histochem J 20: 249-268

Rubin EH and Ferrendelli JA (1977) Distribution and regulation of cyclic nucleotide levels in cerebellum in vivo. J Neurochem 29: 43-51

Schipper J and Tilders FJH (1983) A new technique for studying specificity of immunocytochemical procedures: Specificity of serotonin immunostaining. J Histochem Cytochem 31: 12-18

Schultz JE, Pohl SL and Klumpp S (1986) Voltage-gated $Ca^{2+}$ entry into Paramecium linked to intraciliary increase in cyclic GMP. Nature 322: 271-273

Skagerberg G, Meister B, Hokfelt T, Lindvall O, Goldstein M, Joh T and Cuello AC (1988) Studies on dopamine-, tyrosine hydroxylase and aromatic L-amino acid decarboxylase-containing cells in the rat diencephalon: comparison between formaldehyde-induced histofluorescence and immunofluorescence. Neuroscience 24: 605-620

Somjen GG (1988) Nervenkitt: notes on the history of the concept of neuroglia. Glia 1: 2-9

Steinbusch HWM, De Vente J and Schipper J (1986) Immunohistochemistry of monoamines in the central nervous system. In Panula P, Paivarinta H and Soinila S (Eds) Neurohistochemistry today, Neurology and Neurobiology, Vol.20, Alan Liss, New York, 75-105

Steinbusch HWM and De Vente J (1987) Immunocytochemistry of cyclic GMP in the superior cervical ganglion of the rat: A combined quantitative immunofluorescence and pharmacological in-vitro study. Exp Brain Res 16: 355-360.

Steinbusch HWM and Tilders FJH (1987) Immunohistochemical techniques for light- microscopical localization of dopamine, noradrenaline, adrenaline, serotonin and histamine in the central nervous system. In: Steinbusch HWM (Ed) Monoaminergic Neurons at Light and Electron microscopical Levels, IBRO-Handbook Series: Methods in the Neurosciences, Vol. 10, Wiley and Sons, Chicester, UK

Steinbusch HWM, Wouterlood FG, De Vente J, Bol JGJM and Berkenbosch F (1988) Immunohistochemical localization of monoamines and cyclic nucleotides. Their application in quantitative immunofluorescence studies and tracing monoaminergic neuronal connections. Acta Histochem, Suppl, 35: 86-106

Steiner AL, Ong S and Wedner HJ (1972) Cyclic nucleotide immunocytochemistry. Adv Cycl Nucl Res 7: 115-155

Tramu G, Pillez A and Leonardelli J (1978) An efficient method of antibody elution for the successive or simultaneous localization of two antigens by immunocytochemistry. J Histochem Cytochem 26: 322-324

Twort CHC and Van Breemen C (1988) Cyclic guanosine monophosphate-enhanced sequestration of $Ca^{2+}$ by sarcoplasmic reticulum in vascular smooth muscle. Circ Res 62: 961-964

Waldman SA and Murad F (1987) Cyclic GMP synthesis and function. Pharmacol Rev 39: 163-196

Wedner HJ, Hoffer BJ, Battenberg E, Steiner AL, Parker CW and Bloom FE (1972) A method for detecting cyclic adenosine monophosphate by immunofluorescence. J Histochem Cytochem 20: 293-299

Zwiller J, Ghandour MS, Revel MO and Basset P (1981) Immunohistochemical localization of guanylate cyclase in rat cerebellum. Neurosci Lett 23: 31-36

# TRANSMITTER-RELATED RETROGRADE LABELING AND IMMUNOHISTOCHEMISTRY TO LOCALIZE AMINO ACID TRANSMITTERS IN NEURAL PATHWAYS AND CELLS

P. Streit, P. Grandes*, P. Morino**, Ph. Tschopp and M. Cuénod
Brain Research Institute
University of Zürich
August-Forel-Strasse 1
CH-8029 Zürich
Switzerland

Chemical neuroanatomy has emerged as a relatively new field of intense research activity stimulated by the development of powerful histochemical methods (see Björklund and Hökfelt, 1983, 1984, 1985). Progress has been most impressive in studies on the distribution of neurotransmitters, neuromodulators and receptors for such endogenous neuroactive compounds. The fact that the known neuroactive compounds are numerous and that they often coexist in various combinations has led to a concept superimposing a new, chemically defined level of complexity onto the already extremely complex system described in merely morphological terms. To work out such a concept, methods able to detect chemical specificities even in very small morphological entities are required. A better knowledge of the anatomo-chemical organization of the nervous system should provide further insights into its functional organization if only by establishing a firm basis for detailed physiological and pharmacological investigations.

With respect to methodology in chemical neuroanatomy, our group has mostly been involved in the development and application of transmitter-related retrograde tracing of neural pathways and, more recently, in work on certain aspects of transmitter immunohistochemistry. Thus, we will discuss these two methods, their prospects and limitations, as well as experiences made in using them to attribute neurotransmitters or transmitter candidates, particularly amino acids, to neural pathways and cells.

## Transmitter-related retrograde tracing of neural pathways

Transmitter-related retrograde tracing, a neuronal tracing method involving the retrograde migration of labeled transmitter-related compounds has been proposed as a tool for delineating neural pathways according to their transmitter specificity, particularly those using amino acids and acetylcholine. This approach is based on the general hypothesis that some transmitters or related molecules following their selective uptake into nerve terminals normally utilizing them migrate (possibly in a modified form) towards the perikaryon where the material is accumulated. Selective uptake has been documented for amino acid transmitters, for choline, and for biogenic amines (Snyder et al., 1973; Hökfelt and Ljungdahl, 1975; Iversen et al., 1975). Retrograde, somatopetal migration of material from the terminal to the cell body has been established for various endogenous and exogenous molecules (Grafstein and

Present addresses: *Department of Neurosciences, Faculty of Medicine, Basque Country University, 699-48080 Bilbao, Spain; **The Institute of Physical and Chemical Research (RIKEN), Laboratory for Neural Systems, Wako, Saitama, Japan

NATO ASI Series, Vol. H 58
Neurocytochemical Methods
Edited by A. Calas and D. Eugène

Forman, 1980; Weiss, 1982; Laduron, 1987; Castel et al., 1990). Transmitter-related retrograde tracing of neural pathways has been reviewed by Cuénod and Streit (1983) and with respect to excitatory amino acid neurotransmission by Ottersen and Storm-Mathisen (1986); in the present contribution, some of the later developments and applications of the method will be briefly considered. Very recently, uptake and retrograde axonal transport have also been demonstrated for a neuropeptide, namely for neurotensin (Castel et al., 1990). However, neurotensin uptake appears to be based on a receptor dependent mechanism while amino acids, choline and biogenic amines are taken up by transporter systems. Furthermore, the retrograde labeling observed in the nigrostriatal pathway following application of iodinated neurotensin does not seem to be transmitter-related. In the present short update, we will focus mostly on aspects observed and results obtained with amino acid tracers, [$^3$H]glycine, [$^3$H]GABA and particularly D-[$^3$H]aspartate (for tritiated dopamine, serotonin and choline see Cuénod and Streit, 1983; more recently for [$^3$H]choline: Jones et al., 1986; Jones and Beaudet, 1987; Bernays et al., 1988).

The method of transmitter-related retrograde tracing relies on the fact that a number of neurotransmitters and/or related molecules have been reported to be selective retrograde markers. Thus, in contrast to nonselective retrograde tracers such as horseradish peroxidase (HRP) or wheat germ agglutinin (WGA) which label non-differentially all neurons with terminals exposed to them, transmitter-releated markers retrogradely label only a subset of the afferents to an area in which they are applied in a radioactively labeled form, the labeling pattern being revealed by means of autoradiography. This retrograde labeling, therefore, has been hypothesized to be transmitter-specific - namely to occur only following the administration of a molecule related or identical to the neurotransmitter of the labeled pathway(s) (for review: Cuénod and Streit, 1983).

The experimental procedures underlying transmitter-related retrograde tracing have been described extensively by Wiklund (see appendix in Cuénod and Streit, 1983). Briefly, the radioactively labeled transmitter or transmitter-related compound should be obtained at a high specific activity and with a long-lived isotope, two conditions favoring the use of tritiated tracers. In fact, appropriate tracer compounds are easily available in most cases. Availability may, however, become a limiting factor if newly discovered endogenous neuroactive compounds or related molecules have to be tested as selective retrograde tracers. 30-60 nl containing 15-30 $\mu$Ci of the concentrated labeled compound are microinjected over a period of 10-20 minutes. In appropriate sites, the radioactive substances may also be applied by superfusion in the range of high affinity uptake concentrations (Beaudet et al., 1981; Wiklund et al., 1982, 1984). To prevent false positive results, the concentration of the injected compound should be kept as low as possible. In fact, the use of experimental series in which various concentrations are used is strongly recommended, particularly in the case of tritiated D-aspartate (Behzadi et al., 1990; Wiklund, personal communication). The application of non-selective retrograde tracers in parallel experiments should help to determine the level of anatomical selectivity. The level of chemical selectivity, on the other hand, should be assessed in a further series of experiments in which another transmitter or transmitter-related compound is tested as retrograde tracer. In all experiments involving radioactively labeled compounds, moreover, the actual concentration of the compound injected should be calculated from the specific activity of the labeled material and from the radioactivity measured in an aliquot that has been delivered from the injection device immediately before applying the tracer in the tissue. To obtain spatially restricted uptake sites while making the tracer available for prolonged periods of time, gel particles loaded with the radioactively labeled compound can be implanted (Fischer et al., 1986). After a survival time of a few hours to a few days, depending on the pathway's length, the experimental animal is in most cases perfused with a fixative containing glutaraldehyde, and the tissue sections are processed for light-microscopic autoradiography by conventional techniques.

[$^3$H]Glycine: The first observations suggesting a relation between selective retrograde labeling and transmitter specificity were made by Hunt in our laboratory in 1975 (Hunt et al, 1975, 1977). Following injection of tritiated glycine, serine, alanine and choline into the pigeon optic tectum, he observed an accumulation of autoradiographic silver grains over a small group of perikarya in the nucleus isthmi, pars parvocellularis (Ipc). These and further observations at the levels of light and electron microscopy (Streit et al., 1980 ) were best explained by assuming a selective uptake and retrograde migration of radioactivity. As a result of electrical stimulation in the Ipc-tectal pathway, an increase in the concentration of exogenous as well as of endogenous glycine could be measured in perfusates collected in push-pull cannula experiments (Reubi and Cuénod, 1976; Wolfensberger et al., 1981; Canzek et al., 1981). These findings have been considered as strong indications for the glycinergic nature of this pathway. - Recently, tritiated glycine has been used successfully in retrograde tracing experiments to identify the cell bodies of descending glycinergic projections terminating in the cochlear nucleus of the guinea pig (Benson and Potashner, 1990).

[$^3$H]GABA: Tritiated GABA has also been used as a retrograde tracer. Thus, following its injection into the rat substantia nigra, medium-sized neuronal perikarya in the caudoputamen were overlaid with autoradiographic silver grains (Streit et al., 1979; Streit, 1980). GABA is well established as transmitter in a population of striatal neurons projecting to the substantia nigra. Perikaryal labeling was absent in projections to the substantia nigra involving other transmitters, such as those originating in the dorsal raphé nucleus or the cerebral cortex. Several additional examples of selective retrograde tracing of pathways with tritiated GABA as tracer have been described (for review: Cuénod and Streit, 1983; more recently: Bagnoli et al., 1983; Somogyi et al., 1983; Hall et al., 1984; Künzle and Schnyder, 1984; Matute and Streit, 1985; Kisvarday et al., 1986, 1987; Christie et al., 1987; Beart et al., 1990). On the other hand, injections of tritiated GABA into Deiter's nucleus failed to label retrogradely Purkinje cells in the cerebellar cortex (Wiklund et al., 1983). This surprising negative result in a well established GABAergic system can probably be explained by the apparent lack of a high affinity uptake system for GABA at the terminals of these neurons (Storm-Mathisen, 1975, 1976).

D-[$^3$H]Aspartate: Nerve terminals of excitatory pathways for which glutamate and/or aspartate are transmitter candidates are often characterized by a mechanism of high affinity uptake for these amino acids. D-Aspartate which is not readily metabolized acts as a false transmitter and is taken up by the same high affinity mechanisms as the L-forms of aspartate and glutamate (for review: Ottersen and Storm-Mathisen, 1986). Indeed, D-[$^3$H]aspartate (D-[$^3$H]Asp) has been used successfully to retrogradely label neuronal pathways - initially the rat corticostriatal connection (Streit, 1980) and thereafter a multitude of neuronal projections (for review: Cuénod and Streit, 1983; Ottersen and Storm-Mathisen, 1986; additional references: Watanabe and Kawana, 1984; Matute and Streit, 1985; Matute and Martinez-Millan, 1985; Kalén et al., 1985; Christie et al., 1987; Fuller et al., 1987; Kisvarday et al., 1987; Matute et al., 1987; Bernays et al., 1988; Fuller and Price, 1988; Wiklund et al., 1988; Somogyi et al., 1989; Beart et al., 1990; Behzadi et al., 1990). Indeed, this method has proved its ability to define precisely the location of neuronal perikarya giving rise to presumed excitatory amino acid projections.

In many cases, in fact, selective retrograde labeling with D-[$^3$H]Asp as tracer has added valuable information on pathways for which excitatory amino acids have been suggested as transmitters on the basis of neurochemical, physiological, pharmacological or histochemical evidence. Even minor projections have been revealed, thus demonstrating the method's sensitivity. In contrast, if methods involving lesions are used, minor afferents to a certain region in the CNS might easily be missed. It should be pointed out, however, that D-[$^3$H]Asp does not distinguish between neuronal elements using either glutamate or aspartate as transmitter: The method's specificity is determined by the specificity of the uptake system which in the case of dicarboxylic amino acids is a common one for L-glutamate, L- and D-aspartate.

In certain experiments involving D-[$^3$H]Asp as tracer, 'false negative' observations have been made. Thus, administration of D-[$^3$H]Asp in the cerebellar cortex leads to the retrograde labeling of the inferior olive via the climbing fiber system (Fig. 1) while the granule cells remain mostly unlabeled, with the exception of a few which may have been mechanically injured (Künzle and Wiklund, 1982; Wiklund et al., 1982, 1984; Matute et al., 1987). The cerebellar granule cells represent one of the best documented examples of CNS neurons using glutamate as transmitter (for review: Ottersen and Storm-Mathisen, 1986; Somogyi et al., 1986). The lack of retrograde labeling in cerebellar mossy fiber systems should be considered as another 'false negative' result since immunohistochemical studies demonstrate strong glutamate-like immunoreactivity in mossy fiber terminals (Somogyi et al., 1986; Liu et al., 1989) and since recent pharmacological evidence indicates an excitatory amino acid as the principal mossy fiber transmitter (Garthwaite and Brodbelt, 1990). The 'false negative' result obtained with D-[$^3$H]Asp in mossy fiber afferents may probably be explained by the lack of an efficient uptake system as this tracer labels mossy fiber terminals only poorly in uptake experiments performed at the level of electron microscopy (Garthwaite and Garthwaite, 1988). In contrast to the findings in other corticofugal projections which are readily traced, retrograde labeling with D-[$^3$H]Asp in the cat corticocollicular pathway has initially been reported to be absent (Baughman and Gilbert, 1980, 1981). Lesion of this pathway, on the other hand, has been shown to decrease the collicular Glu/Asp uptake (Fosse et al., 1984). It should be emphasized, however, that later experiments with this tracer have demonstrated labeling in various cortical areas of the cat, some also in the visual cortex, and prominent cortical labeling in rat and rabbit (Matute and Streit, 1985).

Over the past decade, tests in various pathways for which transmitters have been established or suggested by the results of neurochemical, physiological, pharmacological or histochemical investigations have shown selective retrogade labeling of these pathways following application of the corresponding transmitter or a related compound. The evidence is to a large extent compatible with the proposal that transmitter-related retrograde labeling occurs in pathways known - according to other methods - to use the tested transmitter, whereas it is absent in other pathways. As discussed above, discrepancies are observed in some cases and lead to the conclusion that 'false positive' or 'false negative' results have been obtained. However, it often remains open whether the 'fault' is with the specific retrograde labeling or with the other available evidence (for more detailed discussion: Cuénod and Streit, 1983). Furthermore, labeling in pathways for which the transmitter has not been defined can be seriously considered as an indication for their transmitter specificity and should stimulate further investigations by means of different methodological approaches.

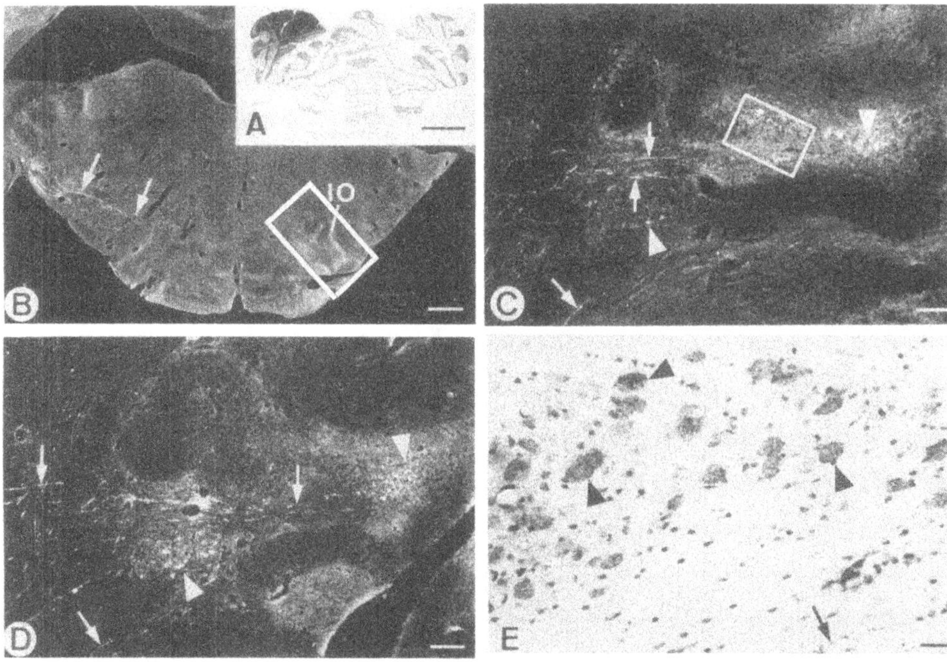


Figure 1A-E: Autoradiographic labeling patterns following injection of D-[$^3$H]aspartate in a monkey cerebellum. Selective retrograde labeling of olivocerebellar projection. A: Injection site. B: Retrogradely labeled cell bodies in inferior olive (IO) and axons of olivocerebellar pathway entering cerebellar peduncle (arrows). Frame: area in D. C,D: Autoradiograms at different levels in IO. Clusters of labeled perikarya (arrowheads) and axons (arrows). Frame in C: area enlarged in E. E: Labeled perikarya (arrowheads) and axons (arrow). Neuropil labeling in surrounding of labeled neurons. Bars: 5 mm in A, 1 mm in B, 200 μm in C and D, 30 μm in E. (Matute et al., 1987; reproduced with permission)

Figure 2A-E (next page): Immunohistochemical labeling patterns in semithin sections from perfusion-fixed brain tissue. A: Staining pattern in part of rat cerebral cortex obtained with monoclonal 'anti-GABA' antibody. Labeled terminal-like elements surrounding perikaryon of pyramidal neuron in layer 5 (arrowheads). Immunoreactive cell bodies of interneurons (arrows). B,C: Patterns of immunoreactivity obtained with monoclonal 'anti-glutamate' antibody in pigeon optic tectum ipsi- (B) and contralateral (C) to retinal ablation. B: Immunoreactive terminal-like elements, partly arranged in clusters (arrows), numerous in layers IIb(3) and IId(5) containing most retinal afferent terminals. C: Retinal ablation causes drastic loss of glutamate-immunoreactive terminal-like elements in contralateral optic tectum; adult retinotectal projection is completely crossed. Remaining elements, particularly beaded processes (arrowheads), resemble afferents from subtectal nucleus isthmi, pars parvocellularis. D: Pattern of glutamate-like immunoreactivity in rat cerebellar cortex. Labeled parallel fibers and their terminals in molecular layer (m) and immunoreactive mossy fiber terminals (arrowheads) in granule cell layer (g). E: Polyclonal 'anti-homocysteate' antibodies stain glial elements in rat cerebellar cortex. Labeled radial processes (arrows) and varicosities (arrowheads) along major Purkinje cell dendrites in molecular layer represent parts of Bergmann glia. Homocysteate-immunoreactive elements also surround Purkinje cell perikarya (P). Immunoreactive dots, probably astrocytic endfeet, in granule cell layer - partly close to blood vessels. Bars: 25 μm in A, B, D; 50 μm in E; same magnification in B and C.

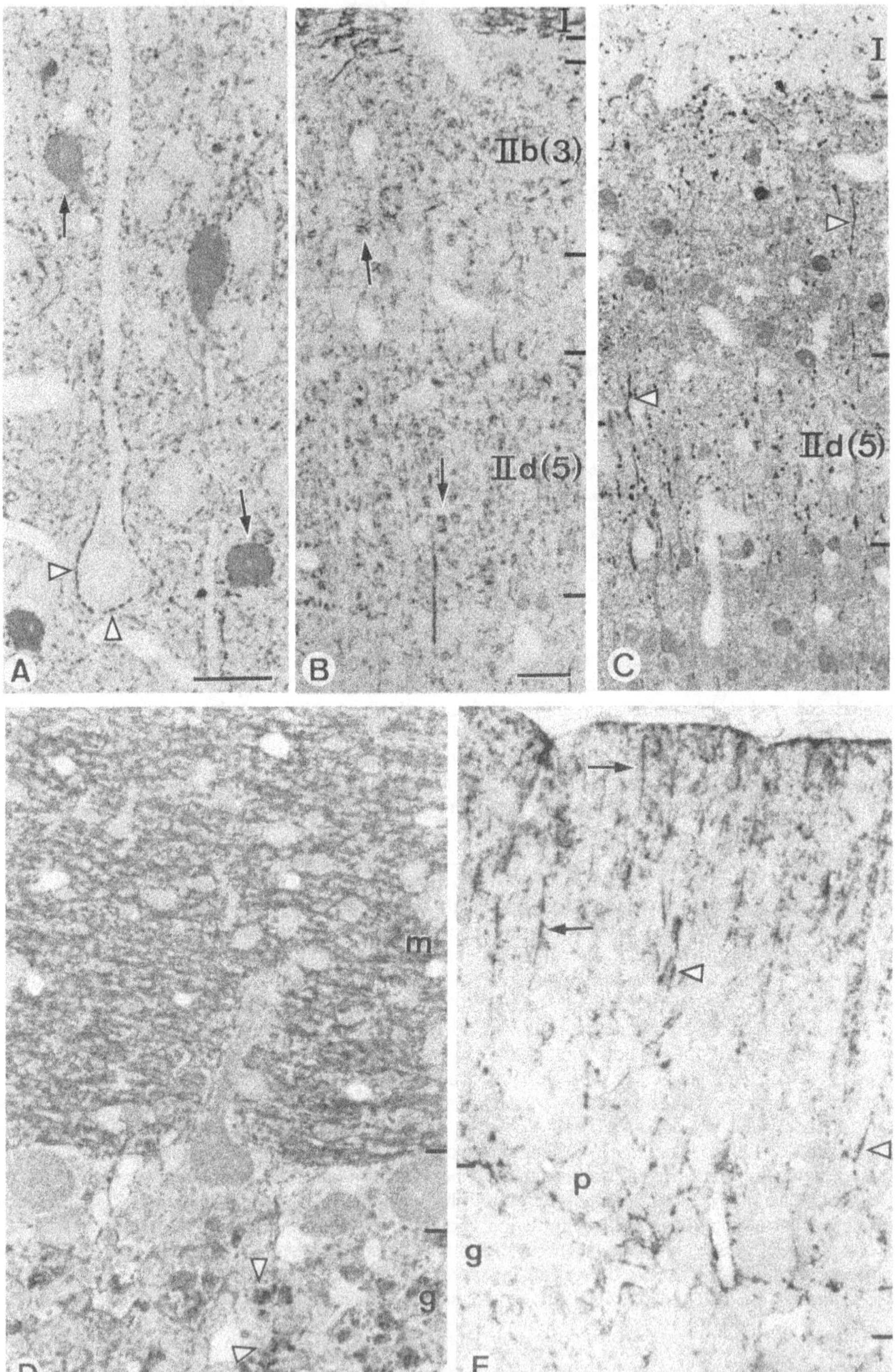

I
IIb(3)
IId(5)
I
IId(5)
m
g
p
g
A
B
C
D
E

Of all the transmitter-related compounds used for retrograde tracing, D-[$^3$H]Asp has probably yielded the most interesting information. Most importantly, this information has expanded our knowledge on presumed excitatory amino acid connections, a system which is notoriously difficult to define by histochemical methods. Indeed, the immunohistochemical localization of enzymes involved in the formation of glutamate and aspartate such as aspartate aminotransferase and phosphate-activated glutaminase failed to define the key enzymes responsible for the transmitter pools of these amino acids (for review: Ottersen and Storm-Mathisen, 1986; Wenthold and Altschuler, 1986). Moreover, because of their involvement in various processes of general cellular metabolism (Fonnum, 1984), the immunohistochemical localization of glutamate- or aspartate-like immunoreactivities in neuronal perikarya does not seem to be a useful marker for transmitter specificity. In fact, relatively high levels of glutamate-like immunoreactivity have been shown in cell bodies of cholinergic and dopaminergic neurons of the brain stem (Ottersen and Storm-Mathisen, 1984, 1985; Storm-Mathisen and Ottersen, 1986). Retrograde labeling with D-[$^3$H]Asp as tracer, therefore, has become a very useful method to point out pathways that may use glutamate and/or aspartate as transmitter(s). Since excitatory amino acids are generally believed to play a decisive role in excitatory chemical neurotransmission in the vertebrate CNS it will be important to obtain detailed and comprehensive information on the neural elements using such compounds as transmitters.

## Development and immunohistochemical application of antibodies to localize endogenous neuroactive amino acids

The recent literature in the field of chemical neuroanatomy has made it clear that established and putative neurotransmitters even if they are low molecular weight substances can be localized at least as immunoreactivities by means of immunohistochemical methods (for review: Ottersen and Storm-Mathisen, 1984, 1987a,b; Geffard et al., 1985; Ottersen et al., 1986; Steinbusch et al., 1986; Storm-Mathisen and Ottersen, 1986; Ottersen, 1989b); the potential of this method has already been indicated by the pioneering immunohistochemical work of Steinbusch and colleagues (1978) on serotonin and of Storm-Mathisen and collaborators (1983) on GABA and glutamate. The methodology to localize amino acids has been reviewed by Ottersen and collaborators (1986), Ottersen and Storm-Mathisen (1987a) and Ottersen (1989b). Our group has mainly concentrated, on the one hand, on the development and characterization of monoclonal antibodies to GABA and glutamate and of poly- and monoclonal antibodies to homocysteate, and, on the other hand, on the immunohistochemical localization of glutamate and homocysteate in neural tissue.

Development and characterization of antibodies: Monoclonal antibodies are defined reagents for which the specificity has to be ensured by selecting appropriate cell lines by means of differential screening tests. Inherent to the use of polyclonal antibodies , on the other hand, there are some general problems and limitations such as low or variable titers, lack of reproducibility when a new animal is used, limited supply as well as antibody heterogeneity. We have, therefore, made efforts to produce monoclonal antibodies to various amino acids, including GABA, glutamate and homocysteate. For immunization and for selection tests, antigen preparations should be chosen that are as closely as possible related to the fixative-linked amino acid of interest in the tissue sample. This similarity seems important particularly in the case of monoclonal antibodies because of their limited range of crossreactivity with antigenic determinants. To fix amino acids in tissue for histological processing, glutaraldehyde is commonly used and is expected to lead to the formation of complex, heterogeneous and mostly unknown reaction products (see Ottersen et al., 1986). The problem of matching antigen preparations and reaction products in tissue may account for the observation that monoclonal antibodies recognizing the antigen preparation in the selection procedure may not yield immunohistochemical staining patterns. Those monoclonal antibodies demonstrating a certain immunoreactivity in tissue, on the other hand, have proved to be reagents which are by no means less reliable than polyclonal antibodies, at least in the case of our monoclonal antibodies to GABA (Matute and Streit, 1986) and glutamate (Liu et al., 1989). Thus, it seems that our antigen preparations contain antigenic determinants that are similar enough to those present at the sites relevant for the localization of immunoreactivity for a certain amino acid in tissue. The antigen preparations, the immunization schedules, the hybridoma production and the selection procedure involving an ELISA system have been described in detail by Matute and Streit (1986) and by Liu and collaborators (1989).

For the characterization of our antibodies we have applied antibody dilution tests and, more importantly, absorption tests performed as ELISA or as immunohistochemical assays. Recently, however, we have also adopted the test system introduced by Ottersen (1987, 1989a,b) that is used in conjunction with our light microscopic immunohistochemical staining procedure involving the PAP technique and silver intensification of the diaminobenzidine reaction product. From the results obtained in these various test systems, one can estimate the level of a monoclonal antibody's crossreactivity with other and, most importantly, with closely related compounds. However, it is still not clear which level of crossreactivity is acceptable for an immunohistochemical reagent. This issue is of particular relevance since a compound may reach very high concentrations in certain tissue elements and, thus, might be recognized by crossreacting antibodies. The levels of crossreactivity have to be low if the issue of colocalization is to be addressed. Monoclonal antibodies exhibit often distinctly different patterns and levels of crossreactivities. Thus, the use of a panel of monoclonal antibodies raised against a certain fixative-linked amino acid may become a convenient and reliable way to decide whether the patterns of immunoreactivities represent the sites where the amino acid of interest is localized or the sites where other compounds are revealed by crossreactivity.

In the course of our attempts to develop monoclonal antibodies to localize amino acids by immunohistochemistry, we have found that for reasons which we have not determined yet this task is easy in the case of substances such as GABA and taurine but much more difficult in the case of compounds such as glutamate and, in particular, homocysteate. In fact, for the localization of homocysteate-like immunoreactivity, we have obtained our first immunohistochemical results only after changing to the development of polyclonal mouse antibodies. These polyclonal antibodies have been raised as ascitic fluids (Lacy and Voss, 1986), thus allowing us obtain relatively large quantities of antibodies in mice and to adopt the immunohistochemical procedure established in our laboratory for monoclonal mouse antibodies to GABA and glutamate (Liu et al., 1989). Even without prior purification, the polyclonal antibodies to homocysteate show a high level of specificity. Thus, in the test system introduced by Ottersen (1987), immunoreactivity is found with these antibodies only for a glutaraldehyde-linked amino acid-protein conjugate containing HCA amongst 25 conjugates assayed.

Immunohistochemical localization: The application of various types of 'anti-GABA' antibodies has made it clear that synaptic terminals containing GABA-like immunoreactivity can easily be visualized even at the level of light microscopy (Fig. 2A), a point that can also be noted with our monoclonal 'anti-GABA' antibodies (Matute and Streit, 1986; Meyer et al., 1986; Braun et al., 1988; Domenici et al., 1988; Bagnoli et al., 1989; Hendry et al., 1989; Liu et al., 1989; Streit et al., 1989). In the case of glutamate, the localization of synaptic terminals containing glutamate-like immunoreactivity (Glu-LI) has proved to be a more difficult task. However, such terminals have indeed been described at the level of electron microscopy in several neural systems (Somogyi et al., 1986 .; Ottersen, 1987, 1989a,b; De Biasi and Rustioni, 1988; Montero and Wenthold, 1989; Liu et al., 1989). Furthermore, the distribution of glutamate-immunoreactive (Glu-ir) terminal-like elements has been analyzed in rinsed and immersion fixed slices (Storm-Mathisen et al., 1983; Ottersen and Storm-Mathisen, 1984; Storm-Mathisen and Ottersen, 1986) as well as in perfusion-fixed tissue (Liu et al., 1989; Grandes and Streit, 1989; Yingcharoen et al., 1989; Walberg et al., 1990) from various regions of the vertebrate central nervous system.

In our group, we have focused on the development of a light microscopic method to map Glu-ir terminal-like elements in neural tissue. We use monoclonal antibodies to glutaraldehyde-linked glutamate in combination with a sensitive postembedding staining procedure for semithin sections from animals rapidly perfusion-fixed with aldehydes (Liu et al., 1989). We have tested this method in several systems. Thus, the pattern of Glu-LI in the rat cerebellar cortex (Fig. 2D) characterized by labeling of parallel fibers and mossy fiber terminals seems to be in line with other information on a transmitter role of Glu in these excitatory systems (Liu et al., 1989; Garthwaite and Brodbelt, 1990). In the hippocampal formation, the pattern of Glu-LI is characterized by large numbers of immunoreactive terminal-like elements in the CA1 stratum radiatum and the dentate molecular layer as well as by labeling of mossy fiber terminals (Liu et al., 1989). The perforant path is generally accepted as one of the best-documented examples of a pathway using Glu as transmitter (Liu et al., 1989; Ottersen and Storm-Mathisen, 1984, 1986). Transections of the perforant path result in the virtual disappearance of Glu-ir puncta in the parts of the dentate molecular layer, related in extent and topography to the lesion (Grandes and Streit, 1988; Grandes and Streit, in prep.). Punctate Glu-LI, thus, is dependent on the presence of the glutamatergic input system. Glu-immunoreactive terminals in the pigeon optic tectum (Fig. 2B,C) are mostly of retinal origin as determined by lesion experiments and also by electron microscopic immunostaining (Bahro et al., 1988; Morino et al., in prep.). This finding is in line with previous indications for the use of an excitatory amino acid (EAA), particularly Glu, in the pigeon retino-tectal pathway (Cuénod and Streit, 1983). Intriguingly, terminations of the Ipc-tectal pathway are Glu-ir particularly in tecta lacking retinal afferents. As discussed in the beginning, the Ipc-tectal pathway has been considered as a glycinergic projection not the least because of the retrograde labeling with tritiated glycine, a result stimulating the development of the method of trasnmitter-related retrograde labeling. Further studies will be necessary to determine the relationship between the results obtained with the two different methods. Nevertheless, the patterns of Glu-LI described are largely compatible with information on neuronal systems using Glu as transmitter in these brain areas. However, studies on the localization of Glu-ir terminals are still relatively rare and, thus, should be extended to additional neural systems and pathways in which indications for a transmitter role of glutamate have been obtained. This undertaking seems to be particularly important since glutamate is generally considered as the major excitatory transmitter in the vertebrate central nervous system.

Only recently, the potent EAA homocysteate (HCA) has been proposed as transmitter candidate (Do et al., 1986b). Indeed, L-HCA activates EAA receptors, preferentially the N-methyl-D-aspartate (NMDA) type wherever present (Mewett et al., 1983; Mayer et al., 1984; Do et al., 1986b; Knöpfel et al., 1987; Thomson, 1989) and it is released in a $Ca^{2+}$-dependent manner by depolarization of brain slices (Do et al., 1986a,b; Vollenweider et al., 1990). To localize homocysteate-like immunoreactivity (HCA-LI), we have used the polyclonal antibodies mentioned above in combination with the light microscopic immunohistochemical staining procedure developed for the localization of Glu-LI (Liu et al., 1989; Streit et al., 1991). In recent immunohistochemical material from rat cerebellar cortex (Fig. 2E), HCA-LI is found in the molecular layer most frequently as fibrous elements with a radial orientation, but varicosities can also be seen in close association with large Purkinje cell dendrites. Large immunoreactive dots are observed at the level of Purkinje cell perikarya. HCA-immunoreactive (HCA-ir) dots of various sizes are also present in the surrounding of blood vessels, in the granule cell layer and in the white matter. The pattern of HCA-LI can be explained most easily by the presence of immunoreactive material in astrocytic endfeet and in parts of Bergmann glial cells (Streit et al., 1991; Morino et al., 1990). Very recently, the same type of staining patterns have also been obtained with a panel of newly developed monoclonal 'anti-HCA' antibodies. The use of alternating ultrathin and semithin sections, the former for electron microscopy and the latter for immunostaining, have allowed us to definitively identify immunoreactive profiles as glial elements. In the molecular layer, thus, varicose immunoreactive profiles in association with large Purkinje cell dendrites can be defined as parts of the glial sheath enclosing these dendrites. Furthermore, the fibrous immunoreactive elements show the typical characteristics of Bergmann glia. Further evidence for a glial localization of HCA-LI has been obtained in the pigeon cerebellar cortex. In contrast to the rat, however, HCA-LI is not only present in Bergmann glial processes but also in cell bodies. In the optic tectum, moreover, astrocytic endfeet as well as cell bodies are HCA-ir. HCA-LI in the hippocampal formation vary in intensity between different animals. In the hippocampal formation, HCA-ir dots are most densely arranged in the infragranular zone of the dentate gyrus, are numerous between the cell bodies of the dentate granule cell layer and of the CA3 pyramidal cell layer and frequent in the dentate molecular layer, in the surrounding of blood vessels and in the outer part of the CA1 stratum radiatum. This pattern of HCA-LI seems to correspond to labeling of certain astrocytic endfeet. Very rarely, however, HCA-LI has also been localized in some neuronal perikarya in the pyramidal layer of the CA1 and CA3 regions (Grandes et al., 1990). Very recently, the pattern of HCA-LI has been studied in the rat retina (Ortega et al., 1990). In this tissue, immunoreactive perikarya resembling those of bipolar cells are regularly observed. This finding is consistent with the suggestion of a transmitter role for HCA in bipolar cells (Neal and Cunningham, 1989).

Although some neuronal localization of HCA-LI has been noticed, it seems to be found predominantly in glial elements. In view of this surprising finding, the following two hypotheses should be considered: (A) HCA (or possibly a precursor) originally contained in neurons is released, and the compound is then taken up by glial elements, where HCA is finally localized as immunoreactivity. The neuronal compound could have been released in the beginning of perfusion-fixation possibly by aldehydes at a low tissue concentration. Following efficient uptake into glial endfeet and processes, it would be fixed by the increasing aldehyde concentration in a somewhat later phase of perfusion-fixation. Preliminary results obtained in experiments on tissue slices indicated that Bergmann glia and hippocampal astrocytes take up HCA as revealed by immunohistochemistry. Occasional HCA-ir neuronal perikarya in hippocampus and retina suggest that certain neurons contain HCA. According to this hypothesis, however, HCA containing neurons would be particularly sensitive to a releasing effect of perfusion-fixation. (B) HCA is synthesized and stored by glial cells from which it can be released. The localization of HCA-LI in parts of Bergmann glial processes and in astrocytic endfeet but not in the entire cell nor in all astrocytic cells is difficult to explain at present. The $Ca^{2+}$-dependency of release as determined in experiments on brain slices (Do et al., 1986a,b; Vollenweider et al., 1990) is generally taken as an indication for a release from a neuronal compartment. A glial release of HCA could have been induced via functional EAA receptors on glial cells (Usowicz et al., 1989) or in exchange for an EAA such as glutamate released in a $Ca^{2+}$-dependent manner from a neuronal compartment and taken up into the glia. It should be mentioned that the HCA release from cerebellar slices is completely abolished in animals in which the olivocerebellar climbing fiber pathway has been destroyed by treatment with 3-acetylpyridine (3-AP; Vollenweider et al., 1990). HCA-LI in glial elements, however, is unaffected in 3-AP treated animals. These findings indicate a special relationship between climbing fibers and HCA release. Furthermore, they seem to strengthen the hypothesis postulating a glial origin of HCA. In the rat cerebellar cortex, HCA release from glial cells may be under the control of climbing fiber activity. In view of our observations in hippocampus and retina, however, the possibility of both a glial and a neuronal localization cannot be excluded at present for HCA in certain parts of the central nervous system. Further investigations are needed to decide between these hypotheses and to determine the role of the potent excitant HCA in physiological or pathophysiological processes. These studies may, on the other hand, lead to a new concept on excitatory chemical transmission postulating a glial influence on their surrounding by means of neuroactive compounds.

## Conclusions

Transmitter-related retrograde labeling seems to be able to point out neural pathways as well as their transmitter specificity. The dependence of this method on presence and specificity of the uptake systems appears to be its major limitation. Immunohistochemistry represents the method of choice to localize accumulations of amino acids at a high degree of spatial resolution and at a high degree of immunochemical specificity. However, tracing or lesion experiments are needed to determine the relation of labeled elements and neural pathways. The results obtained using these two methods should be considered as important pieces in a puzzle. They have to be supplemented by physiological, pharmacological and neurochemical data to shape our concept on the anatomochemical organization of neural systems.

Acknowledgements The authors thank B. Stierli, L. Heeb, R. Emch and R. Schoeb for technical assistance. Supported by Swiss National Foundation (grant 3.389.86), E. Slack-Gyr Foundation and Proyecto del Gobierno Vasco (grant 8916).

Bagnoli P, Burkhalter A, Streit P, Cuénod M (1983) [$^3$H]-GABA selective retrograde labeling of neurons in the pigeon thalamo-wulst pathway. Arch Ital de Biol 121: 47-53

Bagnoli P, Fontanesi G, Streit P, Domenici L, Alesci R (1989) Changing distribution of GABA-like immunoreactivity in pigeon visual areas during the early posthatching period and effects of retinal removal on tectal GABA-ergic systems. Vis Neurosci 3: 491-508

Bahro M, Born-Kaufmann M, Cuénod M, Streit P (1988) Glutamate-immunoreactive terminals in the pigeon optic tectum. Eur J Neurosci Suppl.1: 53

Baughman RW, Gilbert CD (1980) Aspartate and glutamate as possible neurotransmitters of cells in layer 6 of the visual cortex. Nature 287: 848-850

Baughman RW, Gilbert CD (1981) Aspartate and glutamate as possible neurotransmitters in the visual cortex. J Neurosci 1: 427-439

Beart PM, Summers RJ, Stephenson JA, Cook CJ, Christie MJ (1990) Excitatory amino acid projections to the periaqueductal gray in the rat: A retrograde transport study utilizing D-[$^3$H] aspartate and [$^3$H] GABA. Neuroscience 34: 163-176

Beaudet A, Burkhalter A, Reubi JC, Cuénod M (1981) Selective bidirectional transport of [$^3$H] D-aspartate in the pigeon retino-tectal pathway. Neuroscience 6: 2021-2034

Behzadi G, Kalén P, Parvopassu F, Wiklund L (1990) Afferents to the median raphe nucleus of the rat: retrograde cholera toxin and wheat germ conjugated horseradish peroxidase tracing, and selective D-[$^3$H] aspartate labelling of possible excitatory amino acid inputs. Neuroscience 37: 77-100

Benson CG, Postashner SJ (1990) Retrograde transport of [$^3$H] glycine from the cochlear nucleus to the superior olive in the guinea pig. J Comp Neurol 296: 415-426

Bernays RL, Heeb L, Cuénod M, Streit P (1988) Afferents to the rat red nucleus studied by means of D-[$^3$H] aspartate, [$^3$H] choline and non-selective tracers. Neuroscience 26: 601-619

Björklund A, Hökfelt T (1983, 1984, 1985) Handbook of chemical neuroanatomy. Vol 1-4, Elsevier, Amsterdam

Braun K, Scheich H, Zuschratter W, Heizmann CW, Matute C, Streit P (1988) Postnatal development of parvalbumin-, calbindin- and adult GABA-immunoreactivity in two visual nuclei of zebra finches. Brain Res 475: 205-217

Canzek V, Wolfensberger M, Amsler U, Cuénod M (1981) In vivo release of glutamate and aspartate following optic nerve stimulation as determined by a mass fragmentographic technique. Nature 293: 572-574

Castel MN, Malgouris C, Blanchard JC, Laduron PM (1990) Retrograde axonal transport of neurotensin in the dopaminergic nigrostriatal pathway in the rat. Neuroscience 36: 425-430

Christie MJ, Summers RJ, Stephenson JA, Cook CJ, Beart PM (1987) Excitatory amino acid projections to the nucleus accumbens septi in the rat: A retrograde transport study utilizing D-[$^3$H] aspartate and [$^3$H] GABA. Neuroscience 22: 425-439

Cuénod M, Streit P (1983) Neuronal tracing using retrograde migration of labeled transmitter related compounds In: Björklund A, Hökfelt T (eds) Handbook of chemical neuroanatomy, vol 1, Methods in chemical neuroanatomy. Elsevier, Amsterdam, pp 365-397

De Biasi S, Rustioni A (1988) Glutamate and substance P coexist in primary afferent terminals in the superficial laminae of spinal cord. Proc Natl Acad Sci USA 85: 7820-7824

Do KQ, Mattenberger M, Streit P, Cuénod M (1986a) In vitro release of endogenous excitatory sulfur-containing amino acids from various rat brain regions. J Neurochem 46: 779-786

Do KQ, Herrling PL, Streit P, Turski WA, Cuénod M (1986b) In vitro releaase and electrophysiological effects in situ of homocysteic acid, an endogenous N-methyl-(D)-aspartic acid agonist, in the mammalian striatum. J Neurosci 6: 2226-2234

Domenici L, Waldvogel HJ, Matute C, Streit P (1988) Distribution of GABA-like immunoreactivity in the pigeon brain. Neuroscience 25: 931-950

Fischer BO, Ottersen OP, Storm-Mathisen J (1986) Implantation of D-[$^3$H] aspartate loaded gel particles permits restricted uptake sites for transmitter-selective axonal transport. Exp Brain Res 63: 620-626

Fonnum F (1984) Glutamate: a neurotransmitter in mammalian brain. J Neurochem 42: 1-11

Fosse VM, Heggelund E, Iversen E, Fonnum F (1984) Effects of area 17 ablation on neurotransmitter parameters in efferents to area 18, lateral geniculate body, pulvinar and superior colliculus in the cat. Neurosci Lett 52: 323-328

Fuller TA, Price JL (1988) Putative glutamatergic and/or aspartatergic cells in the main and accessory olfactory bulbs of the rat. J Comp Neurol 276: 209-218

Fuller TA, Russchen FT, Price JL (1987) Sources of presumptive glutamergic/aspartergic afferents to the rat ventral striatopallidal region. J Comp Neurol 258: 317-338

Garthwaite G, Garthwaite J (1988) Electron microscopic autoradiography of D-[$^3$H] aspartate uptake sites in mouse cerebellar slices shows poor labelling of mossy fibre terminals. Brain Res 440: 162-166

Garthwaite J, Brodbelt AR (1990) Glutamate as the principal mossy fibre transmitter in rat cerebellum: pharmacological evidence. Eur J Neurosci 2: 177-180

Geffard M, Henrich-Rock AM, Dulluc J, Seguela P (1985) Antisera against small neurotransmitter-like molecules. Neurochem Int 42: 1-11
Grafstein B, Forman DS (1980) Intracellular transport in neurons. Physiol Rev 60:1167-1283
Grandes P, Streit P (1988) Disappearance of glutamate immunoreactive terminal-like elements from dentate molecular layer following perforant path lesion. Eur J Neurosci Suppl. 1: 54
Grandes P, Streit P (1989) Glutamate-like immunoreactivity in calyces of Held. J Neurocytol 18: 685-693
Grandes P, Cuénod M, Streit P (1990) Immunohistochemical localization of homocysteate in rat hippocampal formation. Eur J Neurosci Suppl 3: 118
Hall K, Brauth SE, Kitt CA (1984) Retrograde transport of [$^3$H] GABA in the striatotegmental system of the pigeon. Brain Res 310: 157-163
Hendry SHC, Jones EG, Emson PC, Lawson DEM, Heizmann CW, Streit P (1989) Two classes of cortical GABA neurons defined by calcium binding protein immunoreactivities. Exp Brain Res 76: 467-472
Hunt SP, Künzle H, Cuénod M (1975) The retrograde transport of amino acids and a glycinergic system within the optic lobe. Exp Brain Res S23: 189
Hunt SP, Streit P, Künzle H, Cuénod M (1977) Characterization of the pigeon isthmo-tectal pathway by selective uptake and retrograde movement of radioactive compounds and by Golgi-like horseradish peroxidase labeling. Brain Res 129: 197-212
Hökfelt T, Ljungdahl AA (1975) Uptake mechanisms as a basis for the histochemical identification and tracing of transmitter-specific neuron populations. In: Cowan WM, Cuénod M (eds) The use of axonal transport for studies of neuronal connectivity. Elsevier, Amsterdam, pp 249-305
Iversen LL, Dick F, Kelly JS, Schon F (1975) Uptake and localization of transmitter amino acids in the nervous system. In: Berl S, Clark DD, Schneider D (eds) Metabolic compartmentation and neurotransmission. Plenum Press, New York, pp 65-87
Jones BE, Beaudet A (1987) Retrograde labeling of neurons in the brain stem following injections of [$^3$H]choline into the forebrain in the rat. Exp Brain Res 65: 437-448
Jones BE, Paré M, Beaudet A (1986) Retrograde labeling of neurons in the brain stem following injections of [$^3$H] choline into the rat spinal cord. Neuroscience 18: 901-916
Kalén P, Karlson M, Wiklund L (1985) Possible excitatory amino acid afferents to nucleus raphé dorsalis of the rat investigated with retrograde wheat germ agglutinin and D-[$^3$H] aspartate tracing. Brain Res 360: 285-297
Kisvarday ZF, Cowey A, Hodgson AJ, Somogyi P (1986) The relationship between GABA immunoreactivity and labelling by local uptake of [$^3$H] GABA in the striate cortex of monkey. Exp Brain Res 62: 89-98
Kisvarday ZF, Martin KAC, Friedlander MJ, Somogyi P (1987) Evidence for interlaminar inhibitory circuits in the striate cortex of the cat. J Comp Neurol 260: 1-19
Knöpfel T, Zeise ML, Cuénod M, Zieglgänsberger W (1987) L-homocysteic acid but not L-glutamate is an endogenous N-methyl-D-aspartic acid receptor preferring agonist in rat neocortical neurons in vitro. Neurosci Lett 81:188-192
Künzle H, Schnyder H (1984) The isthmus-tegmentum complex in the turtle and rat/ A comparative analysis of its interconnections with the optic tectum. Exp Brain Res 56: 509-522
Künzle H, Wiklund L (1982) Identification and distribution of neurons presumed to give rise to cerebellar climbing fibers in turtle. A retrograde axonal flow study using radioactive D-aspartate as a marker. Brain Res 252:146-150
Lacy MJ, Voss EW (1986) A modified method to induce immune polyclonal ascites fluid in BALB/c mice using Sp2/O-Ag14 cells. J Immunol Meth 87:329-339
Laduron PM (1987) Axonal transport of neuroreceptors: possible involvement in long-term memory. Neuroscience 22: 767-779
Liu CJ, Grandes P, Matute C, Cuénod M, Streit P (1989) Glutamate-like immunoreactivity revealed in rat olfactory bulb, hippocampus and cerebellum by monoclonal antibody and sensitive staining method. Histochemistry 90: 427-445
Matute C, Martinez-Millan L (1985) Selective retrograde labeling in some afferents to the rabbit lateral geniculate nucleus following injections of tritiated neurotransmitter-related compounds. Neurosci Lett 53: 9-14

Matute C, Streit P (1985) Selective retrograde labeling with D-[$^3$H] aspartate in afferents to the mammalian superior colliculus. J Comp Neurol 241: 34-49

Matute C, Streit P (1986) Monoclonal antibodies demonstrating GABA-like immunoreactivity. Histochemistry 86: 147-157

Matute C, Wiklund L, Streit P, Cuénod M (1987) Selective retrograde labeling with D-[$^3$H] aspartate in the monkey olivocerebellar projection. Exp Brain Res 66:445-447

Mayer ML, Westbrook GL (1984) Mixed-agonist action of excitatory amino acids on mouse spinal cord neurones under voltage clamp. J Physiol 354: 29-53

Mewett KN, Oakes DJ, Olverman HJ, Smith DAS, Watkins JC (1983) Pharmacology of the excitatory actions of sulphonic and sulphinic amino acids. In: Mandel P, DeFeudis FV (eds) CNS Receptors from Molecular Pharmacology to Behavior. Raven Press, New York, pp 163-174

Meyer EP, Matute C, Streit P, Nässel DR (1986) Insect optic lobe neurons identifiable with monoclonal antibodies to GABA. Histochemistry 84: 207-216

Montero VM, Wenthold RJ (1989) Quantitative immunogold analysis reveals high glutamate levels in retinal and cortical synaptic terminals in the lateral geniculate nucleus of the macaque. Neuroscience 31: 639-647

Morino P, Grandes P, Cuénod M, Streit P (1990) Immunohistochemical localization of homocysteate in developing and adult rat cerebellum. Eur J Neurosci Suppl 3: 178

Neal MJ, Cunningham JR (1989) L-homocysteic acid - a possible bipolar cell transmitter in the rabbit retina. Neurosci Lett 102: 114-119

Ortega F, Grandes P, Hennequet L, Gondra J, Donate-Oliver F, Cuénod M, Streit P (1990) Immunohistochemical localization of HCA-like immunoreactivity in rat retina. Eur J Neurosci Suppl 3:119

Ottersen (1987) Postembedding light- and electron microscopic immunocytochemistry of amino acids: description of a new model system allowing identical conditions for specificity testing and tissue processing. Exp Brain Res 69: 167-174

Ottersen OP (1989a) Quantitative electron microscopic immunocytochemistry of amino acids. Anat Embryol 180: 1-15

Ottersen OP (1989b) Postembedding immunogold labelling of fixed glutamate: an electron microscopic analysis of the relationship between gold particle density and antigen concentration. J Chem Neuroanat 2: 57-66

Ottersen OP, Storm-Mathisen J (1984) Neurons containing or accumulating transmitter amino acids. In: Björklund A, Hökfelt T, Kuhar MJ (eds) Handbook of chemical neuroanatomy, vol 3, Classical transmitters and transmitter receptors in the CNS, part II. Elsevier, Amsterdam, pp 141-246

Ottersen OP, Storm-Mathisen J (1986) Excitatory amino acid pathways in the brain. In: Schwarcz R, Ben-Ari Y (eds) Advances in experimental medicine and biology, Vol 203: Excitatory amino acids and epilepsy. Plenum Press, New York, pp 263-284

Ottersen OP, Storm-Mathisen J (1987a) Localization of amino acid neurotransmitters by immunocytochemistry. TINS 10: 250-255

Ottersen OP, Storm-Mathisen J (1987b) Immunocytochemical visualization of glutamate and aspartate. In: Hicks TP, Lodge D, McLennan H (eds) Excitatory amino acid transmission. Alan R Liss, New York, pp 131-138

Ottersen OP, Storm-Mathisen J, Madsen S, Skumlien S, Strömhaug J (1986) Evaluation of the immunocytochemical method for amino acids. Med Biol 64: 147-158

Reubi JC, Cuénod M (1976) Release of exogenous glycine in the pigeon optic tectum during stimulation of a midbrain nucleus. Brain Res 112: 347-361

Snyder SH, Yamamura HI, Pert CB, Logan WJ, Bennet JP (1973) Neuronal uptake of neurotransmitters and their precursors in studies with "transmitter" amino acids and choline. In: Mandell AJ (ed) New concepts in neurotransmitter regulation. Plenum Press, New York, pp 195-222

Somogyi P, Cowey A, Kisvarday ZF, Freund TF, Szentagothai J (1983) Retrograde transport of gamma-amino [3H] butyric acid reveals specific interlaminar connections in the striate cortex of monkey. Proc Natl Acad Sci USA 80: 2385-2389

Somogyi P, Halasy K, Somogyi J, Storm-Mathisen J, Ottersen OP (1986) Quantification of immunogold labelling reveals enrichment of glutamate in mossy and parallel fibre terminals in cat cerebellum. Neuroscience 19: 1045-1050

Somogyi P, Minson JB, Morilak D, Llewellyn Smith I, McIlhinney JR, Chalmers J (1989) Evidence for an excitatory amino acid pathway in the brainstem and for its involvement in cardiovascular control. Brain Res 496: 401-407

Steinbusch HWM, Verhofstad AA, Joosten HWJ (1978) Localization of serotonin in the central nervous system by immunohistochemistry: description of a specific and sensitive technique and some applications. Neuroscience 3: 811-819

Steinbusch HWM, De Vente J, Schipper J (1986) Antibodies against small neuroactive substances: immunohistochemistry of monoamines in the central nervous system. In: Panula P, Päivärinta H, Soinila S (eds) Neurology and Neurobiology, vol 16, Neurohistochemistry: modern methods and applications. Alan R Liss, New York, pp 75-105

Storm-Mathisen J (1975) High-affinity uptake of GABA in presumed GABAergic nerve endings in rat brain. Brain Res 84: 409-427

Storm-Mathisen J (1976) Distribution of the components of GABA systems in neuronal tissue: Cerebellum and hippocampus - effect of axotomy. In: Roberts E, Chase TN, Tower DB (eds) GABA in nervous system function. Raven Press, New York, pp 149-168

Storm-Mathisen J (1981) Autoradiographic and microchemical localization of high affinity glutamate uptake. In: Roberts PJ, Storm-Mathisen J, Johnston GAR (eds) Glutamate: Transmitter in the Central Nervous System. John Wiley, Chichester, pp 89-115

Storm-Mathisen J, Leknes AK, Bore AT, Vaaland JL, Edminson P, Haug FMS, Ottersen OP (1983) First visualization of glutamate and GABA in neurones by immunocytochemistry. Nature 301: 517-520

Storm-Mathisen J, Ottersen OP (1986) Antibodies against amino acid neurotransmitters. In: Panula P, Päivärinta H, Soinila S (eds) Neurology and Neurobiology, vol 16, Neurohistochemistry: modern methods and applications. Alan R Liss, New York, pp 107-136

Streit P, Knecht E, Cuénod M (1979) Transmitter specific retrograde labeling in the striatonigral and raphe-nigral pathways. Science 205: 306-308

Streit P, Knecht E, Cuénod M (1980) Transmitter-related retrograde labeling in the pigeon optic lobe; a high resolution autoradiographic study. Brain Res 187: 59-67

Streit P (1980) Selective retrograde labeling indicating the transmitter of neuronal pathways. J Comp Neurol 191: 429-463

Streit P, Grandes P, Morino P, Tschopp P, Cuénod M (1991) Immunohistochemical localization of glutamate and homocysteate. In: Moroni F, Meldrum BS, Simon R, Woods JH (eds) Excitatory amino acids 1990. Fidia Research Foundation Symposium Series, Vol 5. Raven Press, New York

Thomson AM (1989) Glycine modulation of the NMDA receptor/channel complex. TINS 12: 349-353

Usowicz MM, Gallo V, Cull-Candy SG (1989) Multiple conductance channels in type-2 cerebellar astrocytes activated by excitatory amino acids. Nature 339: 380-383

Vollenweider FX, Cuénod M, Do KQ (1990) Effect of climbing fiber deprivation on release of endogenous aspartate, glutamate and homocysteate in slices of rat cerebellar hemispheres and vermis. J Neurochem 54: 1533-1540

Walberg F, Ottersen OP, Rinvik E (1990) GABA, glycine, aspartate, glutamate and taurine in the vestibular nuclei: an immunocytochemical investigation in the cat. Exp Brain Res 79: 547-563

Watanabe K, Kawana E (1984) Selective retrograde transport of tritiated D-aspartate from the olfactory bulb to the anterior olfactory nucleus, pyriform cortex and nucleus of the lateral olfactory tract in the rat. Brain Res 296: 148-151

Weiss D (1982) Axoplasmic transport. Raven Press, New York

Wenthold RJ, Altschuler RA (1986) Immunocytochemical localization of enzymes involved in the metabolism of excitatory amino acids. In: Roberts PJ, Storm-Mathisen J, Bradford HF (eds) Excitatory amino acids. Macmillan, London, pp 85-101

Wiklund L, Toggenburger G, Cuénod M (1982) Aspartate: possible neurotransmitter in cerebellar climbing fibers. Science 216: 78-80

Wiklund L, Künzle H, Cuénod M (1983) Failure to demonstrate retrograde labeling of cerebellar Purkinje cells after injection of [$^3$H] GABA in Deiter's nucleus. Neurosci Lett 38: 23-28

Wiklund L, Toggenburger G, Cuénod M (1984) Selective retrograde labelling of the rat olivocerebellar climbing fiber system with D-[$^3$H] aspartate. Neuroscience 13: 441-468

Wiklund L, Behzadi G, Kalén P, Headley PM, Nicolopoulos LS, Parsons CG, West DC (1988) Autoradiographic and electrophysiological evidence for excitatory amino acid transmission in the periaqueductal gray projection to the nucleus raphé magnus in the rat. Neurosci Lett 93: 158-163

Wolfensberger M, Reubi JC, Canzek V, Redweik U, Curtius HC, Cuénod M (1981) Mass fragmentographic determination of endogenous glycine and glutamic acid released in vivo from the pigeon optic tectum. Effect of electric stimulation of a midbrain nucleus. Brain Res 224: 327-336

Yingcharoen K, Rinvik E, Storm-Mathisen J, Ottersen OP (1989) GABA, glycine, glutamate, aspartate and taurine in the perihypoglossal nuclei: an immunocytochemical investigation in the cat with particular reference to the issue of amino acid colocalization. Exp Brain Res 78: 345-357

# ULTRASTRUCTURAL IDENTIFICATION AND QUANTIFICATION OF NEUROACTIVE AMINO ACIDS

Ole P. Ottersen
Department of Anatomy
Institute of Basic Medical Sciences
University of Oslo
P.O.B. 1105 Blindern
N-0317 Oslo 3, Norway

## INTRODUCTION

The growing realization that amino acids may be responsible for fast signalling in a large majority of the synapses in the central nervous system (Fonnum 1984, Iversen and Goodman 1986), has stimulated efforts to elucidate the roles of these amino acids in normal as well as in pathologically altered synaptic transmission. Students of neuroactive amino acids are, however, confronted with the problem that the amino acids in question are also involved in protein synthesis and diverse metabolic processes, as is the case in non-neural tissues. It has proved difficult to distinguish between amino acid pools engaged in synaptic functions, and amino acid pools engaged in other tasks.

The present review deals with recent methodological advances in the field of EM postembedding immunocytochemistry, that have made it possible to analyze the cellular and subcellular distribution of neuroactive amino acids in a semiquantitative, and, tentatively, in a quantitative manner. We will show how these advances can be exploited in attempts to discriminate between different amino acid pools in brain tissue.

## SPECIFICITY

It is generally accepted that testing of specificity should be

NATO ASI Series, Vol. H 58
Neurocytochemical Methods
Edited by A. Calas and D. Eugène

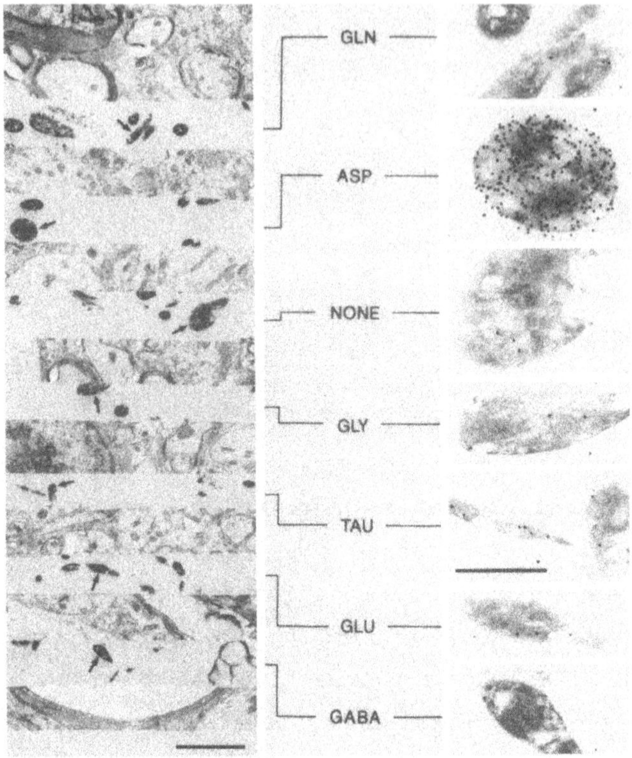


Figure 1. Ultrathin section of test "sandwich" incubated in an antiserum to aspartate. Note selective gold particle labelling of aspartate conjugate. Arrows indicate bodies enlarged in right panel (see Fig. 2 for quantitative analysis). Consult text for further explanation. Bars: 2 um (left), 0.5 um (right). - Taken with permission from Zhang et al. (1990).

carried out under conditions that are identical or closely similar to the conditions of the immunocytochemical procedure. This is because the specificity of an immunocytochemical reaction not only depends on the quality of the antiserum, but also on the incubation parameters, tissue preparation, and many other factors. With respect to postembedding immunogold analysis, it becomes particularly important to determine whether the embedding procedure has altered tissue antigens in such a way as to cause crossreactivities that do not appear under pre-embedding conditions.

Figure 1 shows the type of model system that we routinely use for monitoring the specificity of our postembedding immunogold procedure. The model system consists of a stack of sections containing different amino acid conjugates, alternating with brain sections used merely as spacers (for details, see Ottersen 1987). The different test conjugates form dense clumps that can be regarded as artificial tissue profiles containing a single fixed amino acid. Thus, the preparation of the test conjugates includes the removal, by extensive dialysis, of all small molecules from a total brain extract, followed by glutaraldehyde coupling of a single amino acid to this extract. One would expect that the test antigens that are formed in this way, resemble those formed in the tissue during glutaraldehyde fixation.

Ultrathin sections cut transversely through the stack of alternating conjugate-containing and spacer sections are incubated in the same drops of immunoreagents as the tissue sections, a strategy that should increase further the reliability of the test results. The test section shown in Fig. 1 was incubated with an aspartate antiserum (Zhang et al. 1990), but equally good selectivity has been achieved with our antisera to glutamine, glycine, taurine, glutamate, and GABA (review: Ottersen 1989b). The use of specially designed computer programs (Blackstad et al. 1990) permits the test results to be expressed in a semiquantitative manner (Fig. 2).

## RELATIONSHIP BETWEEN GOLD PARTICLE DENSITY AND ANTIGEN CONCENTRATION

The postembedding immunogold procedure is ideally suited for semiquantitative studies (e.g., Somogyi et al. 1986), inasmuch as the gold particles can be easily detected and counted in electron micrographs. The interpretation of the semiquantitative data would, however, be easier if information was available

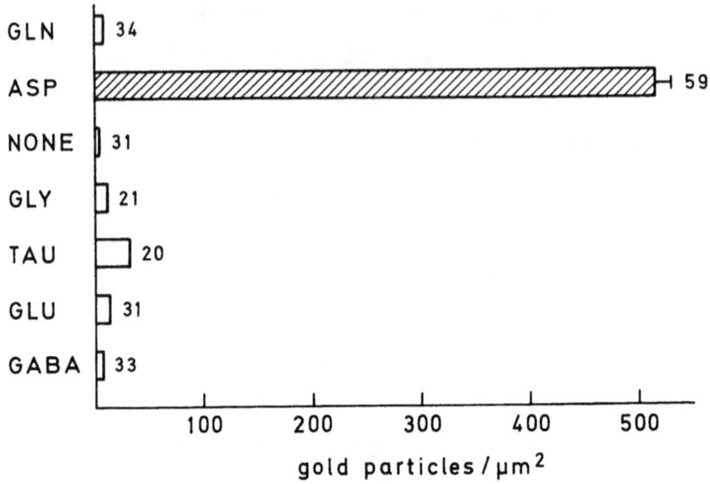


Figure 2. Semiquantitative analysis of test section shown in Fig. 1. S.E.M. (for aspartate) and number of observations are indicated.

concerning the relationship between gold particle density and antigen concentration. To explore this issue, we designed a model system similar to that used for specificity testing in Fig. 1, but with different concentrations of a single amino acid (Fig. 3). Trace amounts of radiolabeled amino acid were used to calculate the concentration of fixed amino acid in the test conjugates. Initial studies of glutamate conjugates in the range of 8.6mM and above (Fig. 3) revealed a nearly linear relationship between particle density and mM concentration of fixed glutamate (Ottersen 1989a,b). Subsequent studies, employing glutamate conjugates in a lower (and more biologically relevant) concentration range, provided evidence for a non-linear relationship at concentrations below 2mM of fixed glutamate (Bramham et al. 1990).

Based on our model system, we have calculated that the concentration of fixed glutamate in putative glutamatergic terminals in perfusion-fixed tissue is about 20 to 30mM (Ottersen 1989a,b, Storm-Mathisen and Ottersen 1990), i.e., 2-3 times the average tissue concentration of free glutamate. It is possible,

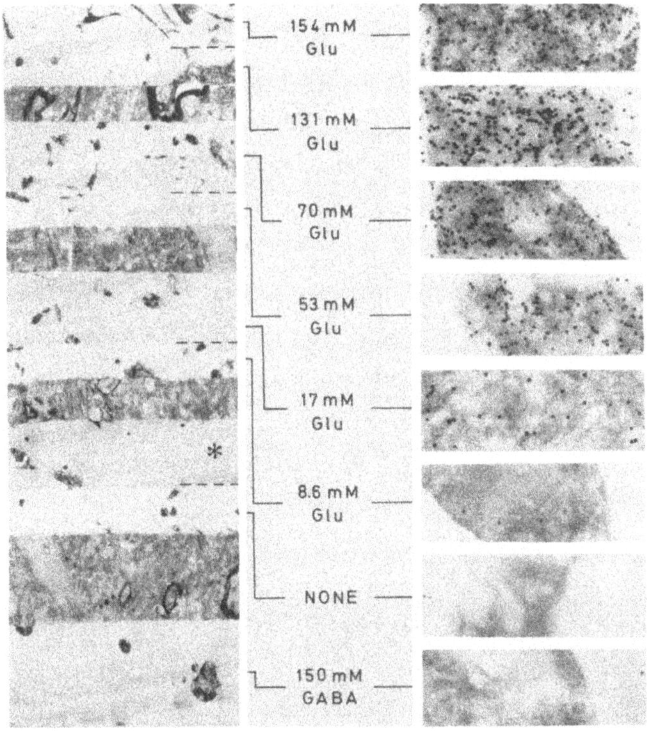


Figure 3. Ultrathin section containing different concentrations of fixed glutamate (NONE = 0 mM glutamate). The section was incubated in glutamate antiserum no. 13. - Taken with permission from Ottersen (1989a), which should be consulted for details.

however, that these values underestimate the real *in vivo* concentration, since a small proportion of the free amino acids in the brain will inevitably escape fixation during the perfusion procedure (Madsen et al. 1990). It is likely that a greater proportion of amino acid will be lost from compartments where the amino acid occurs in high concentrations, than from compartments with low amino acid concentrations.

To allow extrapolation from the model system to the tissue, the model sections must be incubated in the same drops of immunoreagents as the tissue sections, thus avoiding inaccuracies due to inadvertent fluctuations in the incubation parameters.

Even with this precaution, a precise translation of gold particle density into mM concentration of antigen remains difficult or impossible to achieve, for various reasons discussed in detail previously (Ottersen 1989a,b). We feel, nevertheless, that the present model system is accurate enough to facilitate the interpretation of semiquantitative immunogold data.

The system also gives some clues as to the sensitivity of the immunogold procedure. Thus, concentrations of fixed glutamate and glycine as low as 0.1mM are detected by the procedure used (Bramham et al. 1990, Ottersen et al. 1990a). These values represent about 1% and 10%, respectively, of the average brain tissue concentrations of the two amino acids.

## CELLULAR AND SUBCELLULAR DISTRIBUTION OF NEUROACTIVE AMINO ACIDS

Figure 4 shows examples of glutamate and glycine immunolabeling from the hippocampus and spinal cord, respectively. In the stratum radiatum of the hippocampal field CA1, glutamate-like immunoreactivity is enriched in terminals establishing asymmetric contacts with spines (Bramham et al. 1990, Ottersen et al. 1990b, also see Liu et al. 1989). Most of the terminals of this type are known to belong to the Schaffer collateral/commissural system, which, on the basis of extensive neurochemical, neurophysiological and pharmacological investigations, has been claimed to be excitatory and to depend on an excitatory amino acid (Cotman et al. 1987). Motor neurons, on the other hand, are known to be under glycinergic inhibitory control, and the high gold particle density in Fig. 1B probably identifies one of these glycinergic terminals (t1). Note that the adjacent terminal (t2) contains only scattered gold particles, which is likely to reflect background labelling plus labelling due to the presence of small amounts of metabolic glycine. Judged from these pictures, one would be tempted to assume that amino acid-using terminals can be identified on the basis of high amino acid

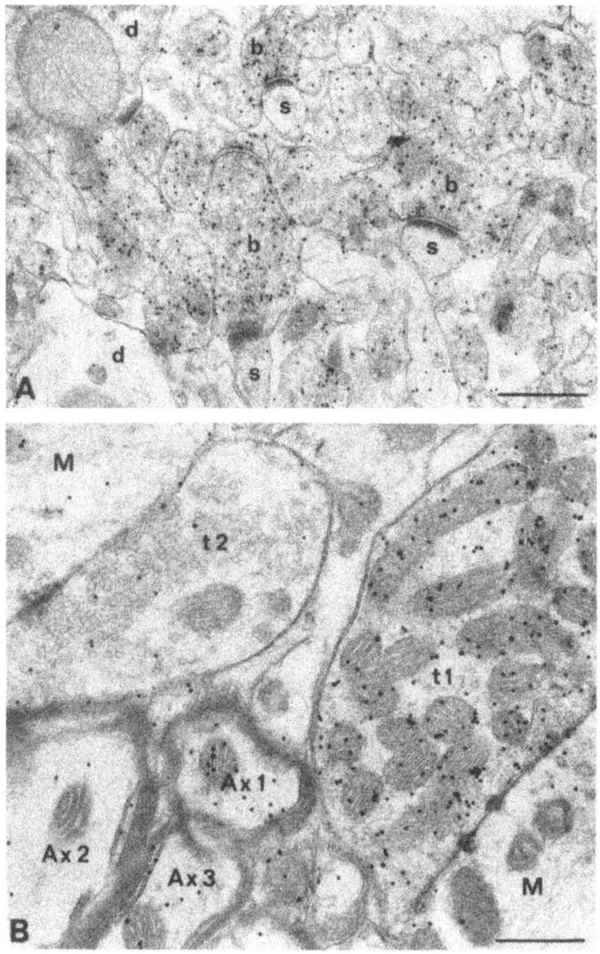


Figure 4. A: Glutamate immunolabelling of nerve terminals (b) in stratum radiatum of CA1 (rat hippocampal slice incubated in physiological $K^+$ concentration and immersion fixed). s, postsynaptic spines; d, dendrites. B: Glycine immunolabelling in nerve terminal (t1) contacting a motor neuron (M) in the ventral horn of the rat spinal cord. t2, weakly labelled terminal; Ax1-3, myelinated axons. Bars: 0.6 um (A); 0.3 um (B).

concentrations alone. However, one cannot exclude that the levels of metabolic glutamate and glycine show considerable variation among neurons.

It is therefore fortunate that further clues as to the distinction between metabolic and transmitter pools can be gained

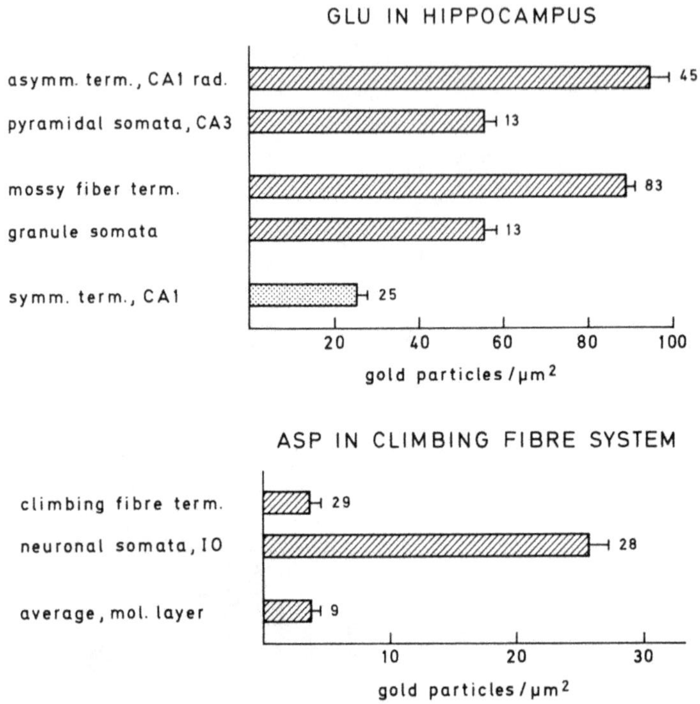


Figure 5. Semiquantitative analysis of immunogold preparations from rat hippocampus (glutamate immunolabelling; A) and rat climbing fibre system (aspartate immunolabelling; B). Perfusion fixed material. S.E.M. and number of observations are indicated. Stippled column represents terminals establishing symmetric contacts with cell bodies and dendritic stems in CA1 (most of these terminals are probably GABAergic). Bottom column indicates average gold particle density in the molecular layer of the cerebellum. IO, inferior olive (source of climbing fibres).

by analyzing the subcellular distribution of amino acids. For example, the finding that terminals of Schaffer collateral/ commissural fibers and of mossy fibers in the hippocampus show an enrichment of glutamate-like immunoreactivity relative to their parent cell bodies (CA3 pyramidal somata and granule somata, respectively; Fig. 5) strengthens the notion that glutamate is a transmitter in these terminals. In contrast, our observation of a higher level of aspartate-like immunoreactivity in the parent cell bodies than in the terminals of the cerebellar climbing fibre system (Fig. 5) is difficult to reconcile with a transmitter role of aspartate in these fibres. Rather, the latter data suggest that aspartate only serves metabolic roles in these neurons (Zhang et al. 1990).

Evidently, pronounced amino acid gradients may exist within individual neurons. In other words, the level of amino acids in the nerve terminals cannot be predicted by an assessment of the labelling intensity in the parent cell bodies. This implies that one should avoid drawing conclusions about a neuron's transmitter identity on the basis of the cell body labelling alone.

The high resolution of the postembedding immunogold procedure also permits an analysis of the distribution of amino acids within the individual nerve terminals. Studies of this kind have revealed a close correlation between the level of glutamate-like immunoreactivity and vesicle density in certain excitatory terminals (Montero and Wenthold 1989, Ottersen et al. 1990b, Ji et al. 1991). This is shown in Fig. 6 for spinocerebellar mossy fiber terminals, identified by anterograde transport of horseradish peroxidase-wheat germ agglutinin.

The strong association of glutamate-like immunoreactivity to synaptic vesicles naturally speaks in favour of glutamate being a transmitter in these terminals. The possibility that this strong correlation is artifactual and merely reflects a preferential retention of amino acids to the synaptic vesicles is excluded since the level of glutamine-like immunoreactivity in the same terminals is not similarly correlated to synaptic vesicle density (Ji et al. 1991). Glutamine is a precursor of glutamate, and is assumed to be entirely extravesicular.

## COMBINATION OF IMMUNOCYTOCHEMISTRY AND RELEASE STUDIES

If the nerve terminal labelling for an amino acid indeed reflects a transmitter pool, it should be possible to abolish or strongly

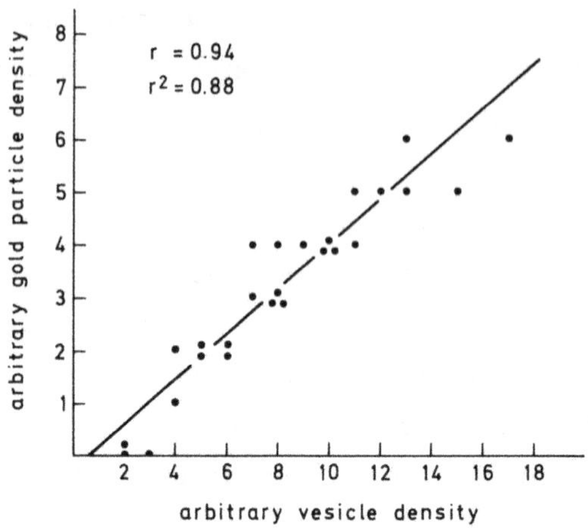


Figure 6. Correlation between gold particle density and synaptic vesicle density (arbitrary values) in identified spinocerebellar mossy fiber terminals following immunolabelling with a glutamate antiserum. The correlation and determination coefficients are given. (For detailed procedure, see Ji et al. 1991).

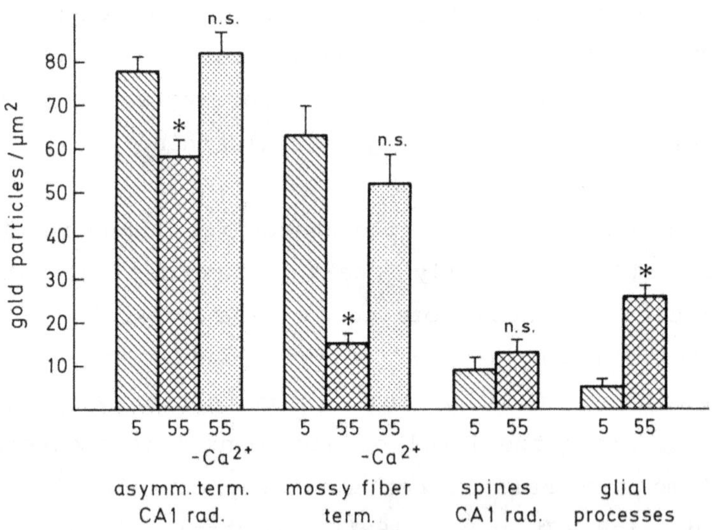


Figure 7. High $K^+$-induced redistribution of glutamate-like immunoreactivity in slices of rat hippocampus. Values below columns indicate the mM concentration of $K^+$ in the incubation medium. * Significantly different from first value in each group ($P<0.001$; Student's t-test). - Taken with permission from Ottersen et al. 1990b.

reduce the immunolabelling by experimental procedures that evoke synaptic release. Based on this line of reasoning we incubated slices of hippocampus in high concentrations of $K^+$. This led to a reduction in the immunolabelling intensity in all major categories of excitatory fiber terminals in the hippocampus, although different categories differed in their sensitivity to depolarization (Fig. 7). Corresponding results have been obtained in the cerebellum (Ottersen et al. 1990c).

The depolarization-induced decrease in immunolabelling at the level of the nerve terminals was associated with an increased immunolabelling intensity in the glial cells (Fig. 7), probably reflecting uptake of glutamate in the latter cells.

Matching the exocytotic nature of the release process, the $K^+$-induced depletion of glutamate was abolished by omission of $Ca^{2+}$ and addition of $Mg^{2+}$ to the medium (Fig. 7).

## CONCLUSION

We have presently shown that postembedding immunogold procedures allow us to study, in a semiquantitative manner, the cellular and subcellular distribution of neuroactive amino acids in tissue fixed while in a physiological state as well as in tissue fixed after experimental manipulations *in vitro*.

Studies of this kind serve as valuable guides for distinguishing between different amino acid pools and have proved fruitful in investigations aimed at studying the synaptic handling of the different neuroactive amino acids in preparations with intact microstructure.

## REFERENCES

Blackstad TW, Karagülle T, Ottersen OP (1990) MORFOREL, a computer program for two-dimensional analysis of micrographs of biological specimens, with emphasis on immunogold preparations. Comput Biol Med 20:15-34

Bramham CR, Torp R, Zhang N, Storm-Mathisen J, Ottersen OP (1990) Distribution of glutamate-like immunoreactivity in excitatory hippocampal pathways: a semiquantitative electron microscopic study in rats. Neuroscience. In press

Cotman CW, Monaghan DT, Ottersen OP, Storm-Mathisen J (1987) Anatomical organization of excitatory amino acid receptors and their pathways. Trends Neurosci 10:273-280

Fonnum F (1984) Glutamate: a transmitter in mammalian brain. J Neurochem 42:1-11

Iversen LL, Goodman EC (eds) (1986) Fast and slow chemical signalling in the nervous system. Oxford University Press, Oxford New York Tokyo

Ji Z, Aas J-E, Laake J, Walberg F, Ottersen OP (1991) Subcellular distribution of fixed glutamate and glutamine in a class of presumed excitatory terminals - the spinocerebellar mossy fiber boutons. A combined anterograde transport-immunogold study in rat with reference to the distribution of fixed aspartate, GABA, and taurine. Submitted

Liu CJ, Grandes P, Matute C, Cuénod M, Streit P (1989) Glutamate-like immunoreactivity revealed in rat olfactory bulb, hippocampus and cerebellum by monoclonal antibody and sensitive staining method. Histochemistry 90:427-445

Madsen S, Ottersen OP, Storm-Mathisen J, Sturman JA (1990) Immunocytochemical localization of taurine: methodological aspects. In: H. Pasantes-Morales et al (eds) Taurine: Functional neurochemistry, physiology and cardiology. Wiley-Liss

Montero VM, Wenthold RJ (1989) Quantitative immunogold analysis reveals high glutamate levels in retinal and cortical synaptic terminals in the lateral geniculate nucleus of the macaque. Neuroscience 31:639-647

Ottersen OP (1987) Postembedding light- and electron microscopic immunocytochemistry of amino acids: description of a new model system allowing identical conditions for specificity testing and tissue processing. Exp Brain Res 69:167-174

Ottersen OP (1989a) Postembedding immunogold labelling of fixed glutamate: an electron microscopic analysis of the relationship between gold particle density and antigen concentration. J Chem Neuroanat 2:57-66

Ottersen OP (1989b) Quantitative electron microscopic immunocytochemistry of neuroactive amino acids. Anat Embryol 180:1-15

Ottersen OP, Storm-Mathisen J, Laake JH (1990a) Cellular and subcellular localization of glycine studied by quantitative electron microscopic immunocytochemistry. In: OP Ottersen, J Storm-Mathisen (eds) Glycine neurotransmission. Wiley, Chichester

Ottersen OP, Storm-Mathisen J, Bramham C, Torp R, Laake J, Gundersen V (1990b) A quantitative electron microscopic immunocytochemical study of the distribution and synaptic handling of glutamate in rat hippocampus. In: J Storm-Mathisen, J Zimmer, OP Ottersen (eds) Understanding the brain through the hippocampus. Progr Brain Res, Vol 83, Elsevier, Amsterdam

Ottersen OP, Laake JH, Storm-Mathisen J (1990c) Demonstration of a releasable pool of glutamate in cerebellar mossy and parallel fibre terminals by means of light and electron microscopic immunocytochemistry. Arch Ital Biol 128:111-125

Somogyi P, Halasy K, Somogyi J, Storm-Mathisen J, Ottersen OP (1986) Quantification of immunogold labelling reveals enrichment of glutamate in mossy and parallel fibre terminals in cat cerebellum. Neuroscience 19:1045-1050

Storm-Mathisen J, Ottersen OP (1990) Immunocytochemistry of glutamate at the synaptic level. J Histochem Cytochem. In press

Zhang N, Walberg F, Laake JH, Meldrum BS, Ottersen OP (1990) Aspartate-like and glutamate-like immunoreactivities in the inferior olive and climbing fibre system: a light microscopic and semiquantitative electron microscopic study in rat and baboon (*PAPIO ANUBIS*). Neuroscience 38:61-80

# VIDEO-MICROSCOPIC TECHNIQUES TO STUDY AXONAL TRANSPORT, NEURONAL FINE STRUCTURE AND CYTOCHEMICAL PARAMETERS IN LIVE NEURONS

Dieter G. Weiss
Institut für Zoologie
Technische Universität München
Lichtenbergstr. 4
D-8046 Garching
Germany

In the early 1980s the developments in microelectronics and computer design opened up the possibility to digitize and process microscope images in "real time", i.e. at video rates. This led to a variety of ways to dramatically improve the quality of microscopic images and to create new applications of light microscopy. This so called "electronic revolution in light microscopy" allows the additional application of optical techniques and the use of living cells in many cases where previously only electron microscopy of dehydrated material could be used. The basic principles of the new techniques and the type of equipment required, as well as some applications in neurobiology are discussed here.

The most unexpected and surprising improvements of image quality were obtained with the introduction of video-enhanced contrast (VEC) microscopy. It enables one to increase contrast and magnification to such an extent that positions and movements of biological objects smaller than 1/10 of the limit of resolution of light microscopy, e.g. synaptic vesicles and microtubules can be visualized and analysed in the living state. Organelle motion and microtubule assembly were quantitatively described by a number of parameters such as velocity, straightness of path, reversals of direction, and others.

The second group of video-microscopic techniques is based on the visualization and quantitation of weak fluorescence images obtained by monochromatic illumination light microscopy (video-intensified microscopy, VIM). The resulting images are two-dimensional arrays of fluorescence measurements and contain information on the amounts and distribution of intracellular metabolites, dyes, antigens, or exogenous agents. Typical parameters for VIM measurements include $Ca^{2+}$-concentration, pH value, metabolites, and membrane potential. Specific non-toxic dyes are also available to verify the identity of the organelles seen by the VEC techniques and to quantitate their position and abundancy.

NATO ASI Series, Vol. H 58
Neurocytochemical Methods
Edited by A. Calas and D. Eugène

The different video microscopy techniques can be applied as a whole battery of assays at selected time intervals to a given set of cultured neurons, vibratome sections or freshly dissected nerve segments to obtain simultaneous measurements of multiple parameters. But, since quantitative data for these parameters can be calculated from the live images in real time and encoded in the form of gray-shaded or pseudo-colour images, they can also be continuously recorded and yield video films of the intracellular events during and after exposure to physiological stimuli or pharmacological agents. Furthermore, video microscopy allows multiparametric studies to be performed especially with cultured cells which yield a wealth of information that could not be reached in the past. It is therefore to be expected that video microscopy will rapidly find increasing application especially in the neurosciences such as in neuro-physiology, -pharmacology, -toxicology, and experimental neuropathology (Stirling 1986; Weiss et al 1989).

Instrumentation

The electronic equipment required for both types of video microscopy is essentially similar. It needs to be considered that also the microscope may require modifications to allow the use of the new techniques to their full extent. Working at extremely high magnifications for VEC microscopy requires a massive, stable microscope stand, in some cases a vibration isolation table, a very bright illumination system and optical magnification of 2x or 4x in addition to a 100x oil immersion objective (NA 1.25 or better) (Fig. 1). For VEC microscopy one needs a geometrically distortion-free video camera (Chalnicon, Newvicon or Pasecon) or highest precision CCD (charge-coupled device) camera. The cameras need to have externally adjustable offset (black level) and gain to allow for analogue contrast enhancement. Automatic gain control (AGC), if present, must be switched off. For VIM a low light level camera is required such as a silicon intensified target camera (SIT or ISIT) or a microchannel plate-intensified device (see also Aikens et al 1989; Weiss et al 1989) with a light sensivity at the limit of the human vision or below.

After analogue contrast enhancement (Fig. 2) one obtains a high contrast image even for very low contrast or low light objects (Fig. 3a,b) This image is then digitized (usually 512x512 picture elements (pixels) with 256 possible grey levels each) and subjected to various kinds of arithmetic operations in the frame memory of the digital real time processor. For VEC microscopy one needs frame averaging to remove pixel noise and all fast moving objects from the video scenes. Continuous background subtraction removes unwanted fixed pattern noise (mottle, see Fig. 3c-e) or, if required, non moving image details (Lichtscheidl and Weiss 1988). Digital contrast enhancement (stretching of the grey level histogram) is usually required after image subtractions (Fig. 3f). For the dual wavelength fluorescence ratio imaging techniques (e.g. $Ca^{2+}$ imaging with FURA2, see Connor and Müller, this volume) an image processor capable of fast division of full video frames is required.

After the suitable images have been derived in real time by video-microscopic image processing and were recorded, they may be subjected to classical image analysis in order to obtain fine morphological parameters. It should be noted that this is usually not done in real-time and would require the long-known typical image analysis systems (Anonymous 1985; Häder

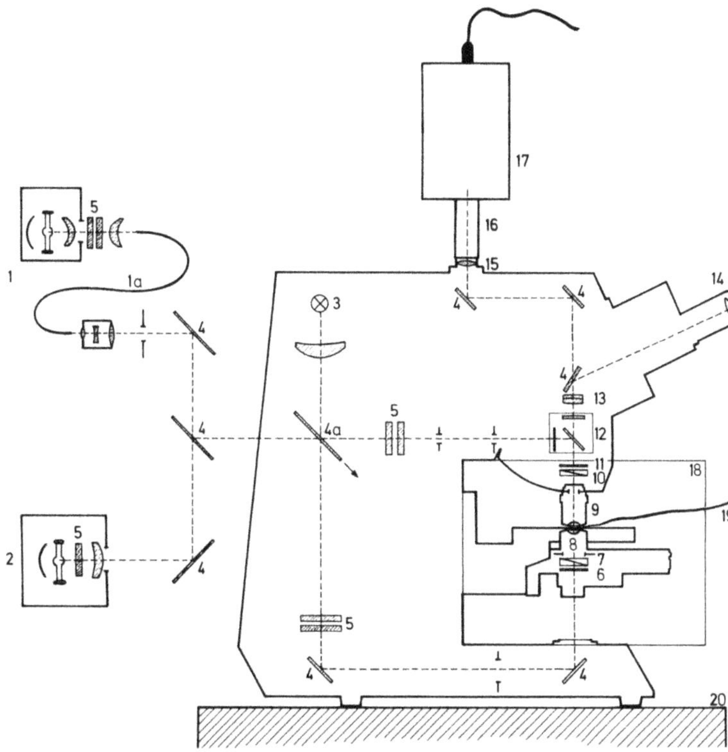


Fig. 1 The video microscope. Schematic representation of a microscope with the modifications required for video-intensified microscopy (VIM) and video-enhanced contrast (VEC) microscopy. 1 Mercury arc lamp with light scrambler (according to Ellis 1985); 1a quartz optical fiber; 1,2 mercury arc lamps (for fluorescence and high magnification differential interference contrast (DIC) microscopy); 3 halogen lamp; 4 mirrors; 4a removable mirror for switching lamps; 5 UV, IR (heat), and green filters; 6 polarizer; 7,10 Wollaston prisms for DIC; 8 high N.A. oil immersion condenser and objective (9); 11 analyser; 12 incident light fluorescence filter cube; 13 additional magnification lens (2x); 14 eyepiece; 15 and 16 projective lens and extension tube, or high magnification eyepiece (16x or 25x) with camera objective (50 or 63 mm); 17 video camera (chalnicon or SIT type); 18 temperature-controlled chamber; 19 thermistor; 20 vibration isolation table. This setup is based on a Polyvar microscope (Reichert/Cambridge Instruments, Vienna). The parts of special importance for video microscopy are nos. 1, 1a, 2, 5, 8, 9, 13, 15, 16, and sometimes 20. From Weiss (1987).

1988; Swedenburg and Conklin 1988; Bradbury 1989), whereas for video microscopy, as is discussed here, we need real-time image processors (Weiss et al 1989; Herman and Jacobson 1990). More details on the equipment required and the detailed steps of the recommended procedures were compiled by Inoué (1986), Weiss et al. (1989), and Spring (1990).

Video-enhanced contrast (VEC) microscopy

When finest details are to be visualized with bright-field, DIC, or polarization optics, then VEC microscopy (Inoué 1986; Shotton 1988; Weiss et al 1989) and especially the variation of it introduced by R.D. Allen (AVEC microscopy) (Allen et al 1981a 1981b; Allen and Allen 1983) should be applied. The required procedure is described in detail (Allen and Allen 1983; Weiss 1986; Weiss et al 1989). As a result resolution of objects of about 150 nm and visualization of even smaller biological objects such as microtubules and vesicles down to 20 nm is achieved with visible light (Fig. 4) and in situations where the limit of resolution of conventional microscopy is about 250nm (Mizushima 1988; Weiss et al 1989; Inoué 1990). Collodial gold particles which are used as tags for antibody molecules can be visualized even in the 5-10 nm range (De Brabander et al 1988).

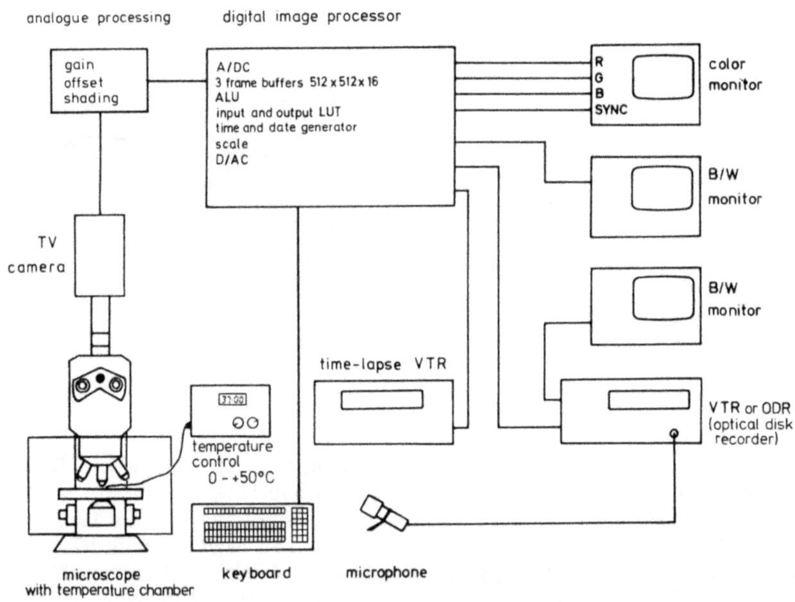


Fig. 2 The video microscopy setup consists of the microscope with camera, an analogue and a digital image processor, the viewing and the recording equipment. Color monitor, raw image monitor and time lapse recorder are not essential but very useful. VTR: video tape recorder, ODR: optical disk recorder, pro image: processed image, A/DC: analog to digital converter, ALU: arithmetic logic unit, LUT: look up table, D/AC: digital to analog converter. RGB: color video signal. From Weiss (1987)

With VEC microscopy we can see for the first time all classes of organelles and even some cytoskeletal filaments (Hayden and Allen 1984; Allen 1985; Weiss 1986; Forscher et al 1987; Travis and Bowser 1988). It should be mentioned that these details become visible only after image processing, but remain invisible when one looks down the microscope or at enlarged photomicrographs. For publication purposes pictures have to be photographed off the video screen (for recommended procedures see Weiss et al 1989). Typical examples of cytoplasmic organelles as seen with AVEC-DIC microscopy are depicted in Fig. 5b,c.

Video-intensified microscopy (VIM)

The imaging of the very weak signals from self-luminous objects typical for fluorescence, luminescence, or dark-field microscopy requires the use of low light level cameras. The technical specifications of low light level cameras and further information on the procedures to be followed and the equipment needed can be found elsewhere (Inoué 1986; Wick 1987; Marcus 1988; Shotton 1988; Aikens et al 1989; Spring and Lowy 1989; Wampler 1989; Weiss et al 1989; Herman and Jacobson 1990; Spring 1990). The fluorescence signals which are of use

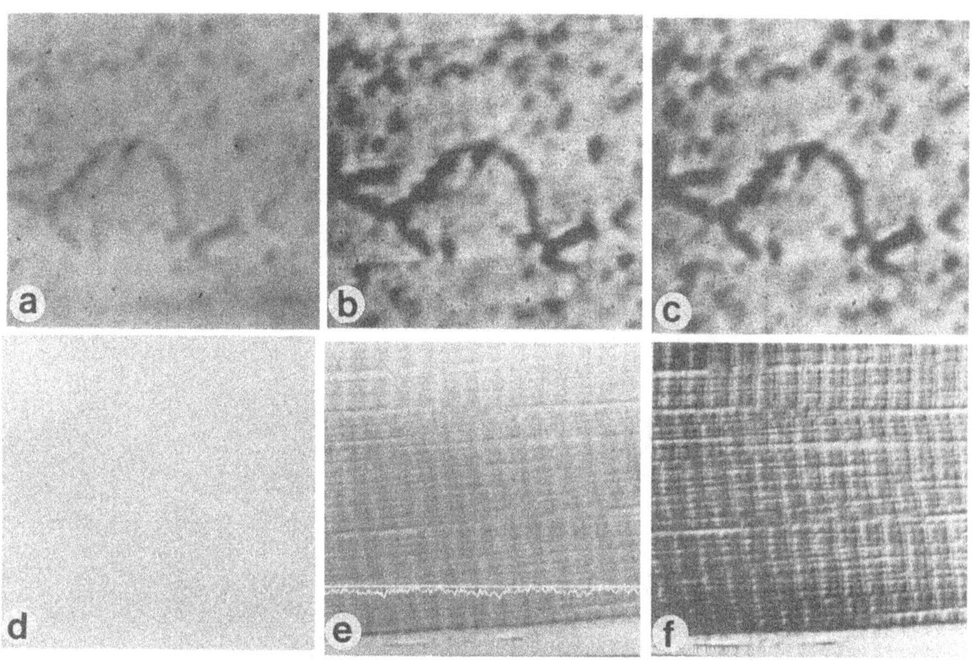


Fig. 3 VEC microscopy of a very weak specimen. Specimen is an unstained electron microscope thin section of striated muscle viewed in DIC. Microscope: Zeiss IM 405, Plan Neofluar x63, NA 1.4; processor: Hamamatsu ARGUS 100; frame width: 42 µm. a. in focus, not enhanced; b. in focus, analogue-enhanced; c. out of focus, with mottle; d. out of focus, mottle subtracted; e. in focus, mottle subtracted; f. digitally enhanced. From Weiss et al. 1989

for neurobiology studies may originate from endogenous compounds (chromophores, cytochromes) from site-specific marker dyes (Fig. 6a,b) (DeBiasio et al 1987; Luby-Phelps et al 1988; Haugland 1989), from exogenously added fluorescent substrates of intracellular enzymes (c.f. Miska and Geiger 1987; Van Noorden 1988), from fluorescent chelators indicating the concentration of intracellular ions (Tsien 1989; Bright et al 1989), or from fluorescently labeled proteins or antibodies added to the cells or organelles under study (Arndt-Jovin et al 1985; Glascott et al 1987; Sammak and Borisy 1988; Harada and Yanagida 1988; McKenna et al 1989; Wang 1989).

A wide variety of specific vital dyes are available as molecular probes which enter the living cell and do not influence cell functions for at least several hours (Haugland 1989). There are dyes which can be used to monitor the concentration of all biologically relevant kations as well as the membrane potential, so that these can be quantitativly determined within living cells (Blasdel and Salama 1986; Grinvald et al 1986; Giuliano, et al. 1990; Loew et al 1990; Prpic et al 1990 ).

Applications in neurobiology

At present the most rapidly expanding application of video microscopy in neurobiology is the use of fluorescent ion chelators. These specially designed dyes cross the plasma membrane in

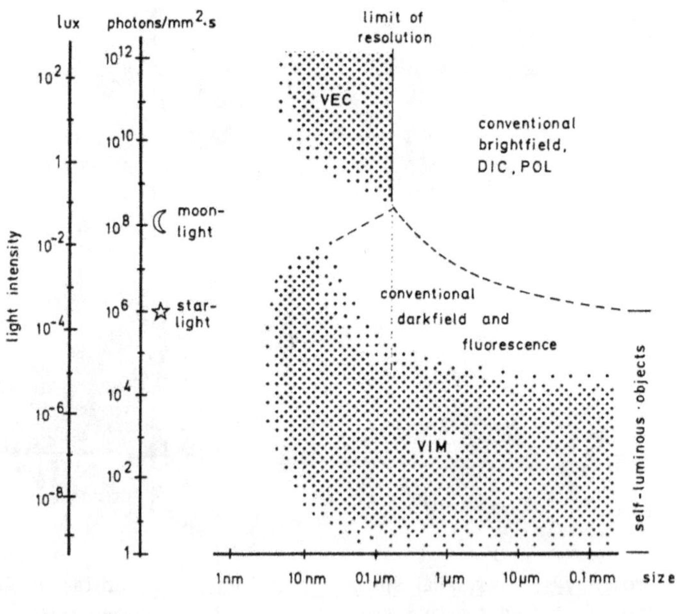


Fig. 4 Video microscopy extends the limits of light microscopy. New applications are opened (dotted areas) both in low light situations (VIM) and when working with very small objects (VEC microscopy). All borders are meant to be approximative. From Weiss et al. (1989)

the form of the acetoxymethyl esters and are trapped in the cell after intracellular cleavage of the ester bond (see however Almers and Neher 1985). The concentration of a specific ion can be calculated from the extent of the shift of the fluorescence spectrum or from the fluorescence intensity (Bright et al 1989; Tsien 1989; Giuliano et al 1990; Prpic et al 1990). The image processors can be programmed to perform automatically the required filter changes and to display the ion concentration in the form of color-coded images. The measurements of $Ca^{2+}$-concentration and of the pH value appear to be the most important ones in both nerve cell physiology and neuropharmacology or toxicology (Bright et al 1987 1989; Forscher et al 1987; Kater et al 1988; Tsien 1989; see also chapter by Connor and Müller this volume).

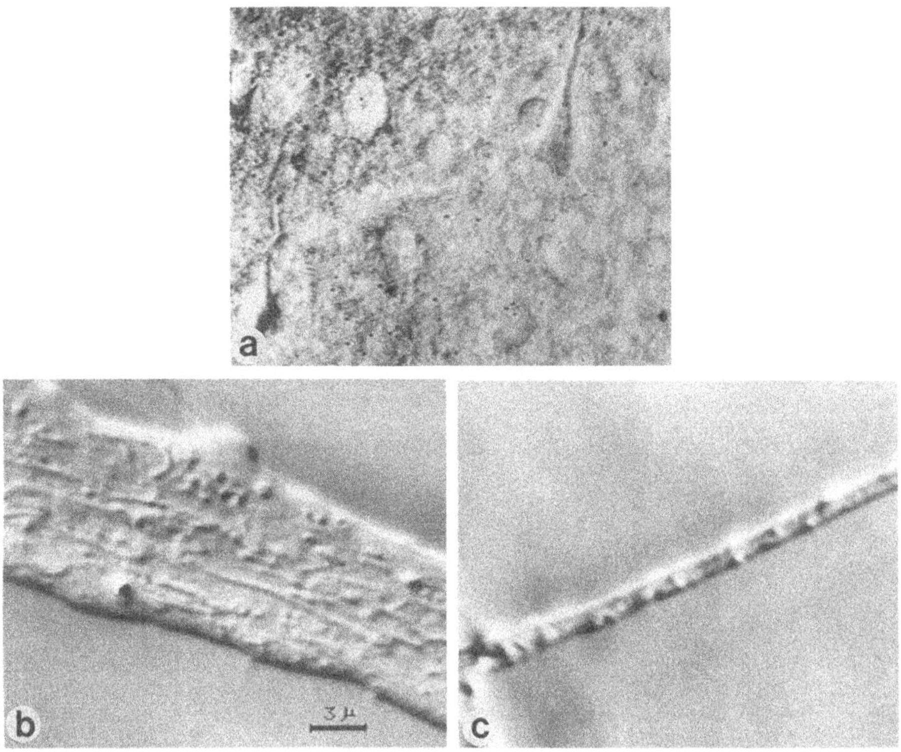


Fig. 5 Video-enhanced contrast microscopy of neurons. a. Live pyramidal cell in a 300 μm thick brain slice. Infrared differential interference contrast VEC microscopy (wavelength 800nm). Zeiss Axiovert, 63x Plan Neofluar NA 1.4, 2.5x optovar; Hamamatsu C2400 Newvicon camera with analogue contrast enhancement only. Width of field 150 μm. Reproduced with permission from Dodt and Zieglgänsberger (1990). b. Neurite of a N18 neuroblastoma cell culture. Most of the organelles move either anterogradely or retrogradely with velocities of 0.5-2 μm/s. AVEC-DIC; Polyvar microscope, Reichert (Vienna Austria), 100x Planapo, NA 1.3, 4x additional magnification; processor: Hamamatsu C1966 Photonic Microscope System with Chalnicon camera. c. Axonal transport of organelles in cultured primary neurons from mouse cerebral cortex. Technical data as in b.

The membrane potential-sensitive dyes can be applied both in vivo (Blasdel and Salama 1986) to monitor e.g. the electric activity of the cortex or in vitro to follow the activity along the circuitry of neuronal networks in slices or cultured neuronal arrays (Grinwald et al 1986; Wu et al 1988). The greatest advantage is the simultaneous recording capability when neuronal arrays of low or intermediate complexity are used. It should be mentioned, however, that recordings over many hours or days are not possible due to the long-term toxicity of these dyes and due to phototoxic effects which may develop after prolonged illumination even at low intensity and when shutters are used to illuminate the specimen only intermittently while

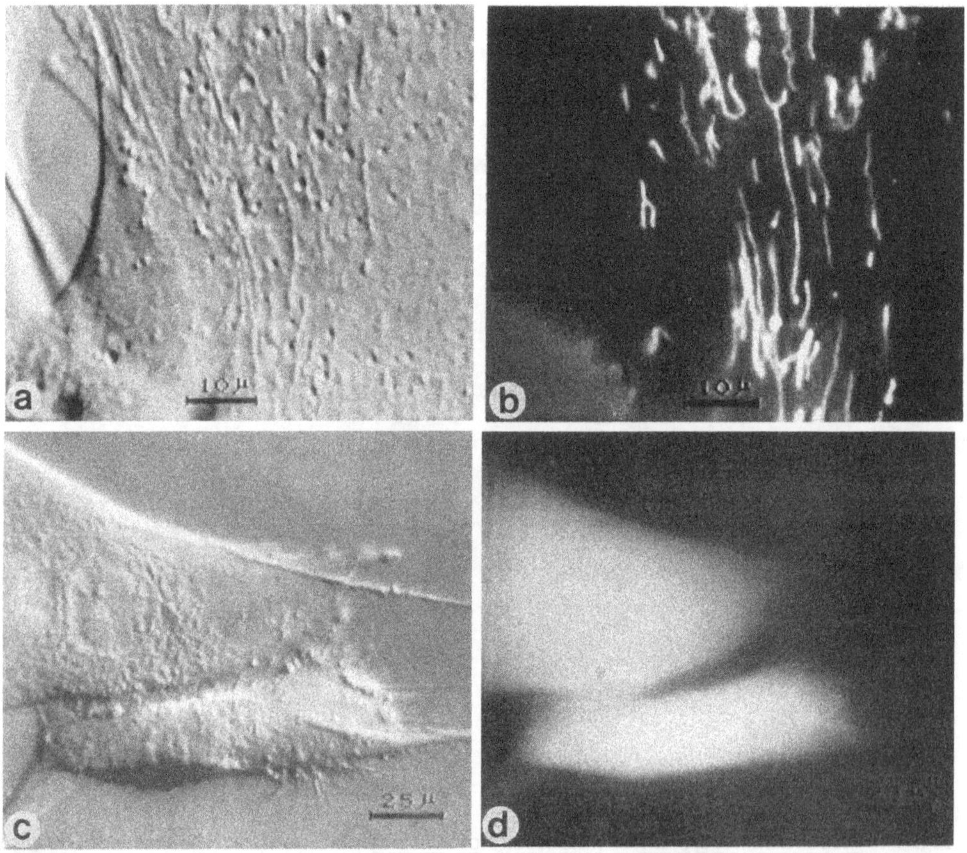


Fig. 6 Combined application of VIM and VEC microscopy a,b. Organelle-specific fluorescence staining. a. Mitochondria and other, spherical organelles in the cytoplasm of a neuroblastoma cell (N18). AVEC-DIC. Technical data as in Fig. 5b. b. Same region after vital staining for mitochondria with rhodamine123 (VIM). All elongated organelles in a. proved to be positive for the mitochondria-specific dye. c,d. Determination of cell thickness as a prerequisite for quantitative intracellular fluorescence cytochemistry. AVEC-DIC image of fresh cultured cells c. and the corresponding fluorescence VIM image after vital staining with fluorescein diacetate d. to image the cytoplasmic volume or optical pathlength. Neuroblastoma cell (N18). From Weiss (1987)

recording. In these cases where long term recording is required multiple microelectrode chips should be used and electrically recorded (Gross and Kowalski 1990).

Similar to the application of the vital dyes, fluorescence-labeled antibodies can be used to detect the location and amount of all kinds of antigens. These are applied after the above observations to fixed cells to give information on e.g. the state of the cytoskeleton in treated cells. Observations on live cells are, however, also possible if microinjection is performed (Hayden et al 1983; De Brabander et al 1988; Sammak and Borisy 1988). The use of 5nm collodial gold-coupled antibodies which is possible with video imaging processing yields non-bleaching images of higher spatial resolution than fluorescence immunocytochemistry (De Brabander et al 1990).

Fluorescence video micrographs are especially well-suited for quantitative evaluation of fine-morphological parameters. Although little work has been done on neurons, typical morphometric parameters can be derived if a suitable software package is applied to images stained with organelle-specific vital dyes for endoplasmic reticulum (ER) networks (Terasaki et al 1984; Lee and Chen 1988), for the filamentous arrays of cytoskeletal elements and mitochondria (Fig. 6b) (Johnson et al 1980) or for spherical organelles such as lysosomes or golgi apparatus (Pagano et al 1989).

Visualization of unstained live neurons with great detail is also possible in brain slices by infrared DIC-VEC microscopy (Fig. 5a) (Dodt and Zieglgänsberger 1990). In infrared light microscopy brain slices are much less opaque than with shorter wavelength light and axons or dendrites of large neurons can be easily seen. This allows correct positioning of micropipettes for patch clamp recording or topical application of pharmacological agents, antibodies or dyes.

The possibility to directly observe organelle motility in neurons and its relation to the cytoskeleton has led to a revolution in axonal transport research (Allen et al 1982 1985; Brady et al 1982) and subsequently to the discovery of the molecular motors of axonal transport kinesin and cytoplasmic dynein (Vale et al 1985; Vallee et al 1989). Axonal transport of organelles can be studied in olfactory axons that are so thin that they could not be resolved by conventional light microscopy (Weiss and Buchner 1988) or in larger myelinated and unmyelinated axons (Allen et al 1985; Seitz-Tutter et al 1988; Weiss et al 1988a,b; Llinas et al 1989; Weiss et al 1990a,b). Studies of changes in transport to and from regenerating axons and growth cones are equally feasible (Kater et al 1988; Goldberg and Burmeister 1989) as the specific staining and morphological monitoring over months of neuro-muscular endplates (Yoshikami and Okun 1984; Lichtman et al 1987; Herrera and Banner 1990).

Several laboratories have developed powerful soft- and hardware to track moving organelles and to subject the resulting time series of positional data to detailed motion analysis (Cohn and Scholey 1987; Gelles et al 1988; Weiss et al 1990b). Fluorescence labelled or negatively charged latex beads are used with great advantage as tools in the field of axonal transport for both cell biological (Allen et al 1985; Gilbert and Sloboda 1986; Gelles et al. 1988) and for fiber-tracing applications (Katz et al 1984; Holländer et al 1989; Härtig et al 1990). In combination with VIM and VEC microscopy latex beads of 20 nm or less obtainable with encapsulated rhodamine or fluorescein (e.g. from B. Paulke, Inst. Polymer Chemie AdW, D-01530 Teltow-Seehof) can be used with great advantage as these show very little bleaching, no diffusion from the injection site, no toxicity, very low tendency to aggregate and high uptake for retrograde tracing (Härtig et al 1990).

Outlining individually stained neurons is also beautifully achieved by microinjecting membrane potential-sensitive dyes such as DiI and DiO (Honig and Hume 1989). The advantage of video-microscopic technology is that less light is required or that even very weak processes can be visualized. Also, the digitized images can be subjected to advanced 3D reconstruction algorithms that process full grey-tone images (Inoué and Inoué 1987).

Towards video-microscopic measurements in cytochemistry

The transfer function of the video cameras used for quantitative microscopy is linear instead of logarithmic as for the human eye and the photographic process (Inoué 1986). The gray-level information in video images can therefore be directly used to gain information on the relative amounts of fluorescent substances. Absolute values can be obtained by the use of a series of calibrated solutions which are imaged at the same instrumental setting as the unknown samples or by making cells permeable for the external medium containing a known dye concentration (Grynkiewicz et al 1985).

If information on intracellular concentrations is desired we need to know in addition to the intensity value the optical pathlength, i.e. the thickness of the cell at each image point (or pixel). This can be obtained by imaging cells loaded with fluorescent dyes which have a homogeneous and exclusively intracellular distribution, such as fluorescein diacetate (Fig. 6c,d) (Möncke et al 1985) or fluorescein-labelled dextrans (Luby-Phelps and Taylor 1988; Giuliano et al 1990). In this case the image brightness codes for cell thickness in a quantitative way. Storing fluorescence intensity or absorption images obtained with monochromatic light at a suitable wavelength of high absorption and dividing these into the cell thickness (or optical pathlength) images, yields "concentration images". This can be done in principle in real-time (i.e. in video frequency of 25 or 30 frames per second) so that concentration changes over time or space can be recorded on video tape.

In the case of the $Ca^{2+}$ chelator FURA2 and some pH indicators, the binding of the ions causes a concentration-dependent spectral shift of the chelator dye fluorescence. By dividing the two images obtained at the two different wavelengths one obtains the image containing the concentration information (ratio imaging) by the use of a calibration curve (ratio values of a standard series plotted vs. concentration) (see Connor and Müller, this volume; Grynkiewicz et al 1985; Bright et al 1989). This procedure is independent of the optical pathlength so that no knowledge of the cell thickness is required. Besides the concentration of $Ca^{2+}$ also $H^+$ (pH-value), $Mg^{2+}$, $Cl^-$, $K^+$ and $Na^+$ concentrations can be determined with suitable dyes (DeBiasio et al 1987; Bright et al 1987, 1989; Haugland 1989; Giuliano et al. 1990)

Similar to the above-mentioned extraction of coordinates from series of video frames, one can also extract intensity values of individual image points or small image regions from such series, so that concentration changes in individual cells or in subcellular regions can be followed over time (temporal analysis) (Wade et al 1986; Weiss et al 1989; Prpic et al. 1990). This makes studies on intracellular transfer, intracellular exchange and metabolism of compounds possible.

These techniques can also be used for regional and temporal assays of activity of selected enzymes which consume or synthesise fluorescent compounds (Van Noorden 1988). Similarly enzymatic substrate measurements of lactate, glucose and ATP have been developed for tissue sections. The cells have to be permeabilized (e.g. frozen and thawed) and exposed to or microinjected with the coupled enzyme system containing luciferase and suitable bioluminescent substrates (c.f. Miska and Geiger 1987; Müller-Klieser et al 1988; Walenta et al 1990).

## Conclusion

Video microscopy has led to both a revolutionary change and a renaissance of biological light microscopy because it helped to overcome the limitations of conventional light microscopy by increasing contrast, by visualizing and resolving yet smaller objects, and by intensifying extremely low light images. Some applications such as measurement of ions with ion-sensitive dyes and quantitative analysis of organelle motion have already profoundly changed some fields of neurobiological research. The possibility to study living cells and tissue slices will lead in the near future to the development of new imaging techniques closer to biochemistry than to histology. This field is not much exploited yet, but the absence of fixation artifacts and the high spatial, temporal and light intensity resolution are of such great promise that videomicroscopic techniques may soon reach more fields of neurocytochemistry.

Acknowledgments

The video-microscopic research in the author's laboratory was supported by Deutsche Forschungsgemeinschaft, Wilhelm-Sander-Stiftung, Bayer. Staatsministerien für Landesentwicklung und Umweltfragen und für Wissenschaft und Kunst.

REFERENCES

Aikens RS, Agard DA, Sedat JW (1989) Solid-state imagers for microscopy. In: Taylor DL, Wang, YL (eds) Methods in Cell Biology. Vol. 29, Chapter 16, Acad. Press, pp 291-305

Allen RD (1985) New observations on cell architecture and dynamics by video-enhanced contrast optical microscopy. Ann Rev. Biophys. biophys. Chem. 14:265-290

Allen RD, Allen NS (1983) Video-enhanced microscopy with a computer frame memory. J. Microscopy 129:3-17

Allen RD, Allen NS, Travis JL (1981b) Video-enhanced contrast, differential interference contrast (AVEC-DIC) microscopy: A new method capable of analyzing microtubule-related motility in the reticulopodial network of Allogromia laticollaris. Cell Motil. 1:291-302

Allen RD, Metuzals J, Tasaki I, Brady ST, Gilbert SP (1982) Fast axonal transport in squid giant axon. Science 218:1127-1129.

Allen RD, Travis JL, Allen NS, Yilmaz H (1981a) Video-enhanced contrast polarization (AVEC-POL) microscopy: A new method applied to the detection of birefringence in the motile reticulopodial network of Allogromia laticollaris. Cell Motil. 1:275-289.

Allen RD, Weiss DG, Hayden JH, Brown DT, Fujiwake H, Simpson M (1985) Gliding movement of and bidirectional organelle transport along single native microtubules from squid axoplasm: Evidence for an active role of microtubules in cytoplasmic transport. J. Cell Biol. 100:1736-1752

Almers W, Neher E (1985) The Ca signal from fura-2 loaded mast cells depends strongly on the method of dye-loading. FEBS Letters 192:13-18.

Anonymous (1985) Image Analysis. Principles and Practice. Published by Joyce Loebl Ltd., distributed by IRL Press, London

Arndt-Jovin DJ, Robert-Nicoud M, Kaufman SJ, Jovin TM (1985) Fluorescence digital imaging microscopy in cell biology. Science 230:247-256

Blasdel GG, Salama G (1986) Voltage-sensitive dyes reveal a modular organization in monkey striate cortex. Nature 321:579-585

Bradbury S (1989) Micrometry and image analysis. In: Lacey AJ (ed) Light Microscopy in Biology: A Practical Approach. IRL Press London, pp 200-220

Brady ST, Lasek RJ, Allen RD (1982) Fast axonal transport in extruded axoplasm from squid giant axon. Science 218:1129-1131.

Bright GR, Fisher GW, Rogowska J, Taylor DL (1987) Fluorescence ratio imaging microscopy: temporal and spatial measurements of cytoplasmic pH. J. Cell Biol. 104:1019-1033

Bright GR, Fisher GW, Rogowska J, Taylor DL (1989) Fluorescence ratio imaging microscopy. Methods. Cell Biol. 30:157-192

Cohn SA, Ingold AL, Scholey JM (1987) Correlation between the ATPase and microtubule translocating acctivities of sea urchin kinesin. Nature 328:160-1163.

De Brabander M, Nuydens R, Geerts H, Hopkins CR (1988) Dynamic behavior of the transferrin receptor followed in living epidermoid carcinoma (A431) cells with nanovid microscopy. Cell Motil. Cytoskel. 9:30-47

De Brabander M, Nuydens R, Geerts H, Nuyens R, Leunissen J, Jacobson K (1990) Using nanovid microscopy to analyse the movement of cell membrane components in living cells. In: Herman B, Jacobson K (eds) Optical Microscopy for Biology. Wiley-Liss, New York, pp 345-356

DeBiasio R, Bright GR, Ernst LA, Waggoner AS, Taylor DL (1987) Five-Parameter Fluorescence Imaging: Wound Healing of Living Swiss 3T3 Cells. J. Cell Biol. 105:1613-1622

Dodt HU, Zieglgänsberger W (1990) Visualizing unstained neurons in living brain slices by infrared DIC-video microscopy. Brain Res., in press

Ellis GW (1985) Microscope illuminator with fiber optic source integrator. J. Cell Biol. 101:83a

Forscher P, Kaczmarek LK, Buchanan JA, Smith SJ (1987) Cyclic AMP induces changes in distribution and transport of organelles within growth cones of *Aplysia* bag cell neurons. J. Neurosci. 7(11):3600-3611

Gelles J, Schnapp BJ, Sheetz MP (1988) Tracking kinesin-driven movements with nanometre-scale precision. Nature 331:450-453.

Gilbert SP, Sloboda RD (1986) Bidirectional transport of fluorescently labeled vesicles introduced into extruded axoplasm of squid Loligo pealei. J. Cell Biol. 99:445-452.

Giuliano KA, Nederlof MA, DeBiasio R, Lanni F, Waggoner AS, Taylor DL (1990) Multi-mode light microscopy. In: Herman B, Jacobson K (eds) Optical Microscopy for Biology. Wiley-Liss, New York, pp 543-558

Glascott jr. PA, McSorley KM, Mittal B, Sanger JM, Sanger JW (1987) Stress fiber reformation after ATP depletion. Cell Motil. Cytoskel. 8:118-129

Goldberg DJ, Burmeister DW (1989) Looking into growth cones. Trends in Neurosci. 12:503-506

Grinvald A, Segal M, Kuhnt U, Manker A, Anglister L, Freeman JA, Hildesheim R (1986) Real time optical mapping of neuronal activity in vertebrate CNS in vitro and in vivo. In: de Weer P, Salzberg BM (eds) Optical Methods in Cell Physiology, Wiley, N.Y., pp 165-198.

Gross GW, Kowalski J (1990) Experimental and theoretical analysis of random nerve cell network dynamics. In: Antognetti P, Milutinovic V (eds) Neural Networks: Concepts, Applications, and Implementations. Vol. 3, Prentice-Hall, Englewood Cliffs, in press

Grynkiewicz G, Poenie M, Tsien RY (1985) A new generation of Ca2+ indicators with greatly improved fluorescence properties. J. Biol. Chem. 260:3440-3450

Häder D-P (1988) Computer-assisted image analysis in biological sciences. Proc. Indian Acad. Sci. (Plant Sci.) 98:227-249

Harada Y, Yanagida T (1988) Direct observation of molecular motility by light microscopy. Cell Motil. Cytoskel. 10:71-76

Härtig W, Paulke B, Brückner G (1990) Development of monodisperse rhodamine-labelled latex microspheres for the retrograde tracing of neurons and the demonstration of glial cells. In: Elsner N, Roth G (eds) Brain - Perception - Cognition. Thieme Verlag Stuttgart, pp 348

Haugland RP (1989) Handbook of Fluorescent Probes and Research Chemicals.. Molec. Probes Inc., Eugene OR, 2nd ed., 234 p

Hayden JH, Allen RD (1984) Detection of single microtubules in living cells: Particle transport can occur in both directions along the same microtubule. J. Cell Biol. 99:1785-1793

Hayden JH, Allen RD, Goldman RD (1983) Cytoplasmic transport in keratocytes: Direct visualization of particle translocation along microtubules. Cell Motil. 3:1-19.

Herman B, Jacobson K (eds) Optical Microscopy for Biology. Wiley-Liss, New York, 1990, p 658

Herrera AA, Banner LR (1990) The use and effects of vital fluorescent dyes: observation of motor nerve terminals and satellite cells in living frog muscles. J. Neurocytol. 19:67-83

Holländer H, Egensperger R, Dirlich G (1989) J. Neurosci., Meth. 29:1-4

Honig MG, Hume RI (1989) DiI and DiO: versatile fluorescent dyes for neuronal labelling and pathway tracing. Trends in Neurosci. 12:333-341

Inoué S (1986) Video Microscopy, Plenum Press, NY., 582 p

Inoué S (1990) Whither video microscopy - towards 4-D imaging at the highest resolution of the light microscope. In: Herman B, Jacobson K (eds) Optical Microscopy for Biology. Wiley-Liss, New York, pp 497-512

Inoué S, Inoué TD (1987) Computer-aided stereoscopic video reconstruction and serial display from high-resolution light-microscope optical sections. Ann. N.Y. Acad. Sci. 483:392-404

Johnson LV, Walsh ML, Chen LB (1980) Localization of mitochondria in living cells with rhodamine 123. Proc. Natl. Acad. Sci. USA 77:990-994

Kater SB, Mattson MP, Cohan C, Connor J (1988) Calcium regulation of the neuronal growth cone. Trends in Neurosci. 11:315–321

Katz LC, Burkhalter A, Dreyer WJ (1984) Nature 310:498–500

Lee C, Chen LB (1988) Behavior of endoplasmic reticulum in living cells. Cell 54:37–46

**Lichtman JW, Magrassi L, Purves, D (1987) Visualization of neuromuscular junctions over periods of several months in living mice. J Neurosci 7:1215-1222.**

Lichtscheidl IK, Weiss DG (1988) Visualisation of submicroscopic structures in the cytoplasm of *Allium cepa* inner epidermal cells by video-enhanced contrast light microscopy. Eur. J. Cell Biol. 46:376–382

Llinás R, Sugimori M, Lin J-W, Leopold PL, Brady ST (1989) ATP-dependent directional movement of rat synaptic vesicles injected into the presynaptic terminal of squid giant synapse. Proc. Natl. Acad. Sci. 86:5656–5660

Loew LM, Farkas DL, Wei MD (1990) Membrane potential imaging - ratios, templates, and quantitative confocal microscopy. In: Herman B, Jacobson K (eds) Optical Microscopy for Biology. Wiley-Liss, New York, pp 131–142

Luby-Phelps K, Taylor DL (1988) Subcellular compartmentalization by local differentiation of cytoplasmic structure. Cell Motil. Cytoskel. 10:28–37

Marcus DA (1988) High-performance optical filters for fluorescence analysis. Cell Motil. Cytoskel. 10:62–70

McKenna NM, Wang Y, Konkel ME (1989) Formation and movement of myosin-containing structures in living fibroblasts. J. Cell Biol. 109:1163–1172

Miska W, Geiger R (1987) Synthesis and characterization of luciferin derivates for use in bioluminescence enhanced enzyme immunoessays. New ultrasensitive detection systems for enzyme immunoessays I. J. Clin. Chem. Clin. Biochem. 25:23–30.

Mizushima Y (1988) Detectivity limit of very small objects by video-enhanced microscopy. Applied Optics 27:2587–2594

Möncke E, Lucas C, Weiss DG (1985) Distribution, diffusion and transport of fluorescent dyes in cultured cells. Eur. J. Cell Biol. 36 Suppl. 7:44.

Mueller-Klieser W, Walenta S, Paschen W, Kallinowski F, Vaupel P (1988) Metabolic imaging in microregions of tumor and normal tissue with bioluminescence and photon counting. J. Natl. Cancer Inst. 80:842–848

Pagano RE, Sepanski MA, Martin OC (1989) Molecular trapping of a fluorescent ceramide analogue at the golgi apparatus of fixed cells: interaction with endogenous lipids provides a *trans*-golgi marker for both light and electron microscopy. J. Cell Biol. 109:2067–2079

Prpic V, Cowlen MS, Adams DO (1990) Application of digital imaging microscopy to studies of ion fluxes in murine peritoneal macrophages. In: Herman B, Jacobson K (eds) Optical Microscopy for Biology. Wiley-Liss, New York, pp 337–344

Sammak PJ Borisy GG (1988 ) Detection of single fluorescent microtubules and methods for determining their dynamics in living cells. Cell Motil. Cytoskel. 10:237–245

Sanger JM, Dome JS, Mittal B, Somlyo AV, Sanger J (1989) Dynamics of the endoplasmatic reticulum in living non-muscle and muscle cells. Cell Motil. Cytoskel. 13:301–319

Seitz-Tutter D, Langford GM, Weiss DG (1988) Dynamic instability of native microtubules from squid axons is rare and independent of gliding and vesicle transport. Exptl. Cell Res. 178:504–512

Shotton DM (1988) Video-enhanced light microscopy and its applications in cell biology. J. Cell Sci. 89:129–150

Spring KR (1990) Quantitative imaging at low light levels - differential interference contrast and fluorescence microscopy without significant light loss. In: Herman B, Jacobson K (eds) Optical Microscopy for Biology. Wiley-Liss, New York, pp 513–522

Spring KR, Lowy RJ (1989) Characteristics of low light level television cameras. In: Taylor DL, Wang YL (eds) Methods in Cell Biol., Vol. 29, Chapter 15, Acad. Press, pp 269–

Stirling RV (1986) Video techniques in neurobiology. Trends in Neurosci. 9:145–147

Swedenburg CE, Conklin JJ (eds) (1988) Imaging Techniques in Biology and Medicine. Academic Press 1988

Terasaki M, Song J, Wong JR, Weiss MJ, Chen LB (1984) Localization of endoplasmic reticulum in living and glutaraldehyde-fixed cells with fluorescent dyes. Cell 38:101–108

Travis JL, Bowser SS (1988) Optical approaches to the study of foraminiferan motility. Cell Motil. Cytoskel. 10:126-136

Tsien RY (1989) Fluorescent probes of cell signaling. Ann. Rev. Neurosci. 12:227-253

Vale RD, Reese TS, Sheetz MP (1985) Identification of a novel, force-generating protein, kinesin, involved in microtubule-based motility. Cell 42:39-50.

Vallee RB, Paschal BM, Shpetner HS (1989) Characterization of microtubule-associated protein (MAP) 1C as the motor for retrograde organelle transport and its identification as dynein. In: Cell Movement, Vol. 2: Kinesin, Dynein, and Microtubule Dynamics. Alan R. Liss., pp 211-222

Van Noorden CJF (1988) Enzyme reaction rate studies in tissue sections. Proc. Roy. Micr. Soc. 23:93-97

Wade MH, Trosko JE, Schindler M (1986) A fluorescence photobleaching assay of gap junction-mediated communication between human cells. Science 232:525-528

Walenta S, Dötsch J, Müller-Klieser W (1990) ATP concentrations in multicellular tumor spheroids assessed by single photon imaging and quantitative bioluminescence. J. Cell Biol. 52:389-393

Wampler JE, Kutz K (1989) Quantitative fluorescence microscopy using photomultiplier tubes and imaging detectors. In: Methods in Cell Biology. Vol. 29, Chapter 14, Acad. Press, pp 239-251

Wang Y-L (1989) Fluorescent analog cytochemistry:tracing functional protein components. In: Taylor DL, Wang YL (eds) Methods in Cell Biology. Vol. 29, Chapter 1, Acad. Press, pp 1-21

Weiss DG (1986) Visualization of the living cytoskeleton by video-enhanced microscopy and digital image processing. J. Cell Sci. Suppl. 5, 1-15

Weiss DG (1987) Videomicroscopic measurements in living cells: dynamic determination of multiple end points for in vitro toxicology. Molec. Toxicol. 1:465-488

Weiss DG, Buchner K (1988) Axoplasmic transport in olfactory receptor neurons. In: Margolis FL, Getchell TV (eds) Molecular Neurobiology of the Olfactory System. Plenum Publ. Corp., N.Y., pp 217-236

Weiss DG, Galfe G, Gulden J, Seitz-Tutter D, Langford GM (1990b) Motion analysis of intracellular objects: trajectories with and without visible tracks. In: Alt W, Hoffmann G (eds) Biological Motion. Lecture Notes in Biomathematics, Vol. 89, Springer Verlag, Berlin, pp 95-116

Weiss DG, Langford GM, Seitz-Tutter D, Gulden J, Keller F (1988b) Motion analysis of organelle movements and microtubule dynamics. In: Rousset BAF (ed) Structure and Functions of the Cytoskeleton. Colloque INSERM/John Libbey Eurotext, Vol 171, Paris, pp 363-378

Weiss DG, Langford GM, Seitz-Tutter D, Keller F (1988a) Dynamic instability and motile events of native microtubules from squid axoplasm. Cell Motil. Cytoskel. 10:285-295

Weiss DG, Maile W, Wick RA (1989) Video microscopy. Chapter 8, In: Lacey AJ (ed) Light Microscopy in Biology. A Practical Approach. IRL Press, Oxford, pp 221-278

Weiss DG, Meyer M, Langford GM (1990a) Studying axoplasmic transport by video microscopy and using squid giant axon as a model system. In: Gilbert DL, Adelman jr. WJ, Arnold JM (eds) Squid as Experimental Animals. Plenum Press, pp 303-321

Wick RA (1987) Quantum-limited imaging using microchannel plate technology. Applied Optics 26:3210-3222

Wu J-Y, London JA, Zecevic D, Höpp H-P, Cohen LB, Xiao C (1988) Optical monitoring of activity of many neurons in invertebrate ganglia during behaviors. Experientia 44:369-376

Yoshikami D, Okun LM (1984) Staining of living presynaptic nerve terminals with selective fluorescent dyes. Nature 310:53-56

# INVESTIGATING THE THREE-DIMENSIONAL MORPHOLOGY OF FLUORESCENCE-LABELLED NEURONS USING CONFOCAL LASER SCANNING MICROSCOPY

Peter Wallén, Gonzalo Viana Di Prisco and Réjean Dubuc

Nobel Institute for Neurophysiology, Karolinska Institutet, Box 60400, S-104 01 Stockholm, Sweden

## SUMMARY

The application of confocal laser scanning microscopy to the field of neuroscience is exemplified in this paper by a brief description of the three-dimensional (3D) morphology of fluorescence-labelled neurons in the central nervous system of the lamprey, a lower vertebrate. Intracellular recordings were performed in an *in vitro* preparation of the lamprey CNS with microelectrodes filled with Lucifer Yellow. After intracellular dye-injection of identified, single neurons, the tissue was prepared as a whole-mount or thick section for optical serial sectioning in a PHOIBOS confocal laser scanning microscope. The extensive dendritic tree of large reticulospinal neurons in the brainstem were reconstructed and its 3D structure studied using different projection angles. The 3D morphology of intraspinal stretch receptor neurons (edge cells) was investigated in detail at high magnification, as was the 3D appearance of areas of possible synaptic contact between pairs of neurons, injected in double-microelectrode experiments. As a result of the substantially improved resolution, the strong suppression of stray light from out-of-focus structures, and the possibility to perform detailed 3D reconstructions, confocal laser scanning microscopy has yielded images that contain new and more detailed morphological information than what has been possible to obtain previously, using conventional fluorescence microscopy.

NATO ASI Series, Vol. H 58
Neurocytochemical Methods
Edited by A. Calas and D. Eugène

## INTRODUCTION

Neurons in the vertebrate central nervous system (CNS) often possess rich and complex dendritic trees, extending in many directions towards sources of synaptic input. This is true for neurons in e.g. the cerebral cortex as well as for cells in the spinal cord. In order to gain a better understanding for the functioning of both the individual cell and for the interactions between the many classes of neurons, it is clearly of significant value to have information on the detailed morphology of the neurons with their dendrites, including the thinnest branches. This paper attempts to illustrate that information on the three-dimensional (3-D) structure is often not only very useful but sometimes even may be essential in studies of neuronal function. To do this we will present data on the 3-D morphology of fluorescence-labelled, functionally identified neurons that are involved in the control of locomotor movements in a lower vertebrate, the lamprey. The analysis was made using confocal laser scanning microscopy.

## THE LAMPREY MODEL FOR LOCOMOTION

The neuronal mechanisms underlying the rhythmic locomotor movements of vertebrates can be conveniently studied in a relatively simple model system, - the in vitro preparation of the CNS of the lamprey (Grillner et al., 1987). This jawless, eel-like animal (cyclostome) swims by performing lateral body undulations that pass along the animal from head to tail. The rhythmic muscle activity pattern that gives rise to these movements is being generated by local networks of interneurons in the spinal cord (Grillner et al., 1987). The activity of the spinal networks is controlled from the brainstem via large, reticulospinal neurons (Brodin et al., 1988b).

Two neuron types, each one subserving an important function in the control of locomotion in the lamprey, have here been chosen for detailed study of their 3-D morphology: the reticulospinal neuron, and the edge cell, which is an intraspinal stretch receptor neuron. The edge cells are located close to the lateral margin of the flattened spinal cord (thickness about 300 $\mu$m), and are activated when the cord margin is being rhythmically stretched during swimming movements (Grillner et al., 1981, 1984). The edge cells make synaptic contact with the network interneurons, thereby providing the circuitry with movement-related feedback, which is essential for the modulation of the locomotor movements to meet changing external demands (Wallén & Williams, 1985). Furthermore, we have analyzed areas of possible synaptic contact between edge cells and interneurons by performing detailed 3D reconstructions.

The isolated spinal cord and/or brainstem was kept in vitro in a cooled and oxygenated recording chamber. The different neurons were impaled intracellularly using recording microelectrodes filled with Lucifer Yellow CH (Stewart 1978). In this way electrophysiological data (membrane potentials, activity patterns, synaptic responses etc) could first be collected, and then the cell could be filled with dye, using the same electrode. This procedure has the advantage that the <u>same</u> individual neuron that has been stained for structural analysis has also been characterized physiologically.

## ANALYZING FLUORESCENT NEURONS USING THE CONFOCAL LASER SCANNING MICROSCOPE

After intracellular recording and staining, the lamprey CNS preparation was fixed in formalin, dehydrated in alcohol, and cleared in methylsalicylate. The tissue was then prepared as a whole-mount or, in some cases with the brainstem, as a thick

(0.5-1.0 mm) cross-section. The fluorescence-labelled neurons were then analyzed with a PHOIBOS confocal laser scanning microscope system (Sarastro AB, Stockholm, Sweden). This system performs automatic, microcomputer controlled, optical serial sectioning and 3-D reconstructions with projections from any desired viewing angle through the data volume (Åslund et al., 1983; Carlsson et al., 1985, 1989; Carlsson, 1988; Liljeborg, 1988; Wallén et al., 1988; Carlsson & Liljeborg, 1989; Forsgren, 1989).

In the present analysis the different neuron types were scanned through by performing between about 100 and 300 serial optical sections, with a step increment in depth of 0.2 - 2.0 $\mu$m, and with each section comprising 512 by 512 pixels.

After scanning, the section series volume was transformed to vector format using the 3D analysis software of the PHOIBOS system (Forsgren, 1989), and subsequently projection calculations were made resulting in 3D images of the cell from any selected viewing angle. Stereo pair reconstructions were made by selecting two angles with a difference of 6°. The images were presented either as untreated projections, or after software treatment for e.g. surface extraction or surface shading. The reconstructed images were then sent to a video copy printer that produced high-resolution prints in color or black-and-white, to be used for illustrations.

## THE RETICULOSPINAL NEURON OF THE BRAINSTEM

The locomotor circuitry of the lamprey spinal cord is controlled from higher centers via reticulospinal neurons located in the brainstem. The axons of these neurons descend to the spinal cord where interneurons as well as motoneurons are excited monosynaptically (Brodin et al., 1988b). Analyzing the 3D struc-

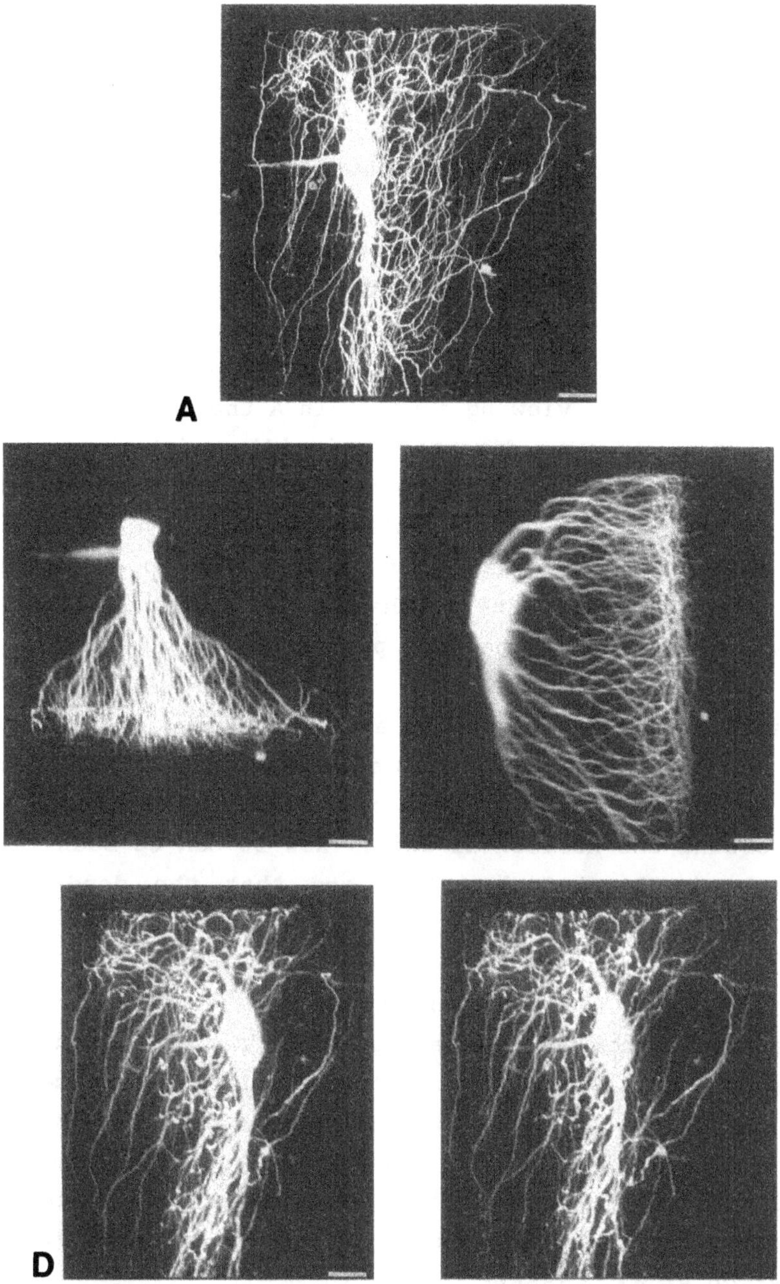

A

B

C

D

Figure 1. Reticulospinal neuron in the lamprey brainstem. A-C: 3D projections from three viewing directions perpendicular to each other, of the same cell. A shows the cell from above, with a 0° projection angle, parallel to the microscope optical axis. In B the volume has been rotated 90° around the horizontal axis, and in C a rotation was performed around the vertical axis (cf. A). D: Stereo pair reconstructions as seen from above at an oblique angle. Scale bars: 50 $\mu$m (objective: 20X/0.75).

ture of the large reticulospinal neurons presented a common problem in microscopy: the need to cover a large enough area that would include the entire cell with its processes, and still to retain enough resolution for reconstruction of the thinnest branches. The dendritic tree of the brainstem neurons often span more than 0.5 mm in the three dimensions and the thinnest processes may be 1 $\mu$m or less in diameter. The enhanced resolution, especially in the depth dimension, achieved in confocal microscopy (Carlsson, 1988) here proved to be of great value. Figure 1 illustrates 3D reconstructions of a brainstem neuron from various viewing angles. In A the cell is shown from above, with the viewing direction coinciding with the microscope optical axis. Note the axon pointing towards the left in the caudal direction. The 20X objective used here covered an area of approximately 0.5 by 0.5 mm; this allowed scanning of almost the entire dendritic tree. Still, the resolution is good enough to allow visualization of the thin processes.

By selecting another projection angle, the neuron can be viewed from other directions. Figure 1B shows the same neuron as viewed from the lateral aspect of the brainstem (90° rotation of the volume around the horizontal axis). Now it can be seen that the cell body is located dorsally, and that the dendrites run ventrally and also spread rostrocaudally. It should be noted that this is not a cross-section of the cell; the whole neuron is still viewed but from another angle. Figure 1C shows the same cell as viewed from the rostral aspect of the brainstem, looking towards the spinal cord (90° rotation around the vertical axis). The large cell body is evident (rostral is here towards the left), with the dendrites forming a dense plexus close to the ventral surface of the brainstem. It is evident that also in the depth dimension, the thin dendrites can be clearly visualized (cf. Wallén et al., 1988).

Performing stereo pair reconstructions from the data volume may also yield valuable information about the structure of the neuron, often giving a striking and immediate impression of the

3D morphology. Figure 1D is a stereo pair of the same brainstem neuron, as seen from above and after rotation (around the vertical axis) by about 20°. The cell body can be seen in the front, with the dendritic processes pointing downwards, away from the viewer.

These reconstructions suggest that this type of reticulospinal neuron receives important synaptic input in a region close to the ventral surface of the brainstem. The 3D structure of the dendritic processes depicts in detail this area of synaptic integration. Information of this kind is obviously of significant value when trying to identify the neurons comprising the source of synaptic input to these brainstem cells.

## THE INTRASPINAL STRETCH RECEPTOR NEURON

The edge cells of the lamprey spinal cord have a unique location and morphology, being specialized for a stretch receptive function during locomotion (Grillner et al., 1984). In order to get a better understanding of how these neurons are being activated during the movement, it is of particular interest to investigate the detailed structure of the lateral, stretch sensitive dendrites that reach out towards the spinal cord margin.

Figure 2A is a stereo pair of an overview reconstruction of an edge cell, where the lateral portion can be seen pointing towards the viewer. A software routine for surface extraction was here utilized. By using a high magnification objective with a high numerical aperture (here 100X/1.4), the detailed structure of the lateral dendrites of the same edge cell was analyzed (Fig. 2B, C). Here a surface shading routine was employed. The stereo pair in B shows the upper part of the lateral dendrites (cf. Fig. 2A), after rotation around the vertical axis such that the stem den-

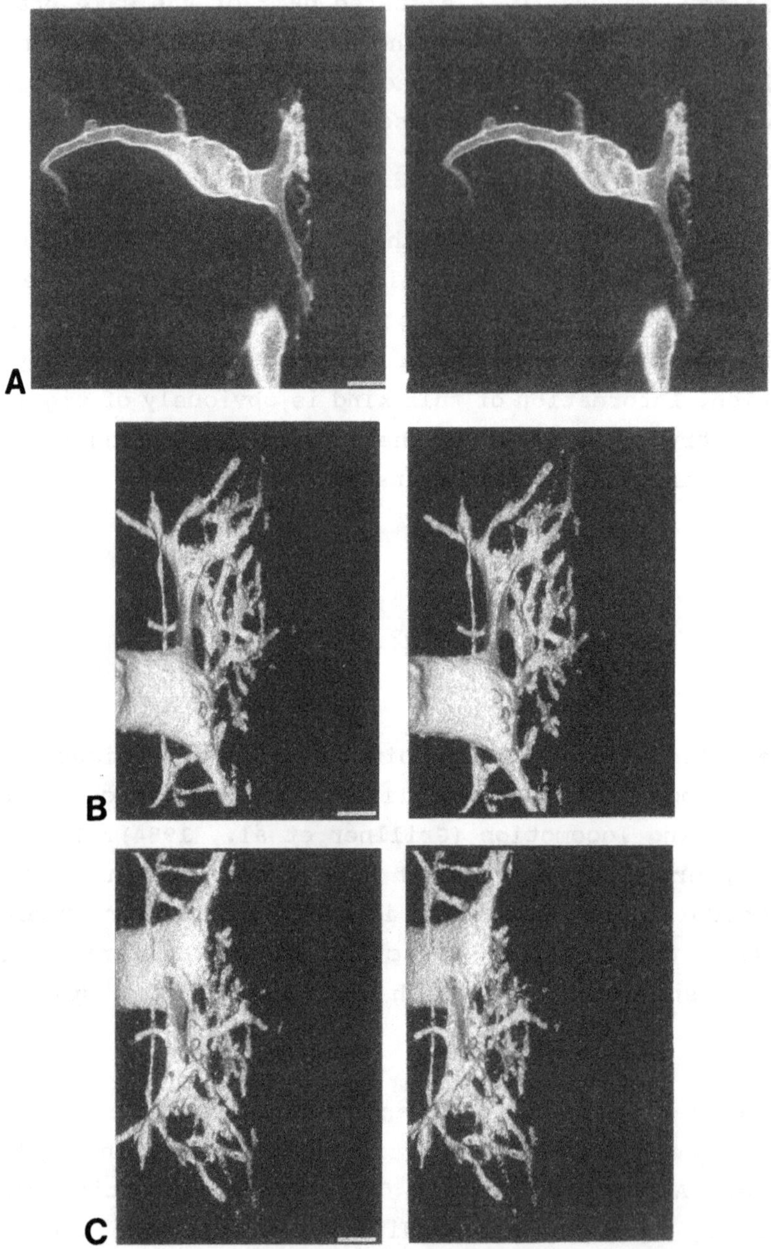


Figure 2. Intraspinal stretch receptor neuron at the lateral margin of the lamprey spinal cord. A: Stereo pair reconstructions employing a surface extraction routine, showing the cell with the lateral dendrites pointing towards the viewer. The structure below the cell is an artefact. Scale bar: 25 μm (objective: 40X/0.95). B,C: Portion of the lateral dendritic tree of the same edge cell (surface shading performed). Stereo pairs. B: View from the cell body location, towards the cord margin. C: View from the cord margin, looking onto the lateral dendritic processes. Scale bars: 10 μm (objective: 100X/1.40).

drite is in front and the lateral processes are in the background, as if the viewer were looking from the cell body location out towards the cord margin. By simply performing a 180° rotation around the horizontal axis (Fig. 2C), the lateral processes are visualized in front, in a view onto the cord margin with the stem dendrite in the background. The thinnest, fingerlike processes are 1-2 μm in diameter.

The possibility to visualize the 3D structure of the edge cell lateral dendrites from different viewing angles, together with the enhanced resolution and stray light suppression achieved with confocal microscopy (Wilson & Sheppard, 1984; Carlsson, 1988), thus has given us information of a level of detail that has not been possible to obtain previously, using conventional fluorescence microscopy.

## RECONSTRUCTING AREAS OF SYNAPTIC CONTACT BETWEEN NEURONS

When investigating the interaction between neurons subserving a specific function, it is of significant value to be able to vizualize the areas where the cells make synaptic contact. Since, however, the dendritic arborizations of the neurons are spread in all three dimensions to receive synaptic input from various sources, conventional two-dimensional reconstructions may not allow conclusive determinations of the points of contact. This is illustrated in Figure 3, where an area of possible synaptic contact between an edge cell (the same cell as in Fig. 2) and an interneuron was analyzed at high resolution (100X/1.4) with the confocal laser scanning microscope, after intracellular injection of Lucifer Yellow into both cells in a double-microelectrode experiment. In A the brightly fluorescing stem dendrite of the interneuron is seen to send a few branches towards the edge cell axon to the left. This 2D image would suggest that there might be two or three points of possible synaptic contact between the two cells in this area. However, if the 3D volume is rotated around the vertical axis (Fig. 3B), it becomes obvious that the lower branch in A does not point in the direction of the edge cell axon (now partly hidden underneath the interneuron dendrite). In the stereo pair reconstruction in Figure 3C, it can be seen that the upper branch appears to make contact with the edge cell axon, indeed the branch seems to terminate on a thin

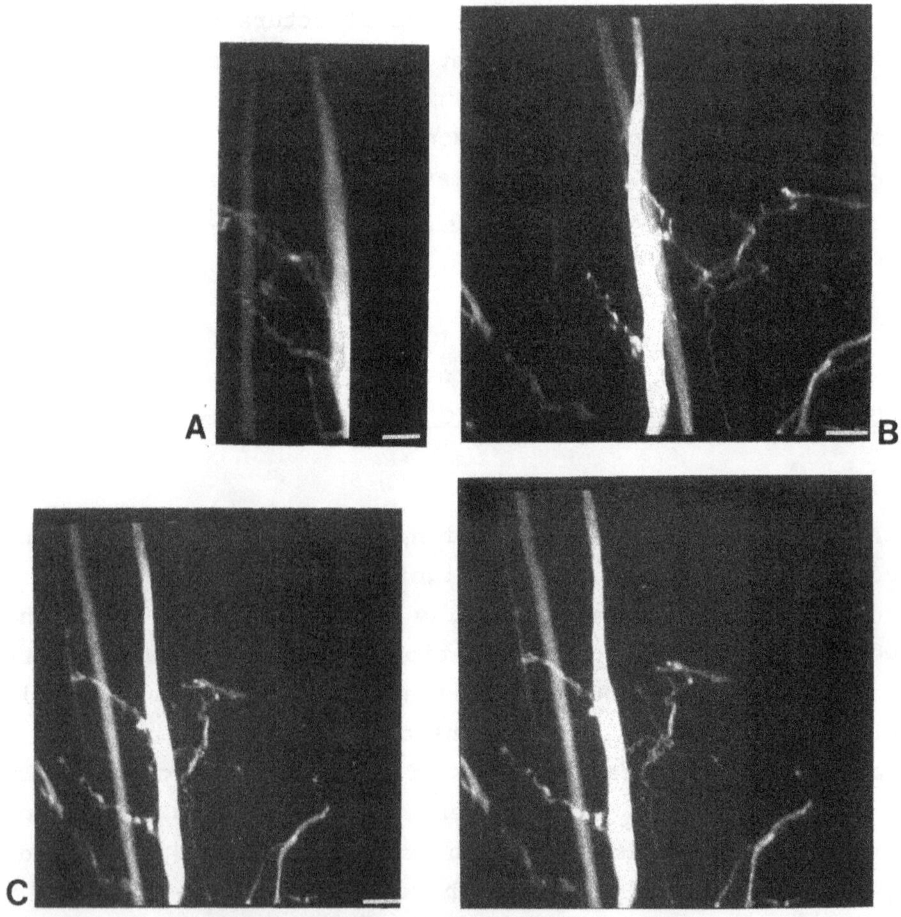


Figure 3. Points of possible synaptic contact between an intraspinal stretch receptor neuron (same as in Fig. 2) and an interneuron in the lamprey spinal cord. A: Reconstruction of an interneuron dendrite (right) and the edge cell axon (left; darker). B: After a 90° rotation around the vertical axis, the lowermost branch is seen not to be in contact with the edge cell axon. C: Stereo pair reconstructions of the same area. Scale bars: 10 $\mu$m (objective: 100X/1.40).

collateral originating from the edge cell axon. The lower interneuron dendrite branch clearly runs above the edge cell axon. It is thus obvious that performing 3D reconstructions of areas of possible synaptic contacts allows a more accurate structural analysis, and determinations based on conventional 2D images may in fact be erroneous.

## CONCLUDING REMARKS

With these examples of 3D reconstructions of fluorescence-labelled neurons of the lamprey CNS we have attempted to illustrate how the unique analyzing capabilities of confocal laser scanning microscopy can be utilized in studies of the nervous system. We have limited the presentation to the structural analysis of single, intracellularly stained neurons, and have not touched upon the usage of confocal microscopy in combination with immunofluorescence techniques. Suffice it here to say that confocal laser scanning microscopy holds promise to become a very powerful and important tool also when analyzing the structure and distribution of immunofluorescence-labelled cell ensemles, e.g. transmitter-identified neurons (cf. Brodin et al., 1988a).

The reconstructions of the large neurons in the lamprey brainstem illustrate the commonly occurring need for low-magnification microscopy with high resolution. Although the enhanced resolution of confocal microscopy yields low-power images in 3D with good resolution, development of better low-magnification objectives with a higher numerical aperture and still with a sufficient working distance would be desirable. This example also illustrates the capability of performing 3D reconstructions from any desired viewing angle. This is an extremely useful and powerful feature when analyzing details of neuronal structures, which is also evident from the example of the contact points in

Figure 3. Finally, when the 3D reconstruction capability and the superior resolution of confocal laser scanning microscopy are combined in high magnification images like in Figures 2B,C and 3, the structures are reconstructed with a level of detail not previously achieved with conventional microscopy, revealing new morphological features that are of significant value in the analysis of neuronal function.

## REFERENCES

Brodin L, Ericsson M, Mossberg K, Hökfelt T, Ohta Y, Grillner S (1988a) Three-dimensional reconstruction of transmitter-identified central neurons by "en bloc" immunofluorescence histochemistry and confocal scanning microscopy, Exp Brain Res 73:441-446

Brodin L, Grillner S, Dubuc R, Ohta Y, Kasicki S, Hökfelt T (1988b) Reticulospinal neurons in lamprey: Transmitters, synaptic interactions and their role during locomotion, Arch Ital Biol 126:317-345

Carlsson K (1988) Development of optics and mechanics for a confocal 3D microscope and a diode array scanner for digital image recording; applications in physiology and medicine, Doctoral thesis, the Royal Institute of Technology, Stockholm

Carlsson K, Danielsson PE, Lenz R, Liljeborg A, Majlöf L, Åslund N (1985) Three-dimensional microscopy using a confocal laser scanning microscope, Opt Lett 10:53-55

Carlsson, K, Liljeborg, A (1989) A confocal laser microscope scanner for digital recording of optical serial sections, J Microsc 153:171-180

Carlsson K, Wallén P, Brodin L (1989) Three-dimensional imaging of neurons by confocal fluorescence microscopy, J Microsc 155:15-26

Forsgren P-O (1989) Development of methods to perform measurements in two and three dimensions using digital image processing techniques; applications to limnology, medicine, and nuclear physics, Doctoral thesis, the Royal Institute of Technology, Stockholm

Grillner S, McClellan A, Perret C (1981) Entrainment of the spinal pattern generators for swimming by mechano-sensitive elements in the lamprey spinal cord in vitro, Brain Res 217:380-386

Grillner S, Wallén P, Dale N, Brodin L, Buchanan JT, Hill R (1987) Transmitters, membrane properties and network circuitry in the control of locomotion in lamprey, Trends Neurosci 10:34-41

Grillner S, Williams T, Lagerbäck P-Å (1984) The edgecell, a possible intraspinal mechanoreceptor, Science 223:500-503

Liljeborg A (1988) Development of electronics and software for a confocal 3D microscope and a diode array scanner for digital image recording; applications in physiology, Doctoral thesis, the Royal Institute of Technology, Stockholm

Stewart WW (1978) Functional connections between cells as revealed by dye coupling with a highly flourescent napthalimide tracer, Cell 14:741-759

Wallén P, Carlsson K, Liljeborg A, Grillner S (1988) Three-dimensional reconstruction of neurons in the lamprey spinal cord in whole-mount, using a confocal laser scanning microscope, J Neurosci Meth 24:91-100

Wallén P, Williams TL (1985) The role of movement-related feedback in the control of locomotion in fish and lamprey, In Feedback and Motor Control in Invertebrates and Vertebrates (eds Barnes WJP, Gladden MH) pp 317-335, Croom Helm

Wilson T, Sheppard CJR (1984) Theory and practice of scanning optical microscopy, Academic Press

Åslund N, Carlsson K, Liljeborg A, Majlöf L (1983) PHOIBOS, a microscope scanner designed for micro-fluorometric applications, using laser induced fluorescence, In Proc Third Scandinavian Conf on Image Analysis, pp 338-343, Studentlitteratur, Lund

# SUMMARY OF THE SESSION ON HODOLOGY

F. Vandesande
Department of Biology
Catholic University of Louvain
Naamsestraat 59
B-3000 Louvain
Belgium

In the first lecture on Hodology, P. Streit discussed in detail transmitter-related retrograde labelling of neural pathways and transmitter immunohistochemistry, two methods that allow the localization of amino acid transmitters in neurons and neuronal pathways.

Transmitter-related retrograde tracing of pathways is based on the presence of a selective uptake mechanism at the axon terminals of neurons, containing this specific transmitter. When a radioactively labelled form of the transmitter, or in certain cases of a related compound, is injected at a restricted site in the central nervous system, only the neurons with a selective uptake mechanism for the given transmitter, are retrogradely labeled. P. Streit demonstrated the applications, perspectives and limitations of the method by showing and discussing his results concerning certain pathways in the rat brain and afferents to the pigeon optic tectum. He warned of false positive and negative results. Especially the use of high transmitter concentrations may result in a non-specific uptake and retrograde labelling. Therefore the results have always to be compared with those of retrograde tracing experiments, in which non-specific tracers, e.g. HRP, have been used.

The speaker demonstrated transmitter immunocytochemistry on the basis of his own results, obtained with monoclonal antibodies to glutamate, GABA and glycine, and with polyclonal antibodies to homocysteate. These monoclonal and polyclonal antibodies were obtained after immunizing mice with a glutaraldehyde conjugate of the respective transmitter with a carrier protein. The main problem with this technique is that

NATO ASI Series, Vol. H 58
Neurocytochemical Methods
Edited by A. Calas and D. Eugène

one has to be sure that the primary antibodies are sufficiently specific. One has to test this specificity in circumstances that mimic, as good as possible, the immunocytochemical situation. Therefore the specificity test developed by O.P. Ottersen, has been used. The results of this specificity test clearly eliminated the possibility that a positive immunocytochemical staining was due to cross-reactions.

In the second lecture O.P. Ottersen demonstrated and discussed the ultrastructural identification and quantification of neuromediators such as GABA, glycine, glutamate, aspartate, and taurine. Since all of these amino acids play an important role in the metabolism of every cell, they are expected to occur in varying amounts in most or all cell compartments. From a neuroanatomical point of view, immunocytochemistry of these transmitters makes only sense, if the observed immunoreactivity can be correlated with their concentration. On the basis of his results the speaker demonstrated that this requirement can be met by using a postembedding immunogold procedure, applied on ultrathin plastic sections of glutaraldehyde fixed brain tissue. The density of gold particles over different cell structures was measured with the aid of a computer program and correlated with particle densities over a series of standards, prepared with different known amino acid concentrations, and incubated together with the tissue sections. Also a specificity test that mimics the immunocytochemical situation was used. The speaker assumed that an immunolabelling of a nerve terminal corresponded to a transmitter pool if he was able to show that : (1) the gold particle density was higher over nerve terminals than over their parent perikarya; (2) the gold particles were enriched over synaptic vesicles relative to organel-free axoplasm; (3) the immunoreactivity could be depleted in a calcium dependent manner, by exposing brain slices to high $K^+$ concentrations _in vitro._

In the third lecture D.G. Weiss demonstrated, with the help of fascinating pictures and a video film of some selected results from his own work, how the possibility to digitize and process microscope images in "real time", has dramatically improved the quality of light microscopy, and how new applications have been created. The speaker discussed the basic principles of two new techniques and the type of equipment required.

With Video Enhanced Contrast (VEC) microscopy, it became possible to study positions and movements of biological objects as small as 15-20 nm, in living cells.

Video Intensified Microscopy (VIM) is based on the measurement of fluorescence intensities by monochromatic light microscopy, and it allows the measurement of different cellular parameters such as $Ca^{2+}$ concentration, pH value, and membrane potentials. Specific non toxic dyes became also available to verify the identity of organels seen by the VEC technique and to quantitate their position and abundancy. These techniques allow researchers to obtain simultaneous measurements of multiple parameters, and may yield video films of intracellular events during and after exposure to physiological stimuli or pharmacological agents.

In the last lecture of the session on Hodology P. Wallen demonstrated the possibilities of 3-D reconstruction of fluorescence-labelled neurons by confocal laser scanning microscopy. The 3-D reconstruction of a neuron, using conventional fluorescence microscopy, has always been a very tedious task. The possibility to generate optical sections by using a confccal laser scanning microscope became recently an efficient alternative. By changing the plane of focus, serial optical sections through the entire cell in a thick preparation, can be obtained and stored into a three-dimensional data volume. The enhanced resolution of this technique, particularly in the depth dimension, and the virtual absence of stray light interference from out of focus structures, results in three-dimensional reconstructions of very high quality. The possibilities of the technique were demonstrated by P. Wallen on the basis of his results, obtained on the central nervous system of the lamprey. It became clear from his lecture, that the confocal microscope is a very useful tool for the neuroanatomist. The application of confocal laser scanning microscopy has yielded new and more detailed morphological information than ever obtained before. It is possible that the present commercial instruments are not ideal, but in a very next future one may expect alternative, and perhaps better designs.

# COMBINED CYTOLABELING METHODS FOR INVESTIGATING CHEMO-SPECIFIC NEURONAL INTERACTIONS AND TRANSMITTER CO-LOCALIZATION

O. Bosler, P. Kachidian, D. André, G. Dusticier and A.M. François-Bellan[1]
Laboratoire de Neurobiologie
C.N.R.S., B.P. 71
13402 Marseille Cédex 9
France

New insights into the functioning of the nervous system have been gained with the emergence of the new discipline of chemo-anatomy in which neuronal circuitry is considered mainly in terms of transmitter-specific interactions. The characterization of the cytological substrate of such interactions called for the development of sensitive combined labeling methods to localize more than one neuronal marker in a single electron microscopic section. Various strategies have been employed to reach this goal, which mainly took advantage of the extensive development of immunocytochemical methods. The present paper will provide an overview of the principles and possible applications of several dual labeling procedures employing immunocytochemical markers, with particular emphasis on their respective advantages and limitations. Specifically, it will be devoted to those methods allowing for a simultaneous detection of two separate molecules either by combined immunocytochemistry and uptake radioautography or by means of dual immunocytochemistry with different constrasting electron dense markers in pre- and/or post-embedding applications.

## COMBINATION OF TRANSMITTER UPTAKE RADIOAUTOGRAPHY AND IMMUNOCYTOCHEMISTRY

Radioautography has long proven to be particularly useful for investigating the cytological features as well as some dynamic properties of

[1] Laboratoire de Neuroendocrinologie Expérimentale, INSERM, U.297, Bd Pierre Dramard, 13326 Marseille Cédex 15, France

NATO ASI Series, Vol. H 58
Neurocytochemical Methods
Edited by A. Calas and D. Eugène

monoaminergic neurons, in applications relying upon the high affinity transport mechanisms by which these neurons can take up and retain their own transmitter (Bosler and Calas, 1982 ; Descarries and Beaudet, 1983). It was also successfully applied to the identification of neurons that take up transmitter aminoacids such as $\gamma$-aminobutyric acid (GABA) (see Descarries and Beaudet, 1983). The coupling of radioautography and immunocytochemistry after *in vivo* or *in vitro* administration of the radioactive transmitter soon appeared as a useful way to push forward the limits of both techniques, by providing a means of characterizing the chemical identity of the radiolabeled elements and their neuronal targets and of investigating transmitter co-localization. It calls, however, for a number of methodological prerequisites which will be briefly discussed first (Bosler *et al.*, 1986).

## Methodological considerations

The specificity of radioautographic and immunocytochemical techniques should not be affected by their being combined on single tissue sections, provided that the antibodies can be shown not to cross-react with the exogenous tracers. Controls are then the same as those commonly used with either technique alone. In contrast, successful coupling of radioautography and immunocytochemistry requires acceptable compromises to be reached between the respective constraints of the two techniques, raising some questions about their respective sensitivities. In particular, brain tissue fixation should allow proper retention of the radioactive accumulated amines in their sites of uptake and storage, together with convenient preservation of the antigenicity of the endogenous molecules to be detected by immunocytochemistry, in adequately preserved tissue.

Glutaraldehyde, which provides for excellent ultrastructural preservation, has long been known also as the agent of choice for cross-linking biogenic amines and aminoacids to tissue proteins. Unfortunately, it reduces the immunoreactivity for most antigens, due either to loss of antigenicity or to failure of the antisera to penetrate deeply glutaraldehyde-fixed tissues. In order to reach a satisfactory compromise that will ensure optimal detection of both labels, glutaraldehyde is therefore usually added with parsimony (0.1% to 0.5%) to the appropriate paraformaldehyde - based solution that will be compatible with immunocytochemical labeling for the endogenous molecule to be simultaneously detected. In some cases, however, successful results have been obtained with fixatives containing concentrations of glutaraldehyde that were superior to 1%. One example is provided by Figure 2B which shows combined labeling for [$^3$H]dopamine (DA) uptake sites and somatostatin (SRIF) immunoreactivity in tissue fixed with not less than 3% glutaraldehyde (Bosler *et al.*, 1982).

The development of antibodies directed against small neurotransmitter molecules after use of the amines coupled to bovine serum albumine and glutaraldehyde as immunogens (Storm-Mathisen *et al.*, 1983 ; Geffard *et al.*, 1985) has further endowed the combined radioautographic and immunocytochemical technique with new potentialities when an exogenous tritiated amine is to be localized together with an endogenous amine. When such antibodies are to be used, indeed, tissue fixation with glutaraldehyde becomes a prerequisite for the corresponding transmitter to be amenable to immunocytochemical detection. In studies aimed at investigating the morphological substrate of catecholamine/serotonin (5-HT) interactions in the dorsomedial nucleus of the hypothalamus, we took advantage of the availability of anti-DA and anti-5-HT antibodies prepared in such a way. Figure 1 and Figure 2A illustrate successful results thus obtained after intraventricular administration of [$^3$H]5-HT or [$^3$H]DA followed by perfusion of the brains with 3% glutaraldehyde and pre-embedding immunostaining for DA and 5-HT, respectively.

Although both pre- and post-embedding immunostaining procedures can be combined with radioautography, as illustrated on Figure 1 and Figure 2, the former approach is generally preferred because of its apparent higher degree of sensitivity and reproducibility for the detection of most neuronal antigens. Incubating free-floating, unembedded tissue sections for relatively long periods of time, however, may entail some loss and/or displacement of the radioactivity from specific sites of accumulation. This is likely to contribute significantly to some underestimation of the results especially since, due to the limited intra-tissular penetration of antibodies in tissue prepared for electron microscopy (in which case detergents are omitted from the incubation media), ultrathin sections to be ultimately analyzed have to be taken from the outermost surface of the tissue blocks, i.e. where tracer molecules have a greater chance of being leaked during immunoprocessing. Then, the areas of optimal immunoreactivity do not correspond to those of optimal radioautographic labeling, implying that the electron microscopic preparations are usually interpretable only within those restricted areas of the sections showing an acceptable compromise. As a whole, it can be assumed that the sensitivity of both amine uptake radioautography and immunocytochemistry is diminished in combined experiments. This limitation, however, did not hamper a considerable amount of information from being obtained with the combined technique, regarding the morphological substrate of transmitter interactions in the central nervous system (CNS) (Bosler *et al.*, 1986).

## Examples of applications

Most double labeling studies having resorted to a combination of radioautography and immunocytochemistry were aimed at identifying the cellular targets of 5-HT neurons in various parts of the brain, such as the hypothalamus (see for example Bosler and Beaudet, 1985 ; Kiss and Halasz, 1985 ; Calas, 1985 ; Guy *et al.*, 1988), the substantia nigra (Hervé *et al.*, 1987), the nucleus raphe dorsalis (Magoul *et al.*, 1986 ; Soghomonian *et al.*, 1988), the nuclei of the solitary tract (Pickel *et al.*, 1984 ; Kachidian *et al.*, 1989), and the spinal cord (Glazer and Basbaum, 1984). In the dorsomedial nucleus of the hypothalamus, the afferent connectivity of those neurons found able to accumulate exogenous 5-HT and presumed to constitute an intrinsic serotoninergic system (Beaudet and Descarries, 1979 ; also see Frankfurt and Azmitia, 1983 ; Becquet *et al.*, 1990) has also been investigated by combined radioautography and immunocytochemistry. Figure 1 provides evidence that these neurons are not dopaminergic in nature, contrary to an initial proposition, while receiving dopaminergic synapses. It also illustrates the specificity of the radioautographic detection of 5-HT elements under the experimental conditions used (see legend), since an interspecific labeling of catecholaminergic neurons, which are the most likely to take up [$^3$H]5-HT non specifically, would have been expected to result in a superimposition of radioautographic grains on the immunopositive fibers.

Such superimpositions have been observed in other applications of the combined method, in which 5-HT/peptide and 5-HT/GABA co-localizations were formally demonstrated in various parts of the brain (Chan-Palay, 1979 ; Armstrong *et al.*, 1984 ; Calas, 1985 ; Harandi *et al.*, 1987). In another more recent study, Magoul *et al.* (1988) have used the combination of peptide immunocytochemistry with radioautography for [$^3$H]5-HT to localize retrogradely labeled 5-HT raphe neurons projecting to specific brain territories and identify their endogenous content. In this way, they were able to demonstrate that raphe 5-HT neurons projecting to the spinal cord, but not those projecting to the olfactory bulb, were co-expressing substance P.

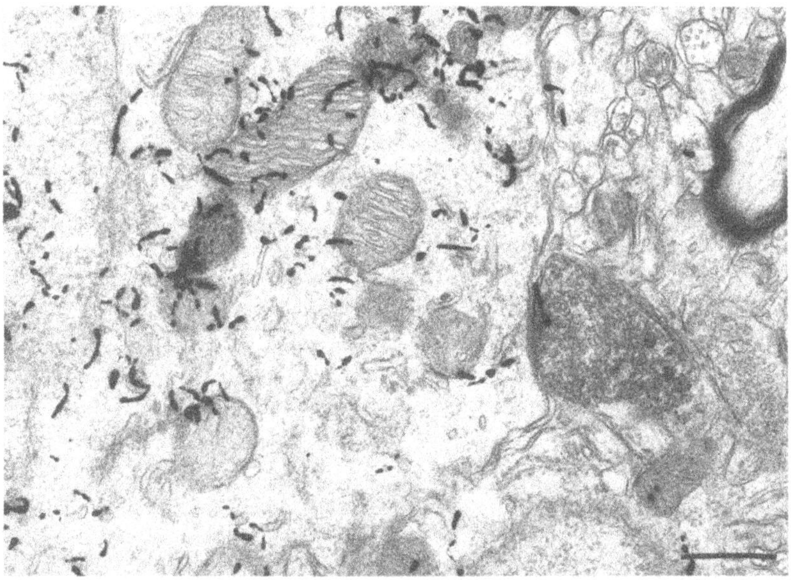

**Fig. 1** - Combined [$^3$H]5-HT uptake radioautography and pre-embedding DA immunocytochemistry (use of a primary antiserum against DA-glutaraldehyde - protein conjugate, Immunotech, Marseille). The tracer ($10^{-4}$M) has been administered intraventricularly with an excess of non-radioactive noradrenaline ($10^{-3}$M) in a pargyline - pretreated rat, and the brain subsequently fixed by perfusion with 3% buffered glutaraldehyde. The figure shows a [$^3$H]5-HT labeled cell body from the dorsomedial nucleus which is synaptically linked to a dopaminergic, peroxidase stained terminal. Scale bar : 0.5 µm.

Relatively fewer electron microscopic studies have been published combining immunocytochemistry and radioautography after administration of [$^3$H]catecholamines or [$^3$H]GABA (see, however, Bosler *et al.*, 1982 ; Pelletier, 1983 ; Nakai *et al.*, 1983 ; Gamrani *et al.*, 1984). This could be accounted for by the fact that the concentrations of glutaraldehyde that are required to adequately retain *in situ* exogenous catecholamines or aminoacids taken up by nerve terminals are greater than those usually required for [$^3$H]5-HT retention, at least in double labeling experimental conditions. In the case of adrenaline, histological requirements for the localization of its uptake sites in the CNS (Bosler and

Descarries, 1983) have even been proved incompatible, in our hands, with any subsequent immunodetection with a pre-embedding technique.

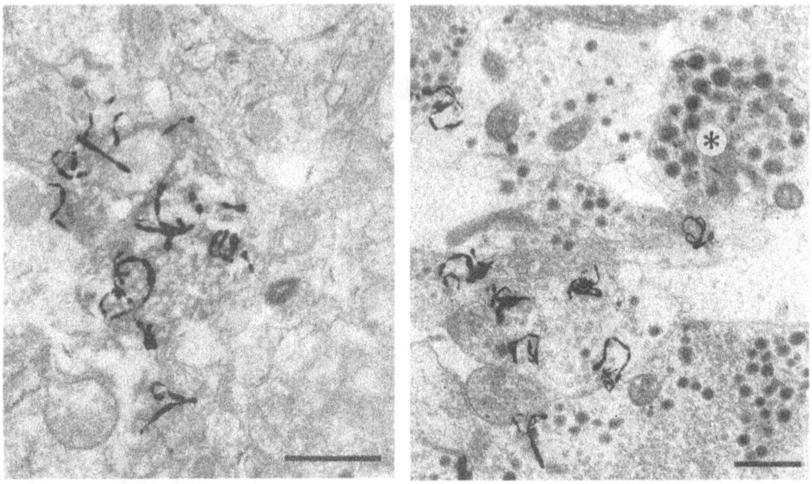

**Fig. 2** - Combination of radioautography and immunocytochemistry after intraventricular administration of $10^{-4}$M [$^3$H]DA and perfusion-fixation with 3% glutaraldehyde. **A** : Pre-embedding PAP staining for 5-HT with a primary antiserum against 5-HT-glutaraldehyde-protein conjugate (Immunotech, Marseille). A doubly labeled axon terminal from the dorsomedial nucleus is shown. **B** : Median eminence. Immunogold labeling for SRIF after osmic acid post-fixation and araldite embedding. Two types of axonal varicosities are separately detected. Note that the SRIF-immunoreactive varicosity (*) exhibits the characteristic features of neurosecretory endings. Reprinted with permission from Editions INSERM, Paris (Bosler *et al.*, 1982). Scale bars : 0.5 µm.

In a series of experiments aimed at investigating 5-HT/DA interactions in the dorsomedial nucleus using combined [$^3$H]DA radioautography and 5-HT immunocytochemistry, we circumvented the difficulty by using anti-5-HT antibodies that required tissue fixation with glutaraldehyde to recognize their corresponding antigen (see above, Geffard *et al.*, 1985). Many radioautographically labeled terminals could be detected in the structure after intraventricular administration of the [$^3$H]DA tracer and fixation of the brain with 3% glutaraldehyde prior to the processing of unembedded sections for 5-HT immunocytochemistry. The presumed 5-HT cell bodies in the nucleus did not exhibit any [$^3$H]DA labeling as expected (see above) but, surprisingly, some of the radiolabeled terminals were found to be also positive for 5-HT (Fig. 2A). Such double labelings could hardly be ascribed to non specific uptake of exogenous DA into serotoninergic terminals, in which case some [$^3$H]DA labeling of 5-HT cells would have been expected. They could indicate, therefore, that some terminals in the dorsomedial nucleus express a serotoninergic and catecholaminergic bipotentiality ([$^3$H]DA was likely to have been accumulated in dopaminergic, noradrenergic and adrenergic axon terminals under the experimental conditions used for radioautographic detection - see Bosler and Calas, 1982). Further support to this suggestion could eventually be provided by means of dual immunocytochemistry (see below), which would also permit specific and separate identification of the three kinds of catecholamines with appropriate antibodies. Another hypothesis would refer to the "dopaminoceptive" rather than true catecholaminergic nature of the [$^3$H]DA-accumulating 5-HT terminals and would then envisage a possible action of DA on 5-HT axonal release. We previously made a similar interpretation to explain the ability of some SRIF terminals to accumulate [$^3$H]DA in the median eminence (Bosler *et al.*, 1982). It has been proposed, indeed, that DA could have an intracellular action on neuropeptide release in that neurohemal region. Most SRIF-immunoreactive terminals in the median eminence, however, did not display any radiolabeling for the exogenous catecholamine (Fig. 2B).

While providing a powerful tool to investigate aminergic neurons in terms of their anatomo-functional organization as well as coexisting transmitters, afferent and efferent connectivity, the combination of radioautography and immunocytochemistry at the electron microscopic level may therefore also be useful to approach some dynamic cellular events such as the uptake of monoamines in chemically identified neuronal elements to be eventually characterized as monoamine targets rather than as true monoaminergic structures. In this sense, it is complementary to the other approaches available for localizing two *endogenous* molecules, namely double labeling immunocytochemical methods. The association of uptake radioautography with such methods should even offer considerable additional potentialities for investigating transmitter interactions (see Fig. 9).

## DUAL IMMUNOCYTOCHEMISTRY

Immunocytochemistry has developed on several fronts over the last few years as it remains the only way to localize any endogenous molecule, and most of all neuropeptides, provided the necessary antibodies and the staining methods used can be proven to be specific by adequate controls. In contrast to combined cytolabeling methods involving uptake radioautography, dual immunocytochemistry can therefore potentially be used to investigate any kind of chemospecific interactions, which explains why they also showed recent extensive developments as a result of an improved sensitivity and increased versatility of immunostaining procedures.

## Immunocytochemical markers

Most staining for electron microscopy has utilized horseradish peroxidase, conjugated to secondary antibodies or to biotin (ABC method) or as a preformed complex with anti-peroxidase antibodies (PAP method), and reaction of sections with diaminobenzidine (DAB) and hydrogen peroxide to localize tissue-bound immunoglobulins (IgG's). More recently, other chromogens have also proved useful for peroxidase staining at the electron microscopic level, namely benzidine dihydrochloride (BDHC) and tetramethylbenzidine (TMB), especially in double-labeling studies (see further). As alternatives to enzymatic markers, particulate tracers such as ferritin and colloidal gold have also been introduced in immunocytochemical research. They are applied on tissue sections as conjugates with IgG's or protein A, a major component of *Staphylococcus aureus* cell walls which selectively and strongly interacts with the Fc portion of most mammalian IgG molecules. Gold particles may be prepared in several homogeneous sizes, providing interesting advantages for multiple labeling. Also, unlike peroxidase reaction products which tend to obscure cytological details, they provide punctate dots of high electron density, allowing semi-quantitative analysis of relative numbers of particles to be eventually performed.

More recently, isotope-based methods, currently referred to as radioimmunocytochemical methods, have also developed as a consequence of the needs for more and more versatile labeling strategies. In its initial sense, radioimmunocytochemistry specifically designated the method devised by Larsson and Schwartz (1977) that involved forming a complex of a radiolabeled antigen bound to one of the two binding sites of its specific antibody, but this generic term is now applied for any immunocytochemical method in which the tissular antigens are detected by radioautography. At least when $^{3}H$ or $^{125}I$ are used as labels, radioimmunocytochemistry offers excellent sensitivity and acceptable resolution.

## Requirements for double immunolabeling

Combined immunocytochemical methods must fulfill at least three major requirements to be potentially applied to the investigation of chemically-defined neuronal interactions. In addition to allowing for clearcut electron microscopic discrimination of the labels for the individual antigens, the immunolabeling methods to be associated will have to retain an acceptable degree of sensitivity in the dual staining procedure and, most of all, will have to result in no cross-reactions between antisera of the first and second stages of the procedure. There are additional requirements when dual immunocytochemistry is to be applied to the demonstration of two neurochemical markers in one same cytological element. The accessibility of antibodies to the antigenic determinants of the molecule to be localized in the second stage of the procedure must not be impeded after the first staining for the co-localized substance while the associated labeling methods must provide sufficient electron microscopic resolution to allow separate localization of both antigens.

## Double enzymatic methods

They can be performed in a two step pre-embedding procedure, without chemical removal of the first antiserum. Although a recent paper has dealt with the association of peroxidase and β-galactosidase as differential enzymatic markers for dual antigen localization in CNS (Kohno *et al.*, 1989), most experimental protocols have resorted to dual peroxidase labels. One of these protocols was based on the consecutive use of the silver-gold intensified and non-intensified forms of the oxidized DAB chromogen (Liposits *et al.*, 1986 ; Gorcs *et al.*, 1986 ; Maegawa *et al.*, 1987). The silver intensification of the DAB reaction product significantly increases the electron density of, and results in metal deposition on, the labeled profiles (under the light microscope, they harbor a black color as opposed to the brown color yielded by the non-intensified DAB endproduct), then allowing for clearcut discrimination of the labels at both the light and electron microscopic levels. Other methods rely on the combined use of DAB as a first chromogen with either BDHC (Lakos and Basbaum, 1986 ; Levey *et al.*, 1986) or TMB (Norgren and Lehman, 1989) which produce a crystalline blue-green reaction product readily distinguishable from the diffuse brown reaction DAB product. All these techniques are difficult in use at the electron microscopic level, requiring the experimental conditions for fixation and staining procedure to be rigourosly controlled to achieve good sensitivity, convenient cytological preservation and reproducibility of the results. It has been stressed, in particular, that the ultrastructural aspect of the BDHC reaction product was highly dependent on both the composition and the acidity of the buffer used and that some loss of BDHC deposits, eventually resulting in a diffuse and brown DAB-like reaction, may occur during tissue processing for electron microscopy (e.g. if dehydration times are not kept to a minimum). This implies that the DAB/BDHC double labeling procedure, and this is also the case for the DAB/TMB and DAB intensified/non intensified method, is not useful for investigating transmitter co-localization since the presence of the two signals within one same cytological element will not necessarily reflect, although possibly providing some indication for, actual co-localization of the two corresponding antigens (François-Bellan *et al.*, 1990).

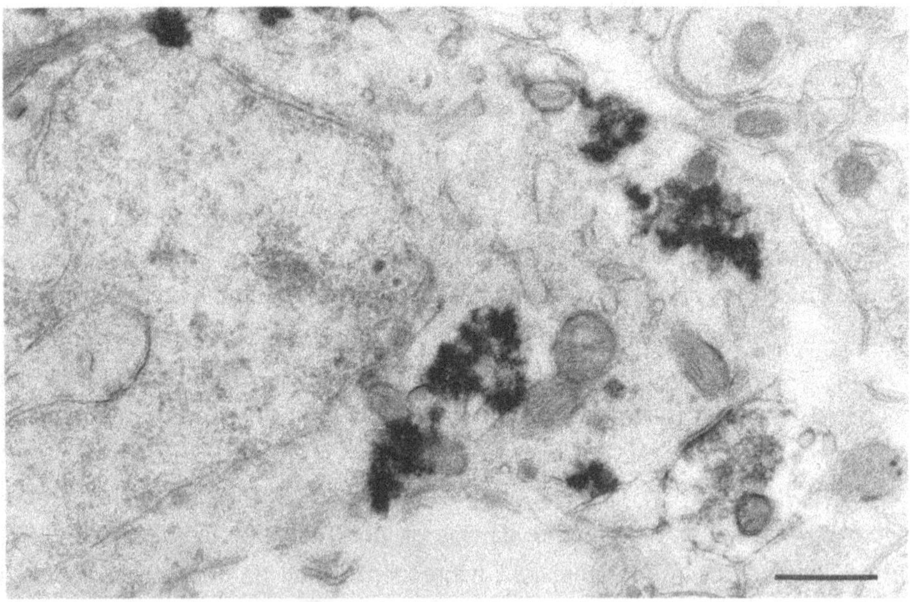

**Fig. 3** - Double PAP staining for GAD and VIP in the rat suprachiasmatic nucleus. Combined DAB/BDHC method carried out with primary antisera from different animal species (sheep anti-GAD and rabbit anti-VIP provided by Dr. M. Tappaz and Dr. M.C. Tonon, respectively). The diffuse DAB reaction product identifies a GAD-positive terminal and BDHC granular deposits a VIP-immunoreactive cell body. These two labeled elements are involved in a synaptic junctional complex. Scale bar : 0.5 µm.

In spite of their own drawbacks, each double peroxidase method possesses specific advantages that make them reliable for investigating chemospecific synaptic connectivity in the brain. Figure 3 illustrates double DAB/BDHC staining obtained with primary antisera raised in different animal species, but control experiments have demonstrated that double enzymatic methods may be used also with primary antisera from one same species without color mixing and antibody crossreactions (Levey *et al.*, 1986). This constitutes an important advantage of such methods, which has been ascribed to the inaccessibility of the immunoreagents used for the detection of the second tissue antigen to the sites of first antigen-antibody complexes, due to the shield that forms around the first antibody sequence after development.

## Combined peroxidase and gold-based methods

The association of peroxidase and gold marker compounds on one same electron microscopic section does not suffer from the same drawbacks, in terms of possible misinterpretation of the results, as those evoked above. However, when used in a double pre-embedding procedure, significant membrane damage must occur to achieve penetration of gold particles in the thickness of sections. This problem has been successfully circumvented by using gold particles of no more than 5 nm in diameter and further increasing their size by silver precipitation around them (Van den Pol, 1985a). In spite of a reduced severity of membrane damage, the pre-embedding silver-intensified gold technique retains some important limitations mainly related to particle penetration and ultrastructural preservation. An alternative of this method involving the use of ferritin instead of colloidal gold, without intensification of the signal, has also been successfully carried out (Leranth *et al.*, 1985).

Immunogold methods, in fact, have mainly resorted to post-embedding labeling procedures, which possess the considerable advantages over pre-embedding procedures of circumventing the problem of antibody penetration and of being easier in use and much less time consuming. Then, for dual labeling purposes, the first antigen may be localized with a peroxidase-based method prior to embedding and the second one by using a post-embedding immunogold method (Van den Pol, 1985b ; Theodosis *et al.*, 1986 ; André *et al.*, 1987). An example is seen in Figure 4 which shows two differentially labeled nerve terminals in the rat red nucleus, one positive for 5-HT (PAP staining) the other positive for GABA (gold-adsorbed IgG labeling). With this double labeling method, there was no apparent problem associated with possible cross-reactivity of the IgG-gold label with IgG's embedded in tissue, even when using primary antisera from the same animal species. However, questions must be raised as to the validity of the method for investigating potential cellular co-localization since the deposit of peroxidase reaction products around the antigenic determinants of the first molecule to be detected and, most of all, subsequent resin embedding might be expected to impede further immunolabeling for any molecule that would be co-localized with the first one.

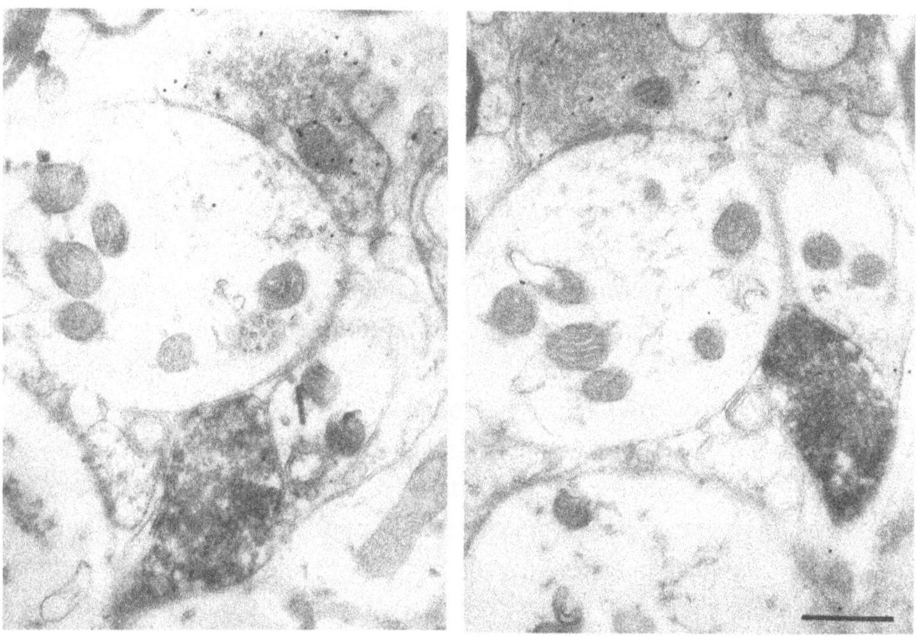

**Fig. 4** - Dual 5-HT/GABA immunocytochemistry. Both primary antisera were from rabbit (Immunotech, Marseille). 5-HT immunoreactivity has been first detected by PAP staining in a pre-embedding procedure and GABA immunoreactivity after application of gold-adsorbed secondary IgG's on ultrathin sections from araldite-embedded specimens (post-embedding labeling). Two immunopositive, differentially labeled endings are illustrated converging onto the same dendritic profile on these two consecutive sections from the rat red nucleus. Scale bar : 0.5 µm. Reprinted with permission from Pergamon Press PLC (André *et al.*, 1987).

## Double post-embedding immunogold methods

They are based on the combined use of different size, small and large, gold particles adsorbed either to secondary anti-species antibodies or to protein A to reveal two distinct antigens on the same ultrathin section. Essentially two main categories of dual post-embedding methods, referred to as simultaneous or sequential methods, have been used so far. With the simultaneous double-labeling methods, the two primary antisera are mixed and applied together on the sections. The sites of binding are subsequently revealed either using gold-labeled corresponding antigens, thanks to the bivalency of antibody molecules (gold-labeled antigen detection-GLAD-method) (Larsson, 1979) or using species-specific labeled secondary antibodies, in which case the availability of primary antisera derived from different species is required (Tapia *et al.*, 1983 ; Merighi *et al.*, 1989) (see Fig. 5). In practise, with the GLAD method, each antigen to be localized requires it own gold-labeled reagent, which makes the GLAD method laborious and expensive, and therefore prohibitive on a high scale in spite of its interesting potentialities for multiple antigen detection.

Sequential techniques have essentially resorted to one of the three following experimental principles : (1) double protein A staining which involved blocking potential unreacted IgG binding sites with an excess of free protein A after application of the first primary antiserum (Geuze *et al.*, 1981), (2) double-face staining, i.e. separate immunogold labeling of each side of the ultrathin section (Bendayan, 1982), which is rendered possible by the fact that antibodies do not penetrate into the plastic sections, and (3) double IgG-gold labeling involving inactivation of free binding sites on the first antigen-antibody complex by formaldehyde vapour treatment or silver enhancement (Wang and Larsson, 1985 ; Bienz *et al.*, 1986). In spite of the specific drawbacks associated with each of them, all those techniques have been proved to offer considerable potentialities for electron microscopic dual labeling, including for investigations on transmitter co-localizations. To date, however, they have been mostly exploited in studies on non-neural tissues. That comparatively fewer applications have been made to the nervous system is to be ascribed to some difficulty in localizing most molecules involved in neuronal communication in osmicated, plastic-embedded specimens. However, a few studies have been recently published demonstrating excellent sensitive dual immunogold stainings for e.g. GABA, tyrosine hydroxylase and/or certain neuropeptides on ultrathin sections from CNS regions (Beauvillain *et al.*, 1984 ; Vuillez *et al.*, 1987 ; Decavel *et al.*, 1989 ; Merighi *et al.*, 1989).

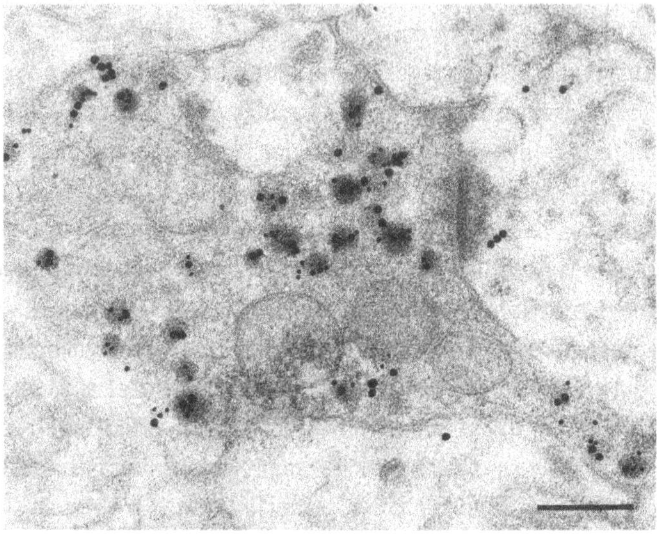

**Fig. 5** - Post-embedding simultaneous double immunogold labeling for substance P (10 nm particles) and calcitonin gene-related peptide (CGRP, 20 nm particles) in the rat dorsal horn. Co-localization of the peptide immunoreactivities in large granular vesicles of a synaptic terminal is demonstrated. Immunolabeling was performed on osmicated, epoxy-embedded tissue by incubating the grid with a mixture of rat anti-substance P and rabbit anti-CGRP antisera, then with appropriate gold-adsorbed secondary IgG's. Scale bar : 0.2 µm. Reproduced, with permission, from Merighi *et al.* (1989). Courtesy of Dr. D.T. Theodosis.

The study by Merighi *et al.* (1989) has evaluated the impact of tissue preparation and embedding on both the sensitivity and accuracy of different double immunogold techniques while re-emphasizing the necessary compromises that must be reached. It has been stressed for example that, in comparison with conventional epoxy embedding, low-temperature embedding in Lowikryl K4M (also see Vuillez *et al.*, 1987) and most of all, cryoultramicrotomy, allowed more sensitive immunogold labelings for various peptide antigens to be obtained in the CNS, but this was detrimental to the cytological preservation and, therefore, to accuracy of the antigen detection.

## Combined immunoperoxidase and radioimmunocytochemical methods

Association of peroxidase and isotope tracers for dual immunocytochemistry has been first introduced by Cuello *et al.* (1982) who reported on the production of internally $^{3}H$-labeled monoclonal antibodies. Since then, the usefulness of combining peroxidase immunocytochemistry with radioimmunocytochemistry for achieving sensitive dual staining together with clearcut differentiation of the labels has also been shown by a number of studies having employed, in pre-embedding procedures, radiolabeled secondary antibodies or biotine (see for example Glazer *et al.*, 1984 ; Hunt and Mantyh, 1984 ; Pickel *et al.*, 1986, and a series of other works by Pickel's group - e.g. Milner *et al.*, 1989) (Fig. 6) or radiolabeled peptide-antibody complexes prepared in vitro according to the technique of Larsson and Schwartz (1977) (Alonso and Siaud, 1989). The advantages of such combinations in terms of the preservation of cellular integrity and/or, above all, applicability for localizing two antigens in the same neuronal element have also been well illustrated (e.g. Cuello *et al.*, 1983 ; Milner *et al.*, 1989 ; Alonso and Siaud, 1989 ; François-Bellan *et al.*, 1990).

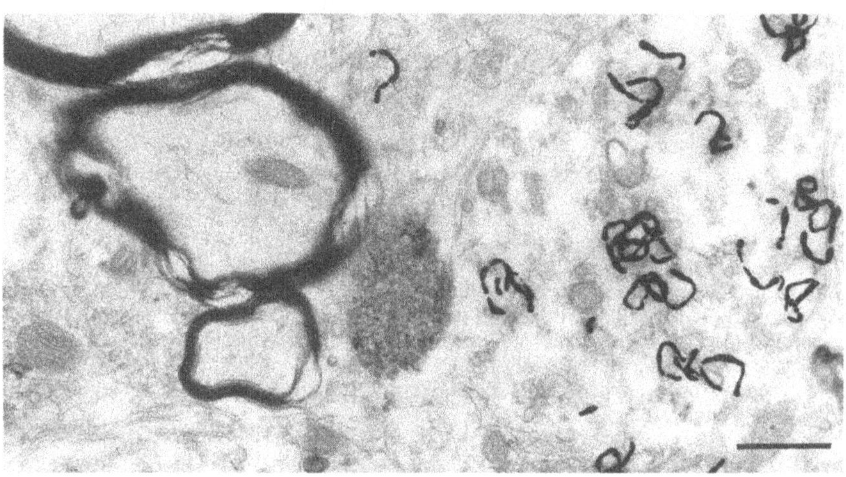

**Fig. 6** - Combined immunoperoxidase staining for phenylethanolamine N-methyl transferase (PNMT) (PAP technique) and radioimmunocytochemical labeling for tyrosine hydroxylase (TH) with $^{125}I$-labeled secondary IgG's in the rat locus coeruleus. Rabbit polyclonal anti-PNMT antibodies (provided by Dr. L. Denoroy) and mouse monoclonal anti-TH antibodies (Boehringer Mannheim) were used. The PNMT-positive (i.e adrenaline-synthesizing) terminal illustrated here forms a synapse with a TH-radiolabeled dendritic shaft (also see Kachidian *et al.*, 1990). Scale bar : 0.5 µm.

Not all of the protocols which have been involved so far, however, have proven valid for selective dual labeling with two primary antibodies raised in the same animal species. We have recently developed a method that definitely circumvents the problem of possible cross-reactions between reagents of the first and second stages of the dual labeling sequence, even when primary antisera from one same species are used. The procedure involves detection of the first antigen using peroxidase-conjugated fragments, and after revelation with intensified DAB, labeling the second one using $^{125}$I-labeled protein A (Fig. 7). It has been designed from the theoretical background that the revelation of the first antigen with Fab fragments in lieu of whole IgG's and in such a way that steric hindrance is significantly enhanced at the site of the antigen-antibody complex (DAB intensification) was likely to minimize any subsequent binding of protein A to Fc fragments other than those belonging to the second primary IgG's.

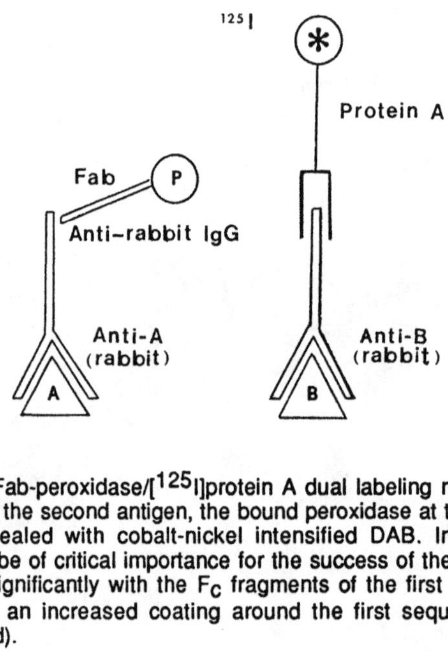


**Fig. 7** - Principle of the Fab-peroxidase/[$^{125}$I]protein A dual labeling method (see text). Prior to protein A radiolabeling for the second antigen, the bound peroxidase at the site of the first antigen-antibody complex is revealed with cobalt-nickel intensified DAB. Intensification of the DAB endproduct was found to be of critical importance for the success of the method, as it did prevent protein A of interacting significantly with the $F_c$ fragments of the first primary IgG's, which was presumed to result from an increased coating around the first sequence of immunoreagents (Kachidian *et al.*, submitted).

That the technique is valid has been assessed by controls having involved either processing sections for dual staining with the second primary antiserum omitted, in which case only a light and diffuse radiolabeling superimposed on the peroxidase staining was observed, or dual staining of some histological sections for two antigens known to be localized in separate neuronal populations in the area tested, e.g. neuropeptide Y (NPY) and tyrosine hydroxylase (TH) in the arcuate nucleus of the hypothalamus. In the latter case, irrespective of the order of application of primary antibodies (peroxidase staining or $^{125}$I radiolabeling), the distributional patterns of NPY and TH positive neurons closely matched previous immunohistochemical localizations (Kachidian *et al.*, submitted).

The Fab/[$^{125}$I]protein A technique can be said to constitute a new reliable and sensitive tool for dual antigen localization, even when the available primary antisera have been generated in one same species. As shown in Figure 8 which provides first direct evidence that adrenaline may be synthesized in central nerve terminals also containing NPY, the technique can be successfully applied to electron microscope co-localization studies. It can then be viewed as offering a possible alternative to other comparable dual labeling methods, e.g. to the ingenious technique involving application of a radiolabeled antigen - antibody complex (Alonso and Siaud, 1989) and whose prerequisite for success is, therefore, that the corresponding tissular antigen to be detected should not be modified by the fixation in such a way that it would not be recognized by antibodies against the native molecule.

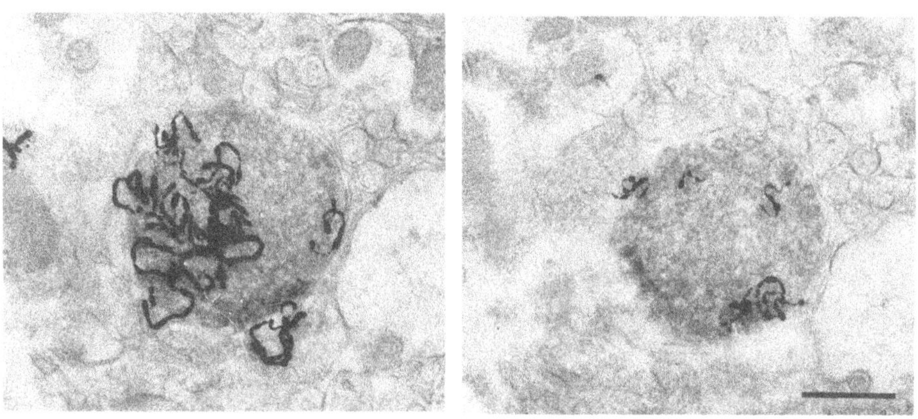

**Fig. 8** - Application of the Fab-peroxidase/[$^{125}$I]protein A method to the demonstration of PNMT/NPY axonal coexistence in the paraventricular nucleus of the hypothalamus, after PNMT peroxidase staining and NPY radiolabeling with two primary antisera raised in rabbit (antisera provided by Dr. L. Denoroy and Dr. G. Pelletier, respectively). Serial sections. Scale bar : 0.5 µm.

## FUTURE GOALS : MULTIPLE LABELING STRATEGIES

Combined cytolabeling methods have now progressed in such a way that the development of new strategies for the routine identification of more than two molecules on one and the same electron microscopic section no longer appears as an unrealistic goal. Successful results have already been obtained in this respect. Figure 9 shows an example in which the radioautographic detection of [$^3$H]5-HT uptake sites was coupled to a dual immunocytochemical staining for VIP and GAD in the hypothalamus (François-Bellan *et al.*, in prep.).

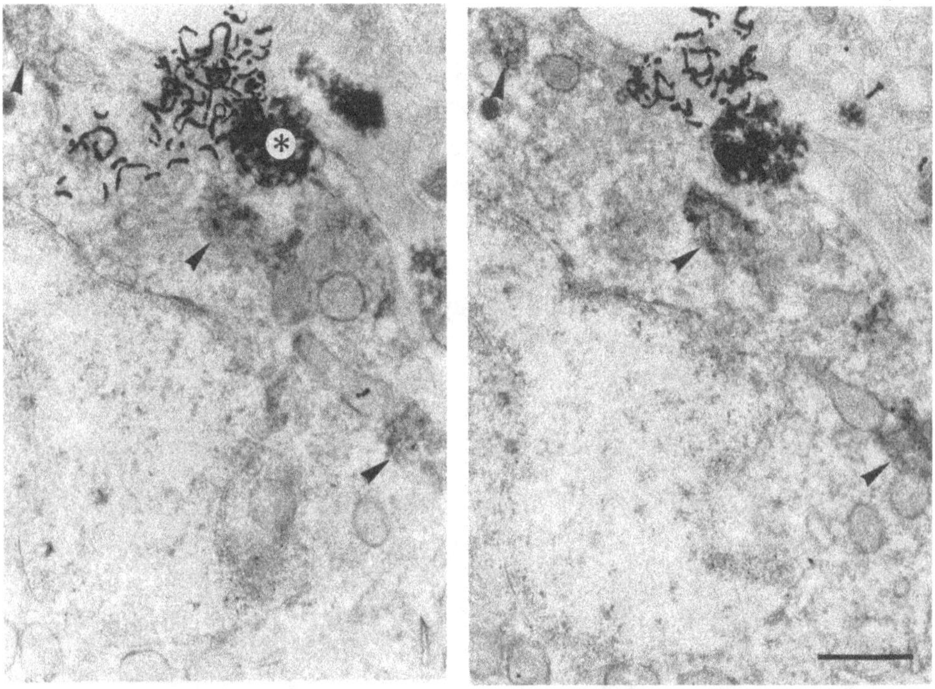

**Fig. 9** - Suprachiasmatic nucleus. Example of triple labeling obtained after radioautographic detection of intraventricularly administered [$^3$H]5-HT in sections that had been previously immunostained for GAD and VIP with the DAB/BDHC method as shown in Fig. 3. In the two consecutive planes of section illustrated here, a GABAergic terminal (DAB staining, asterisk) and a serotoninergic terminal (aggregated silver grains) are identified contacting the same VIP-immunoreactive neuron (BDHC granular deposits, arrowheads). Scale bar : 0.5 µm.

There are a few other works reporting on successful localizations of more than two molecules up to the electron microscopic level, either after application of a triple immunogold staining method (Doerr-Schott and Lichte, 1986 ; Merighi *et al.*, 1990) or after combination of retrograde tracing with double immunolabeling carried out on one same section (Zaborsky *et al.*, 1986) or, according to the so-called "mirror technique" (Kubota *et al.*, 1985), on consecutive sections (Kawai *et al.*, 1987). Combined immunocytochemical and neuronal tracing methods did gain popularity in the last few years, as they provide invaluable tools to reconcile classical neuroanatomy with chemical neuroanatomy (see for examples Gerfen and Sawchenko, 1985 ; Wouterlood *et al.*, 1987 ; Luppi *et al.*, 1988). Their routine incorporation in multiple labeling strategies in the future would be therefore all the less surprising, since such combined methods can be expected to permit significant progress in the understanding of the anatomo-functional organization of the nervous system.

Another major challenge for the next few years will be to design reliable methods for associating such combined cytolabeling tools with receptor localization techniques. Dual labeling electron microscopic results have already been published reporting on the relationships between immunocytochemically identified receptors and terminals (Triller *et al.*, 1988) and on the localization of radiolabeled transmitter-specific receptors on chemically - defined immunopositive neurons (Szigethy *et al.*, 1990). It can be then reasonably hoped that given subsets of neurons will be ultimately amenable to electron microscope characterization not only in terms of their specific transmitter molecule(s) and of the chemical nature and origin of their afferent inputs, but also in terms of the different types of receptors specifically associated with their surface. Moreover, given the considerable recent developments of *in situ* hybridization histochemistry for localizing specific nucleotide sequences of messenger RNAs, it can finally be expected that the association of the potentialities of this important technical breakthrough with those of combined cytolabeling tools and of receptor labeling technology will ultimately permit major advances in the field of chemo-anatomical research.

## REFERENCES

Alonso G, Siaud P (1989) Combined use of immunoperoxidase and radioimmunocytochemistry for double immunocytochemical labeling of neurons at light and electron microscopic level. J Histochem Cytochem 37:1799-1809

André D, Vuillon-Cacciuttolo G, Bosler O (1987) GABA nerve endings in the rat red nucleus. Combined detection with serotonin terminals using dual immunocytochemistry. Neuroscience 23:1095-1102

Armstrong DM, Miller RJ, Beaudet A, Pickel VM (1984) Enkephalin-like immunoreactivity in rat area postrema : Ultrastructural localization and coexistence with serotonin. Brain Res 310:269-278

Beaudet A, Descarries L (1979) Radioautographic characterization of a serotonin-accumulating nerve cell group in the adult rat hypothalamus. Brain Res 160:231-243

Beauvillain JC, Tramu G, Garaud JC (1984) Coexistence of substances related to enkephalin and somatostatin in granules of the guinea-pig median eminence : demonstration by use of colloidal gold immunocytochemical methods. Brain Res 301:389-393

Becquet D, François-Bellan AM, Boudouresque F, Faudon M, Héry F, Guillaume V, Héry M (1990) Serotonin synthesis from tryptophan by hypothalamic cells in serum-free medium culture. Dev Brain Res 54:142-146

Bendayan M (1982) Double immunocytochemical labeling applying the protein A-gold technique. J Histochem Cytochem 30:81-85

Bienz K, Egger D, Pasamontes L (1986) Electron microscopic immunocytochemistry. Silver enhancement of colloidal gold marker allows double labeling with the same primary antibody. J Histochem Cytochem 34, 1337-1342

Bosler O, Beaudet A (1985) VIP neurons as prime synaptic targets for serotonin afferents in rat suprachiasmatic nucleus : a combined radioautographic and immunocytochemical study. J Neurocytol 14:749-763

Bosler O, Beaudet A, Pickel VM (1986) Characterization of chemically defined neurons and their cellular relationships by combined immunocytochemistry and radioautographic localization of transmitter uptake sites. J Electr Microsc Tech 4:21-39

Bosler O, Bloch B, Bugnon C, Calas A (1982) Bases morphologiques des interactions monoamines-peptides dans l'hypothalamus neuroendocrine du rat. Données radioautographiques et immunocytochimiques. *In* A. Tixier-Vidal and P. Richard (eds), Régulations cellulaires multihormonales en neuroendocrinologie, Colloques de l' INSERM, vol. 110, Editions INSERM, Paris, pp. 17-36

Bosler O, Calas A (1982) Radioautographic investigation of monoaminergic neurons : an evaluation. Brain Res Bull 9:151-169

Bosler O, Descarries L (1983) Uptake and retention of [$^{3}$H]adrenaline by central monoaminergic neurons : a light- and electron-microscope radioautographic study after intraventricular administration in the rat. Neuroscience 3:561-583

Calas A (1985) Morphological correlates of chemically specified neuronal interactions in the hypothalamo-hypophyseal area. Neurochem Int 7:927-940

Chan-Palay V (1979) Combined immunocytochemistry and autoradiography after in vivo injections of monoclonal antibody to substance P and $^{3}$H-serotonin : coexistence of two putative transmitters in single raphe cells and fiber plexuses. Anat Embryol 156:241-254

Cuello AC, Milstein C, Galfré G (1983) Preparation and application of monoclonal antibodies for immunohistochemistry and immunocytochemistry. In Cuello AC (ed), Immunohistochemistry, IBRO Handbook Series : Methods in the Neuroscience, Vol 3, John Wiley, Chichester, pp. 215-255

Cuello AC, Priestley JV, Milstein C (1982) Immunocytochemistry with internally labeled monoclonal antibodies. Proc Natl Acad Sci USA, 79:665-669

Decavel C, Dubourg P, Leon-Henri B, Geffard M, Calas A (1989) Simultaneous immunogold labeling of GABAergic terminals and vasopressin-containing neurons in the rat paraventricular nucleus. Cell Tiss Res 255:77-80

Descarries L, Beaudet A (1983) The use of radioautography for investigating transmitter-specific neurons. In A. Björklund and T. Hökfelt (eds), Handbook of Chemical Neuroanatomy, Vol. 1 : Methods in Chemical Neuroanatomy, Elsevier, Amsterdam, pp 286-364

Doerr-Schott J, Lichte CM (1986) A triple ultrastructural immunogold staining method. Application to the simultaneous demonstration of three hypophyseal hormones. J Histochem Cytochem 34:1101-1104

François-Bellan AM, Kachidian P, Dusticier G, Tonon MC, Vaudry H, Bosler O (1990) GABA neurons in the rat suprachiasmatic nucleus : involvement in chemospecific synaptic circuitry and evidence for GAD-peptide colocalization. J Neurocytol, in press.

Frankfurt M, Azmitia EC (1983) The effect of intracerebral injections of 5,7-dihydroxytryptamine and 6-hydroxydopamine on the 5-HT-immunoreactive cell bodies and fibers in the adult rat hypothalamus. Brain Res 261:91-99.

Gamrani H, Harandi M, Belin MF, Dubois MP, Calas A (1984) Direct electron microscopic evidence for the coexistence of GABA uptake and endogenous serotonin in the same rat central neurons by coupled radioautographic and immunohistochemical procedures. Neurosci Lett 48:25-30

Geffard M, Heinrich-Rock AM, Dulluc J, Séguéla P (1985) Antisera against small neurotransmitter-like molecules. Neurochem Int 7:403-413

Gerfen CR, Sawchenko, PE (1985) A method for anterograde axonal tracing of chemically specified circuits in the central nervous system : combined *Phaseolus vulgaris* - leucoagglutinin (PHA-L) tract tracing and immunohistochemistry. Brain Res 343:144-150

Geuze HJ, SLot JW, Van der Ley PA, Scheffer RCT (1981) Use of colloidal gold particles in double-labelling immunoelectron microscopy of ultrathin frozen tissue sections. J Cell Biol 89:653-665

Glazer EJ, Basbaum AI (1984) Axons which take up [$^{3}$H]serotonin are presynaptic to enkephalin immunoreactive neurons in cat dorsal horn. Brain Res 298:386-391

Glazer EJ, Ramachandran J, Basbaum AI (1984) Radioimmunocytochemistry using a tritiated goat anti-rabbit second antibody. J Histochem Cytochem 32 : 778-782

Görcs TJ, Leranth C, MacLusky NJ (1986) The use of gold-substituted silver-intensified diaminobenzidine (DAB) and non-intensified DAB for simultaneous electron microscopic immunoperoxidase labeling of tyrosine hydroxylase and glutamic acid decarboxylase immunoreactivity in the rat medial preoptic area. J Histochem Cytochem 34:1439-1447

Guy J, Pelletier G, Bosler O (1988) Serotonin innervation of neuropeptide Y-containing neurons in the rat arcuate nucleus. Neurosci Lett 85:9-13

Harandi M, Aguera M, Gamrani H, Didier M, Maitre M, Calas A, Belin MF (1987) γ-aminobutyric acid and 5-hydroxytryptamine interrelationship in the rat nucleus raphe dorsalis : combination of radioautographic and immunocytochemical techniques at light and electron microscopy levels. Neuroscience 21:237-251

Hervé D, Pickel VM, Joh TH, Beaudet A (1987) Serotonin axon terminals in the ventral tegmental area of the rat : fine structure and synaptic input to dopaminergic neurons. Brain Res 435:71-83

Hunt SP, Mantyh PW (1984) Radioimmunocytochemistry with [$^3$H]biotin. Brain Res 291:203-217

Kachidian P, Astier B, Renaud B, Bosler O (1990) Adrenergic innervation of noradrenergic locus coeruleus neurons. A dual labeling immunocytochemical study in the rat. Neurosci Lett 109:23-29

Kachidian P, Colin I, Astier B, Renaud B, Bosler O (1989) Are adrenergic neurons subject to a serotoninergic influence in the nucleus tractus solitarii ? A morphological and biochemical study in the rat. Cell Tiss Res 258:603-610

Kawai Y, Takagi H, Kumoi Y, Shiosaka S, Tohyama M (1987) Nigrostriatal dopamine neurons receive substance P-ergic inputs in the substantia nigra : application of the immunoelectron microscopic mirror technique to fluorescent double-staining for transmitter-specific projections. Brain Res 401:371-376

Kiss J, Halasz B (1985) Demonstration of serotoninergic axons terminating on luteinizing hormone-releasing hormone neurons in the preoptic area of the rat using a combination of immunocytochemistry and high resolution autoradiography. Neuroscience 14:69-78

Kohno J, Shinoda K, Kawai Y, Peng Y, Ono K, Shiotani Y (1989) Enkephalin fibers synapse on cholinergic neurons in the rat sacral intermediolateral nucleus : a double-immunostaining at the light and electron microscopic levels. Neuroscience 28:487-494

Kubota Y, Takagi H, Morishima Y, Powell JF, Smith AD (1985) Synaptic interaction between catecholaminergic neurons and substance P-immunoreactive axons in the caudal part of the nucleus of the solitary tract of the rat : demonstration by the electron microscopic mirror technique. Brain Res 333:188-192

Lakos S, Basbaum AI (1986) Benzidine dihydrochloride as a chromogen for single- and double-label light and electron microscopic immunocytochemical studies. J Histochem Cytochem 34:1047-1056

Larsson LI (1979) Simultaneous ultrastructural demonstration of multiple peptides in endocrine cells by a novel immunocytochemical method. Nature 282:743-746

Larsson LI, Schwartz TW (1977) Radioimmunocytochemistry. A novel immunocytochemical principle. J Histochem Cytochem 25:1140-1146

Leranth Cs, Sakamoto H, MacLusky NJ, Shanabrough M, Naftolin F (1985) Application of avidin-ferritin and peroxidase as contrasting electron-dense markers for simultaneous electron microscopic immunocytochemical labelling of glutamic acid decarboxylase and tyrosine hydroxylase in the rat arcuate nucleus. Histochemistry 82:165-168

Levey AI, Bolam JP, Rye DB, Hallanger AE, Demuth RM, Mesulam MM, Wainer BH (1986) A light and electron microscopic procedure for sequential double antigen localization using diaminobenzidine and benzidine dihydrochloride. J Histochem Cytochem 34:1449-1457

Liposits Zs, Sherman D, Phelix C, Paull WK (1986) A combined light and electron microscopic immunocytochemical method for the simultaneous localization of multiple tissue antigens. Tyrosine hydroxylase immunoreactive innervation of corticotropin releasing factor synthesizing neurons in the paraventricular nucleus of the rat. Histochemistry 85:95-106

Luppi PH, Sakai K, Font P, Salvert D, Jouvet M (1988) The nuclei of origin of monoaminergic, peptidergic, and cholinergic afferents to the cat nucleus reticularis magnocellularis : a double-labeling study with cholera toxin as a retrograde tracer. J Comp Neurol 277:1-20

Maegawa M, Hisano S, Tsuruo Y, Katoch S, Nakanishi J, Chikamori-Aoyama M, Daikoku S (1987) Differential immunolabeling for electron microscopy of diverse peptidergic neurons. J Histochem Cytochem 35:251-255

Magoul R, Oblin A, Calas A, Araneda S (1988) Serotonergic projections to the spinal cord but not those to the olfactory bulb also contain substance P. A combined immunocytochemical and autoradiographic study following retrograde axonal transport of [$^3$H]serotonin labeled products. Neuroscience 26:959-969

Magoul R, Onteniente B, Oblin A, Calas A (1986) Inter- and intracellular relationship of substance P-containing neurons with serotonin and GABA in the dorsal raphe nucleus : combination of autoradiographic and immunocytochemical techniques. J Histochem Cytochem 34:735-742

Merighi A, Polak JM, Fumagalli G, Theodosis DT (1989) Ultrastructural localization of neuropeptides and GABA in rat dorsal horn : a comparison of different immunogold labeling techniques. J Histochem Cytochem 37:529-540

Merighi A, Polak JM, Theodosis DT (1990) Ultrastructural visualization of glutamate and aspartate immunoreactivities in the rat dorsal horn, with special reference to the co-localization of glutamate, substance P and calcitonin gene-related peptide. Neuroscience, in press

Milner TA, Pickel VM, Reis DJ (1989) Ultrastructural basis for interactions between central opioids and catecholamines. I. Rostral ventrolateral medulla. J. Neurosci 9:2114-2130

Nakai Y, Shioda S, Ochiai S, Kudo J, Hashimoto A (1983) Ultrastructural relationship between monoamine- and TRH-containing axons in the rat median eminence as revealed by combined autoradiography and immunocytochemistry in the same tissue section. Cell Tiss Res 230:1-14

Norgren RB, Lehman MN (1989) A double-label pre-embedding immunoperoxidase technique for electron microscopy using diaminobenzidine and tetramethylbenzidine as markers. J Histochem Cytochem 37:1283-1289

Pelletier G (1983) Identification of endings containing dopamine and vasopressin in the rat posterior pituitary by a combination of radioautography and immunocytochemistry at the ultrastructural level. J Histochem Cytochem 31:562-564

Pickel VM, Chan J, Milner TA (1986) Autoradiographic detection of [$^{125}$I]-secondary antiserum : a sensitive light and electron microscopic labeling method compatible with peroxidase immunocytochemistry for dual localization of neuronal antigens. J Histochem Cytochem 34:707-718

Pickel VM, Joh TH, Chan J, Beaudet A (1984) Serotoninergic terminals : ultrastructure and synaptic interaction with catecholamine-containing neurons in the medial nuclei of the solitary tract. J Comp Neurol 225:291-301

Soghomonian JJ, Beaudet A, Descarries L (1988) Ultrastructural relationships of central serotonin neurons. In N.N. Osborne and M. Hamon (eds), Neuronal Serotonin, John Wiley, New York, pp. 57-92

Storm-Mathisen J, Leknes AK, Bore AJ, Vocaland JL, Edminson P, Hang FMS, Ottersen OP (1983) First visualization of glutamate and GABA in neurons by immunocytochemistry. Nature 301:517-520

Szigethy E, Leonard K, Beaudet A (1990) Ultrastructural localization of $^{125}$I neurotensin binding sites to cholinergic neurons of the rat nucleus basalis magnocellularis. Neuroscience 36:377-391

Tapia FJ, Varndell IM, Probert L, De Mey J, Polak JM (1983) Double immunogold staining method for the simultaneous ultrastructural localization of regulatory peptides. J Histochem Cytochem 31:977-981

Theodosis DT, Paut L, Tappaz ML (1986) Immunocytochemical analysis of the GABAergic innervation of oxytocin- and vasopressin-secreting neurons in the rat supraoptic nucleus. Neuroscience 19:207-222

Triller A, Cluzeaud F, Korn H (1988) Subcellular localization of glycine receptor : transmembrane organization and relationships with GABA- and serotonine-containing terminals. 11th Annual Meeting European Neuroscience Association, Eur J Neurosci (suppl.), Abstract 85.3

Van den Pol AN (1985a) Silver-intensified gold and peroxidase as dual ultrastructural immunolabels for pre- and postsynaptic neurotransmitters. Science 228:332-335

Van den Pol AN (1985b) Dual ultrastructural localization of two neurotransmitter-related antigens : colloidal gold-labeled neurophysin-immunoreactive supraoptic neurons receive peroxidase-labeled glutamate decarboxylase- or gold-labeled GABA-immunoreactive synapses. J Neurosci 5:2940-2954

Vuillez P, Perez SC, Stoeckel ME (1987) Colocalization of GABA and tyrosine hydroxylase immunoreactivities in the axons innervating the neurointermediate lobe of the rat pituitary : an ultrastructural immunogold study. Neurosci Lett 79:53-58

Wang BL, Larsson LI (1985) Simultaneous demonstration of multiple antigens by indirect immunofluorescence or immunogold staining. Histochemistry 83:47-56

Wouterlood FG, Bol JGJM, Steinbusch HWM (1987) Double-label immunocytochemistry : Combination of anterograde neuroanatomical tracing with *Phaseolus vulgaris* - leucoagglutinin and enzyme immunocytochemistry of target neurons. J Histochem Cytochem 35:817-823

Zaborszky L, Heimer L, Eckenstein F, Leranth C (1986) GABAergic input to cholinergic forebrain neurons : an ultrastructural study using retrograde tracing of HRP and double immunolabeling. J Comp Neurol 250:282-295

**Acknowledgements** - Part of the personal studies referred to in the text were supported by grants 856003 and 886003 from the INSERM (O.B.). We thank Mrs A. Aouane for expert photographic work, Mrs B. Besson for typing and make-up of the manuscript and Mrs T. Sefer for linguistic revision.

# MOLECULAR NEUROANATOMY OF SYNAPSES, CELLS AND SYSTEMS IN THE BRAIN

**PETER SOMOGYI**
Medical Research Council,
Anatomical Neuropharmacology Unit
Oxford University
South Parks Rd.
Oxford OX1 3QT
England

Molecular neuroanatomy involves the identification of the molecular machinery of neurons and glia within the framework of connections and physiological properties of cells. It provides information on the physical constraints imposed by the brain on neuronal circuits in meeting their functional demands. Thanks to tremendous technical advances, molecular neuroanatomy often paves the way for functional questions tackled by other disciplines, and it also helps to integrate information from many disciplines, leading to the explanation of behaviour in terms of specific neuronal networks.

Neural networks are thought to be composed of repeating modules, which can be defined as the basic circuit consisting of several types of neurone and associated glial cells. The operational role of each neurone in a class is stereotyped, and is determined by the temporal and spatial distribution of its input and output as well as by the molecular events underlying the transformation of signals. The regular architecture of some areas of the brain allows the use of simple methods for the prediction of the connections and the chemistry of neurons. However, in other areas such as the neocortex, the identification of neuronal signatures can only be based on synaptic level examination of the identified neurons and their molecular composition. Of particular interest are neurotransmitters, second messengers, receptors and ion channels, i.e. the molecular machinery concerned with neural communication.

More often than not, the structural, chemical and dynamic properties of cells are examined separately for technical reasons. In exceptional cases it is possible to relate all these properties to single <u>identifiable neurons</u>. These cells can be recognised on the basis of structural, dynamic or chemical characteristics from specimen to specimen. In most cases however, both in vertebrates and in invertebrates, a role in the network is carried out by a <u>probabilistically equivalent set of cells,</u> members of which we would like to reexamine from experiment to

NATO ASI Series, Vol. H 58
Neurocytochemical Methods
Edited by A. Calas and D. Eugène

experiment. Therefore catalogues of properties are used to identify individual members of the set of neurons.

In the following brief survey, some aspects of the organization of cortical, hippocampal and cerebellar circuits will be examined, demonstrating how the localization of molecules and the spatial relationships of synaptic connections can be used to identify equivalent sets of neurons and to predict the mode of cellular interactions. I will concentrate on examples which show how different techniques have helped us to answer questions related to the function of the systems that we have studied; I will not attempt to review the literature in the areas discussed below. Technical details of the procedures can be found in recent reviews (Freund and Somogyi, 1989; Somogyi and Freund, 1989; Somogyi 1990).

When trying to explain the operation of a complex neural network on a realistic basis, the first task is to define the basic circuit that carries out the transformation between input and output. With my colleagues I have been studying the basic circuit of the cerebral cortex, establishing its properties with a rigour allowed by current technology.

## Chemical neuroanatomy of the basic cortical circuit

The cortical circuit is an integral part of all sensory-motor transformation. Sensory information arrives at the cortex via the thalamus, and information about different aspects of the outside world are carried in parallel channels in the same modality. For example in the visual system of the cat at least three physiologically distinct channels, the X, Y and W pathways, have been described from retina to cortex (Sherman, 1985). The thalamic axons involved in these pathways terminate in the middle, small cell populated layers of cortex and the monosynaptic target cells are activated by one of the pathways. Cortical cells are largely silent, in spite of considerable spontaneous activity in the thalamus of both anaesthetized and alert animals. When cortical cells respond they are particular for the parameters of the stimulus; e.g. in the visual domain for the orientation and length of stimulus, direction of movement etc. This specificity in the response is generated to a great extent in cortex, already in the middle layers where thalamic axons terminate. Rich intracortical connections, mediated by both intrinsic neurons and by the local collaterals of output cells, distribute the thalamic input to supra- and infragranular layers in a lamina specific and horizontally patchy manner.

## Excitatory amino acids in cortical cells

The output leaves the cortex via the axons of pyramidal cells, which use excitatory amino acid transmitters and produce monosynaptic activation of their targets, as shown by physiological and pharmacological studies. Anatomical evidence for the excitatory amino acid nature of pyramidal cell transmitter(s) comes from the selective retrograde labelling of pyramidal cells by $^3$H-D-aspartate injected into their termination areas (for review see Streit 1991, this volume). Since all pyramidal cells also possess axon collaterals and make synapses within the area where their cell body is located, small intracortical injections of the radiolabelled false transmitter can be used to study their intra- and interlaminar local connections (Kisvarday et al., 1989). This strategy was also used to study the chemical nature of another large population of cortical cells, the spiny stellate cells. Most spiny stellate cells have axons intrinsic to the cortical area where their cell body is located. Physiological and pharmacological methods have not elucidated their chemical nature because it was not possible to activate them selectively without also activating pyramidal cells. The retrograde labelling by $^3$H-D-aspartate, injected into the termination of their axons, showed that spiny stellate cells have the molecular machinery for the selective uptake of acidic amino acids, and this indicates that they use such an amino acid as transmitter (Kisvarday et al., 1989).

The endogenous compound that mediates excitatory amino acid interactions is thought to be glutamate, but it is not at all clear whether glutamate can account for all of the interactions, and other compounds have also been suggested (e.g. Do et al., 1986). The development of antibodies to glutamate and aspartate (Storm-Mathisen and Ottersen, 1986) led to numerous studies showing some degree of immunoreactivity in the somata of cortical cells including pyramidal cells. However, somatic immunoreactivity for amino acids, which are also involved in various metabolic pathways and protein synthesis, has little relevance to transmitter function. Indeed it has been demonstrated that the somata of cholinergic motoneurons is strongly immunopositive for glutamate (Ottersen and Storm-Mathisen, 1984), and the highest aspartate immunoreactivity was localized in the somata of GABA-positive, inhibitory cells in the hippocampus and cerebellum (Ottersen and Storm-Mathisen, 1985, Madl et al., 1987).

The presence of glutamate or other excitatory amino acids in the nerve terminals should have more relevance to a transmitter role, if it can be shown that the identified population of terminals contains distinctly higher levels of immunoreactivity than other neurochemically different populations of terminals. On a regional level, sensitive immunoenzyme methods can provide a first step towards

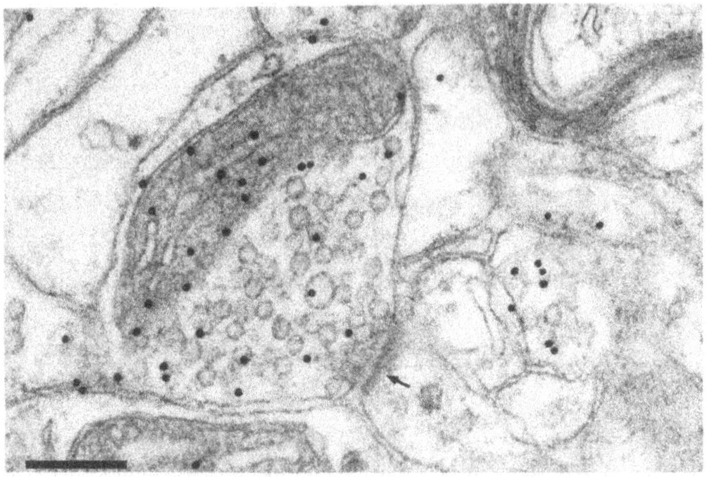

Fig. 1. Glutamate immunopositive bouton establishing a synapse (arrow) with a spine in the striate cortex of cat. Postembedding immunogold method using a rabbit antiserum to glutamate conjugated to bovine serum albumin by glutaraldehyde, and goat anti-rabbit IgG conjugated to colloidal gold. In spite of the impressive accumulation of gold particles over the bouton, suggestions based on such pictures, for the transmitter of terminals remain anecdotal without quantitative comparison of immunogold density over bouton populations, identified by their common origin and/or by their common biochemical character. Scale: 0.2 μm

such a goal (e.g. Liu et al., 1989). However, quantitative comparisons can only be obtained using particulate markers. This was first achieved in the cerebellum using quantitative immunogold localization of glutamate in parallel and mossy fibre terminals, which were shown to contain higher level of immunoreactivity than glial cell processes or GABA-positive terminals (Somogyi et al., 1986). Although individual terminals can be shown to contain immunoreactivity for glutamate in cortex as well (Fig. 1), without quantitative measurement of the immunoreactivity of a population identified with respect to their common origin, such data remain anecdotal. The quantitative postembedding immunogold method, combined with relative measurement of glutamate levels within distinct populations of terminals in the same sections, remains the only valid way to study selective distribution of amino acids which participate in important metabolic pathways as well as in neurotransmission (Ottersen, 1989).

The role of pyramidal cell axon collaterals. In addition to their distant targets pyramidal cells establish hundreds of synapses locally. Depending on the monosynaptic targets of the axons, these collaterals can either activate other excitatory neurons or, by activating local inhibitory cells, can exert disynaptic inhibition. Quantitative studies on the targets of local terminals of pyramidal cells (Kisvarday et al., 1986) showed that most of the postsynaptic elements were the spines of other pyramidal cells. The origin of the pyramidal cell terminals was identified using intracellular HRP injection of physiologically characterised cells. Subsequent electron microscopic (EM) examination of the synapses, and post-embedding immunogold reaction of the postsynaptic elements for GABA, revealed that less than 5% of the targets originated form GABAergic inhibitory neurons. A similar technique was used for the estimation of GABAergic targets in long cortico-cortical pathways (Lowenstein and Somogyi, 1991). However, instead of intracellular HRP, which does not travel well over long distances within the short survival time of a physiological experiment, *Phaseolus vulgaris* leucoagglutinin (PHA-L) was used to label cortico-cortical terminals. Most of the targets were found to be spines of pyramidal cells and only 17% were dendritic shafts, of which about 70% were GABA-positive (Lowenstein and Somogyi, 1991). These quantitative studies combining physiological recording, intracellular marking and anterograde labelling methods with the immunocytochemical characterization of postsynaptic elements, reveal the cortex as a profusely interconnected network of spiny neurons using excitatory amino acid transmitters. The break for this network is provided by local inhibitory cells which must receive highly convergent input from sets of excitatory cells making up the same pathway. The identification of some of these inhibitory cells has been achieved using a battery of combined techniques.

## Identification of cortical GABAergic neurons

Physiological and pharmacological studies demonstrated that GABA is the major inhibitory neurotransmitter in cortex. Immunolocalization of glutamate decarboxylase (GAD), the main synthetic enzyme for GABA, showed that it was present in non-pyramidal cells (Ribak, 1978), many of which, unlike pyramidal and spiny stellate cells, have smooth or sparsely spiny dendrites. The enzyme was also localized in synaptic boutons establishing type 2 (symmetrical) synaptic contacts (Ribak, 1978). However, immunocytochemistry on its own proved insufficient to identify the origin of any particular terminal because, in the extensive immunoreactivity of the neuropil, the terminal axon branches could not be followed back to the somata of cells. Furthermore the dendritic immunoreactivity was insufficient to relate the GAD-positive cells to well known neuronal classes described by Golgi methods or by intracellular labelling.

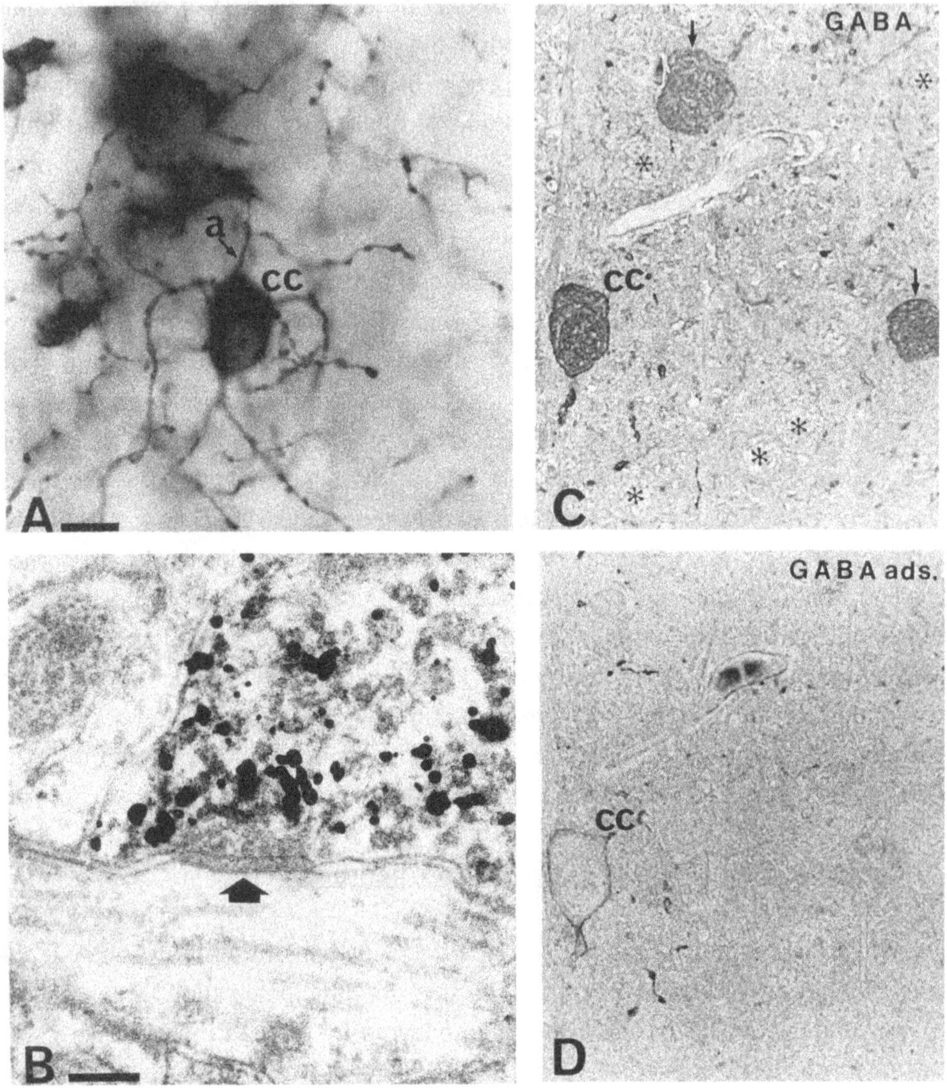


Fig. 2. Immunocytochemistry of a neurone identified as a **clutch cell** (cc) by Golgi impregnation in the striate cortex of monkey. **A.** Resin embedded section (90 μm thick) showing the processes of the cell (axon, a). **B.** One bouton, identified by the gold/silver deposit as originating from this cell, makes a type 2 synapse (arrow). **C.** A 0.5 μm thick section of the somata of the same cell after immunoperoxidase reaction for GABA. Note the peroxidase deposit in the clutch cell and in two more neurons (arrows), but not in others (asterisks). **D.** A consecutive section was reacted in the same way as shown in C, but the anti-GABA serum was preadsorbed with GABA conjugated to polyacrylamide beads. Note that only the thin rim of gold/silver deposit is visible, but the interior of the cell is free of deposit. Data from the work of Kisvarday et al., (1986). Scales: A, C and D, 10 μm ; B, 0.2 μm.

Three combined techniques were developed to identify particular classes of GABAergic cells (for review and detailed methods see Somogyi, 1988; Somogyi and Freund, 1989; Freund and Somogyi, 1989).

1. Pre-embedding immunoenzyme reaction for GAD, followed by Golgi impregnation and gold toning of cells.

2. Golgi impregnation and gold toning of cells, followed by postembedding immunoenzyme demonstration of GABA in semithin sections of their somata (Fig. 2).

3. Intracellular injection of HRP, followed by EM immunogold demonstration of GABA in the terminals of the identified neurons (Fig. 3.).

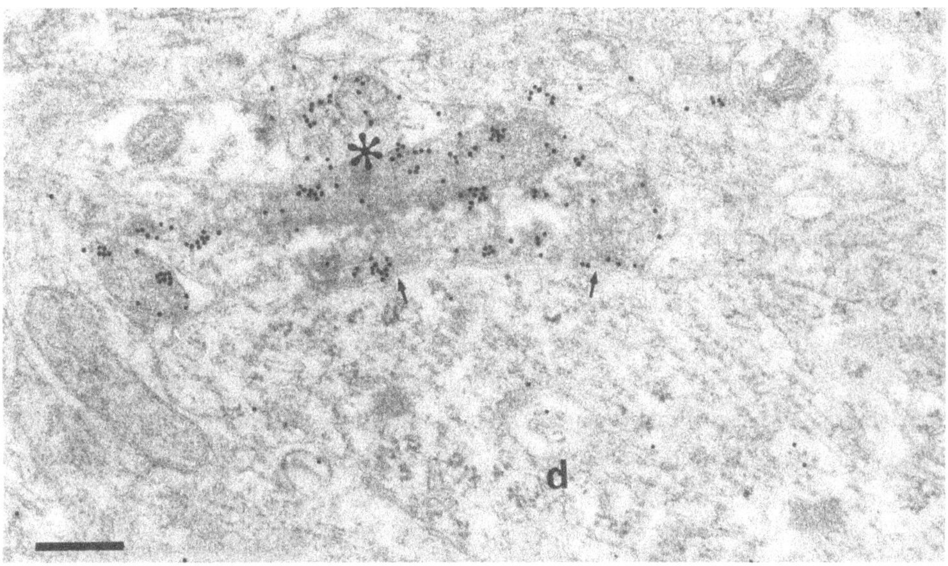


Fig. 3. Postembedding immunogold demonstration of GABA in a bouton (asterisk) of a large basket cell identified by intracellular recording and filling with HRP in the striate cortex of cat (see cell 3 in Somogyi et al., 1983). Note the weak electron density of the peroxidase reaction-product following sodium periodate treatment for the removal of osmium. The bouton makes two synaptic junctions (arrows) with a dendrite (d). For technical details see chapter by Somogyi (1988). Scale: 0.2μm

All three methods result in the localization of a transmitter marker and at the same time provide the opportunity to follow terminals of the axon of the identified cell and sample its postsynaptic targets. The quantitative distribution of postsynaptic targets together with the laminar distribution of the terminals was used as a signature for the delineation of GABAergic neuronal classes (for review see Somogyi, 1989). It was shown that chandelier cells terminate exclusively on the axon initial segment of pyramidal cells. Since the axon initial segment is thought to generate the propagated action potential, it have been suggested that the strategic location of chandelier cell synapses could provide a mechanism for setting the threshold for firing and/or synchronizing populations of pyramidal cells. Further studies revealed at least three classes of basket cells with distinct laminar patterns of their axons: large basket cells in layers 2-3, clutch cells in layer 4, and deep basket cells in layers 5-6. They all terminate on the somata (20-40%), dendritic spines (20-40%) and dendritic shafts of neurons. Further classes, such as neurogliaform and bitufted cells, terminate exclusively on spines and dendritic shafts of neurons. The overall distribution of postsynaptic elements to all GABAergic boutons was established using post-embedding EM immunogold reaction for GABA (Beaulieu and Somogyi, 1990). The average distribution is different from the distribution profile of postsynaptic targets of identified neuronal classes mentioned above, therefore it is clear that none of the classes of cells examined so far picks its synaptic targets randomly.

From these quantitative studies a new picture of inhibitory influences emerges. Different parts of the same postsynaptic cell receive GABAergic input from distinct populations of local circuit neurons. This is a fundamental feature of the basic cortical circuit as shown by the presence of the same organisation in other cortical areas such as the hippocampus. Chandelier cells containing GABA have been discovered in the hippocampus (Somogyi et al., 1985), and the presence of basket cells and local circuit neurons of the stratum radiatum and moleculare of the hippocampus and the dentate gyrus have been demonstrated by classical Golgi studies. The quantitative determination of the distribution of the axon terminals and the neuronal elements postsynaptic to these GABAergic cells awaits further work.

## Selective inputs to cortical GABAergic neurons

The multiplicity of GABAergic inputs converging onto the same postsynaptic cell was suggested to provide a structural basis for different subcortical and cortical pathways to govern the gain of their own effect through the activation of separate sets of GABAergic cells (Somogyi 1989). Evidence is accumulating that in the cortex as well as in the hippocampus different populations of GABAergic cells receive distinct subcortical input.

**Thalamic afferents.** The first evidence was provided by the measurement of perikaryon sizes of neurons which received monosynaptic somatic input from physiologically identified X and Y-type visual thalamic axons in the striate cortex of cat. Freund et al. (1985) found that all such cells were GABA-positive, and the somata innervated by Y-type axons were significantly larger than somata innervated by X-type axons. Therefore the two physiologically distinct streams of afferents innervate different populations of cells.

**Serotonergic axons** from the brain stem have been shown to provide dense innervation of the soma and proximal dendrites of a subpopulation of non-pyramidal cells (Mulligan and Tork, 1988). Both GAD and GABA have been demonstrated in some of the densely innervated cells using dual immunocytochemical methods (Tork et al., 1988). In any area only a small proportion of GABAergic cells were densely innervated indicating that they play a special role in mediating serotonergic effects on cortical neurons which themselves received sparse and diffuse input from other or the same afferents.

**Septo-hippocampal pathway.** A GABAergic pathway from the medial septum has been shown to innervate selectively subpopulations of GABAergic cells in the hippocampus (Freund and Antal, 1988). In this elegant study the septo-hippocampal terminals were identified by the anterograde transport of PHA-L; the targets of the boutons (black in nickel intensified DAB reaction) were revealed first by light microscopic pre-embedding immunohistochemistry for calcium binding proteins (brown after DAB reaction), which were present in high concentration in subsets of GABAergic cells. This two colour localization of presynaptic afferents and postsynaptic cells made it possible to demonstrate the selectivity of innervation. Subsequent electron microscopic demonstration of synaptic contacts and an EM immunogold reaction showed the presence of GABA in both the presynaptic terminals and in the postsynaptic somata and dendrites. Interestingly not all GABAergic cells received such input.

**Cholinergic basal forebrain pathway.** Further evidence for the selective innervation of subpopulations of GABAergic cells by subcortical afferents comes from our recent work on the termination of cholinergic synapses in the cortex of cat (Beaulieu and Somogyi, 1991). The cholinergic innervation of the cortical mantle by the basal forebrain plays an important role in the activation of cortical systems (Buzsaki et al., 1988; Vanderwolf, 1988). Receptors for acetylcholine (ACh) modulate the way in which cortical neurons respond to other inputs, and thus have an influence on all cortical activity from the responses to sensory thalamic input to higher integrative functions such as learning and memory.

In the visual cortex of cat the application of ACh to cortical cells enhances the visually evoked responses in about 60% of cells, but indirect GABA-mediated inhibitory influences have also been reported (Muller and Singer, 1989; Sato et al., 1987; Sillito and Kemp, 1983). As an indicator of the relative importance of direct facilitatory and indirect inhibitory effects we determined the extent to which

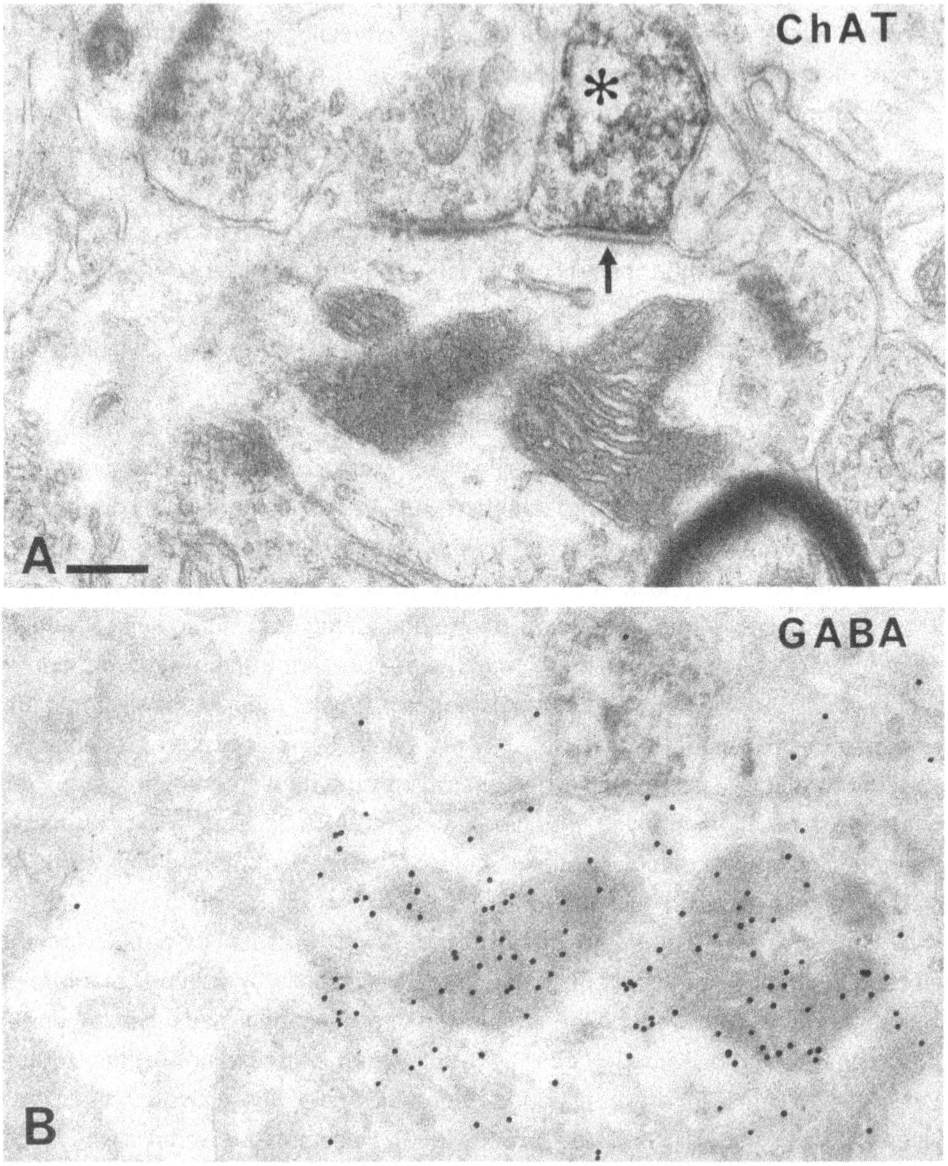


Fig. 4. Immunocytochemical characterization of pre- and postsynaptic elements in the striate cortex of cat. **A.** Pre-embedding immunoperoxidase labelling of a synaptic (arrow) terminal, which is immunopositive for choline acetyltransferase (asterisk). **B.** The postsynaptic dendrite is immunopositive for GABA as shown on a consecutive section by post-embedding immunogold reaction. Data from the work of Beaulieu and Somogyi (1991). Scale for A and B, 0.2 µm

cholinergic synaptic release sites are devoted to the principal cells as compared to the GABAergic local circuit neurons (Beaulieu and Somogyi, 1991). Antibodies to choline acetyltransferase were used in a pre-embedding immunoperoxidase method followed by postembedding EM immunogold demonstration of GABA on ultrathin sections (Fig. 4). This enabled us to determine quantitatively the proportion of elements which contained immunoreactive GABA and therefore originated from local inhibitory cells, and were also postsynaptic to cholinergic terminals .

Most of the cholinergic boutons formed synapses with dendritic shafts (87.3%), much fewer with dendritic spines (11.5%), and only occasional contacts were made on neuronal somata (1.2%). Overall, 27.5% of the postsynaptic elements, all of them dendritic shafts, were immunoreactive for GABA, thus demonstrating that they originated from inhibitory neurons (Fig. 5). This is the highest value for the proportion of GABAergic postsynaptic targets obtained so far for any intra- or subcortical afferents in cortex. The proportion of GABA-positive dendritic shafts was highest in layer IV (49% of targets, two thirds of all dendritic shafts), and lowest in layers V-VI (14%). As seen in fig. 5. the quantitative distribution of targets postsynaptic to choline acetyltransferase-positive terminals is very different from the postsynaptic targets of GABAergic boutons, or from the targets of all boutons in layer IV reported previously (Beaulieu and Somogyi, 1990, Bueno-Lopez et al., 1989). In both cases the proportion of GABA-positive dendrites was only 8-9% of the postsynaptic elements (Fig. 5). These results, obtained with the molecular characterization of both pre- and postsynaptic elements, demonstrate that cortical GABAergic neurons receive a richer cholinergic synaptic input than non-GABAergic cells. The activation of GABAergic neurons by cholinergic afferents and the subsequent fast $GABA_A$ receptor mediated inhibition of their targets could increase the response specificity of cortical cells, which are directly facilitated by slower muscarinic mechanisms during cortical arousal. It has been suggested that the two mechanisms operate synergistically to enhance the signal to noise ratio of responses (Sato et al., 1987; Sillito and Kemp, 1983).

A further unexpected result of the combined immunocytochemical study was that at least 8% of the total population of choline acetyltransferase-positive boutons were also immunoreactive for GABA (Beaulieu and Somogyi, 1991). This raises the possibility of cotransmission at some cholinergic synapses, leading again to a faster $GABA_A$ receptor mediated inhibition followed by slow cholinergic facilitation.

The main message from these four examples of selective innervation of inhibitory neurons is that <u>the operational role of the subcortical afferents can only be explained if the physical constraints i.e. the specificity in the wiring is taken into account.</u> Most, if not all, cortical cells are responsive to GABA, ACh or serotonin.

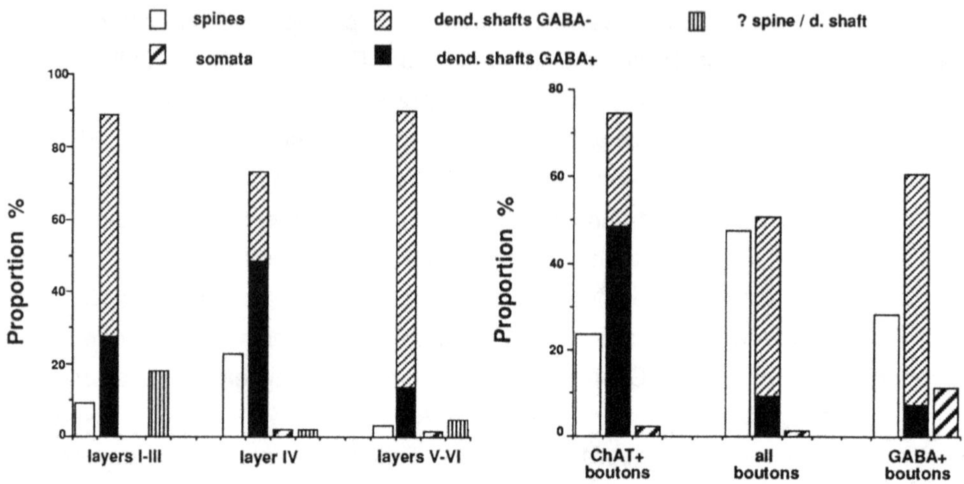


A. Postsynaptic targets of ChAT + boutons in different layers of the striate cortex of cat

B. Postsynaptic targets in layer IV of cat striate cortex

Fig. 5. Quantitative comparison of the synaptic targets of characterized nerve terminal populations in the striate cortex of cat. For the cholinergic terminals a pre-embedding immunoperoxidase method, for the postsynaptic dendrites an EM post-embedding immunogold method was used on the same material (Beaulieu and Somogyi, 1991). For **B** data was recalculated from the above study, and from studies by Beaulieu and Somogyi, (1990) and Bueno-Lopez et al., (1989). It is apparent that cholinergic terminals show preferential termination on dendritic shafts, particularly on those originating from GABA-positive cells.

However, the explanation of the effect of these neurotransmitters can only be as accurate as our understanding of the neuronal network where they are released. It is now clear that subcortical pathways such as the brain stem serotonin or the

cholinergic basal forebrain input to cortex, which have been thought to terminate in a diffuse manner, use subsets of heavily innervated GABAergic cells, which not only multiply their effect, but channel it into the highly selective synaptic circuits of the cortex.

## Subcellular distribution of some neurotransmitter receptors

The above examples show that it is now possible to determine the origin of presynaptic terminals, their putative transmitters and the structural and chemical identity of postsynaptic elements with great precision. However, for each neuroactive substance there are a multitude of receptors which mediate interactions with widely different effects. Therefore, it is essential to localize the receptors with a precision comparable to that of the wiring and distribution of synaptic terminals. This requires EM resolution and the localization of receptors to particular cellular membranes.

**$GABA_A$/benzodiazepine receptor - chloride channel complex.** The $GABA_A$ receptor belongs to the family of ligand gated receptor channel complexes; its activation leads to opening of $Cl^-$ channels. The receptor complex is also the site of action of such clinically important drugs as benzodiazepines and barbiturates. We used monoclonal antibodies (Haring et al., 1985; Vitorica et al., 1988) and immunoenzyme methods to determine the localization of the receptor in relation to synaptic terminals with known chemistry and origin in the cerebellum (Somogyi et al., 1989), the visual thalamus (Soltesz et al., 1990) and in the cortex (Somogyi et al., 1989, Somogyi, 1990). Both the alpha- and the beta-subunits were distributed similarly as revealed by immunoperoxidase methods. Intracellular immunoreactivity was associated with the endoplasmic reticulum, the Golgi apparatus and multivesicular bodies, suggesting the sites of synthesis, glycosylation and degradation of the receptor complex respectively. Extracellular immunoreactivity was associated with the plasma membrane of neurons.

Both the intracellular and the extracellular immunoreactivity for the receptor complex were expressed in a neurone type specific manner, but was stereotyped among members of the same class of cell. Some classes of cells, such as cerebellar granule cells and large thalamic relay neurons, showed strong immunoreactivity on their plasma membrane, but very little intracellular immunoreactivity. Other cells, such as large GABAergic cortical neurons, stellate and basket cells in the cerebellum and a population of small relay cells in the lateral geniculate nucleus, showed strong plasma membrane and intracellular immunoreactivity. Further cell types such as cortical pyramidal cells, many thalamic relay cells and Purkinje cells exhibited moderate immunoreactivity. It was suggested that the degree of intracellular immunoreactivity correlates with the turnover of receptor proteins, while the degree of plasma membrane immunoreactivity reflects the density of receptor/channel complexes. These results predict the differential sensitivity of different neuronal classes to $GABA_A$ receptor agonists.

High resolution electron microscopy showed that the epitopes for all four monoclonal antibodies were on the extracellular surface of the plasma membrane. Immunoreactivity was localized on the somatic and dendritic plasma membrane as well as on dendritic spines, particularly of those of Purkinje cells. Immunoreactivity for the receptor complex was present in the synaptic cleft of boutons known from separate studies to contain GABA or GAD. However, equally strong immunoreactivity was also present at non-junctional plasma membrane in all areas of the brain at sites, such as the somata of cerebellar granule cells and the spines of Purkinje cells, which never receive GABAergic synaptic input. The synaptic clefts of known non-GABAergic terminals were generally immunonegative. Thus, GABA and endogenous substances or drugs that affect $GABA_A$ receptor function could act at both synaptic and non-synaptic sites. The distribution of the receptor complex suggests that the cellular topography of GABAergic influences is not determined by the precise placement of receptors, but by the precise spatial placement of the GABA-releasing terminals, a characteristic of the local circuits described above.

The antibodies (Haring et al., 1985; Vitorica et al., 1988) used in these studies react only with a few of the known receptor subunits. It is possible that some of the more recently discovered subunits, which have been shown to be expressed in a cell type specific manner (Khrestchatisky et al., 1989; Luddens et al., 1990; Malherbe et al., 1990; Schofield et al., 1989; Shivers et al., 1989), have a different or more restricted cell surface distribution. Further high resolution studies using subunit specific antibodies will elucidate the spatial distribution of receptors on the neuronal surface.

**Putative kainate receptor in the cerebellum.** As discussed above, most of the neuronal synaptic interaction in the cerebral cortex as well as in other areas such as the cerebellum takes place via excitatory amino acid receptors. High resolution localization of these receptors has become possible with the preparation of antibodies to some of the purified receptor proteins (Gregor et al., 1988; Klein et al., 1988; Hampson et al., 1989).

The cerebellar cortex of submammalian vertebrates contains a particularly high density of binding sites in the molecular layer for one of the excitatory amino acids, kainate. Since the molecular layer contains a very high density of glutamate releasing nerve terminals originating from granule cells, the receptor was thought to be associated with the synapses predominantly on the Purkinje cells. However, a high resolution light and EM study of the distribution of receptor immunoreactivity in the cerebellum of the chick and fish showed that immunoreactivity was exclusively localized to Bergmann glial cells (Somogyi et al., 1990), which are also restricted to the molecular layer (Fig. 6). Bergmann glial cells surround the dendrites and the spine-parallel fibre terminal synapses of Purkinje cells with thin lamellar processes. Thus, the most likely endogenous compound acting on the putative glial kainate receptor would be glutamate released from parallel fibre terminals (Somogyi et al., 1990). With the cloning of several of the neuronal excitatory amino acid receptors (Keinanen et al., 1990) the visualization of their distribution at identified synapses will soon be possible.

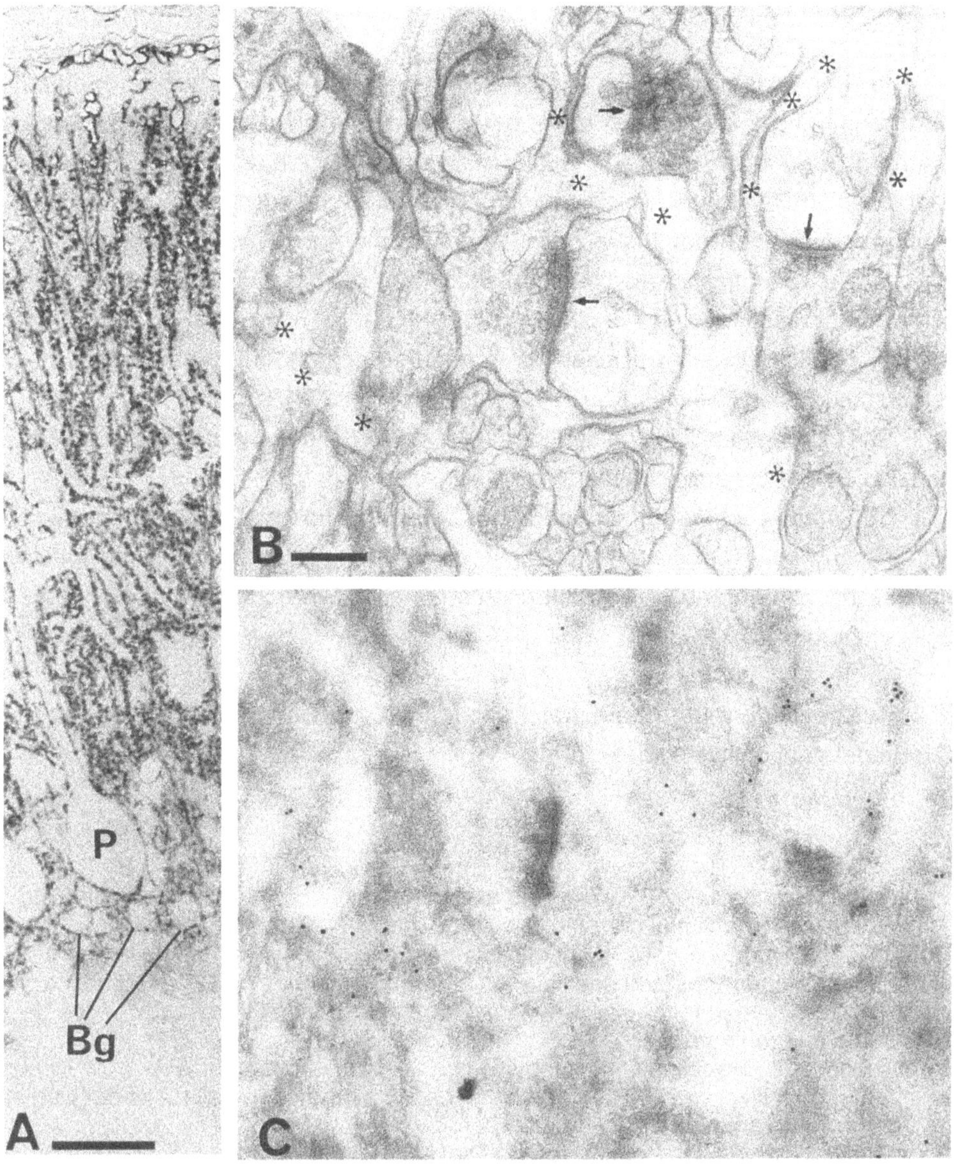


Fig. 6. Immunocytochemical demonstration of a kainate receptor on Bergmann glial cells using the monoclonal antibody IX-50 in the chick cerebellum. **A.** Semithin section, immunoperoxidase method. Note reaction around Bergmann glial cells (Bg), and over their processes enveloping Purkinje cell (P) dendrites. B and C. Serial sections of chick cerebellum treated with uranyl acetate before embedding, but not with osmium tetroxide at any stage. The section in C was etched with ethanolic sodium hydroxide and reacted with the monoclonal antibody followed by colloidal gold conjugated to anti-mouse IgG. Note that the gold particles are found over glial processes (asterisks), but not at the synapses (arrows) of parallel fibres and Purkinje cell spines. (Somogyi et al., 1990). Scales: A, 20μm; B and C, 0.25μm.

## Conclusion

In this brief survey I tried to demonstrate how different molecular neuroanatomical techniques have been used to reveal particular aspects of neuronal organisation. The identified properties of cells can be related to the properties of the network, leading eventually to an explanation of the transformation that takes place between the incoming and outgoing signals. In this process:

1. The identification of structural properties defines the degree to which we understand the connections of cells.

2. The identification of chemical properties defines the degree to which we understand molecular events.

3 The identification of biophysical properties defines the degree to which we can explain the dynamic constraints of cells.

The nature of neuronal operations will only be predicted by information obtained by strict criteria on the same identified neurons. Such predictions will necessarily be of an abstract form based on quantitative data on real networks and cells of the brain.

**Acknowledgement.** I am grateful to Mr. J. D. B. Roberts for his assistance in the preparation of illustrations.

## Literature Cited

Beaulieu C, Somogyi P (1990) Targets and quantitative distribution of GABAergic synapses in the visual cortex of the cat. Europ J Neurosci 2:296-303

Beaulieu C, Somogyi P (1991) Enrichment of cholinergic synaptic terminals on GABAergic neurons and coexistence of immunoreactive GABA and choline acetyltransferase in the same synaptic terminals in the striate cortex of cat. J Comp Neurol in press

Bueno-Lopez JL, Beaulieu C, Somogyi P (1989) Proportions of GABA-immunopositive and GABA-negative synaptic elements in layer 4 of cat's striate cortex. Europ J Neurosci Suppl.2:106

Buzsaki G, Bickford R.G, Ponomareff G, Thal LJ, Mandel,R, Gage F.H (1988) Nucleus basalis and thalamic control of neocortical activity in the freely moving rat. J Neurosci 8:4007-4026

Do KQ, Herrling PL, Streit P, Turski WA, Cuenod M. (1986) In vitro release and electrophysiological effects in situ of homocysteic acid, an endogenous N-Methyl-(D)-aspartic acid agonist, in the mammalian striatum. J Neurosci 6:2226-2234

Freund TF, Antal M (1988) GABA-containing neurons in the septum control inhibitory interneurons in the hippocampus. Nature 336:170-173

Freund TF, Martin KAC, Somogyi P, Whitteridge D (1985) Innervation of cat visual areas 17 and 18 by physiologically identified X- and Y-type thalamic afferents.

II. Identification of postsynaptic targets by GABA-immunocytochemistry and Golgi impregnation. J Comp Neurol 242 : 275-291.

Freund TF, Somogyi P (1989) Synaptic relationships of Golgi impregnated neurons as identified be electrophysiological or immunocytochemical techniques. In : Heimer, Zaborszky (eds) Neuroanatomical Tract-Tracing Methods. II : Recent Progress. Plenum Publishing Corporation, New York, p 201.

Gregor P, Eshhar N, Ortega A, Teichberg VI (1988) Isolation, immunochemical characterization and localization of the kainate sub-class of glutamate receptor from chick cerebellum. Eur. Mol Biol Org J 7 : 2673-2679.

Hampson DR, Wheaton KD, Dechesne CJ, Wenthold RJ (1989) Identification and characterization of the ligand binding subunit of a kainic acid receptor using monoclonal antibodies and peptide mapping. J Biol Chem 264 : 13329-13335.

Haring P, Stahli C, Schoch P, Takacs B, Staehelin T, Mohler H (1985) Monoclonal antibodies reveal structural homogeneity of g-aminobutyric acid/benzodiazepine receptors in different brain areas. Proc Natl Acad Sci USA 82 : 4837-4841.

Keinanen K, Wisden W, Sommer B, Werner P, Herb A, Verdoorn TA, Sakmann B, Seeburg PH (1990) A family of AMPA-selective glutamate receptors. Science 249 : 556-560.

Khrestchatisky M, MacLennan AJ, Chiang MY, Xu W, Jackson B, Brecha N, Sternini C, Olsen RW, Tobin AJ (1989) A novel subunit in rat brain GABA-A receptors. Neuron 3 : 745-753.

Kisvarday ZF, Cowey A, Smith AD, Somogyi P (1989) Interlaminar and lateral excitatory amino acid connections in the striate cortex of monkey. J Neurosci 9 : 667-682.

Kisvarday ZF, Cowey A, Somogyi P (1986a) Synaptic relationships of a type of GABA-immunoreactive neuron (clutch cell), spiny stellate cells and lateral geniculate nucleus afferents in layer IVC of the monkey striate cortex. Neuroscience 19 : 741-761.

Kisvarday ZF, Martin KAC, Freund TF, Magloczky ZS, Whitteridge D, Somogyi P (1986b) Synaptic targets of HRP-filled layer III pyramidal cells in the cat striate cortex. Exp Brain Res 64 : 541-552.

Klein AU, Niederoest B, Winterhalter KH, Cuenod M, Streit P (1988) A kainate binding protein in pigeon cerebellum : purification and localization by monoclonal antibody. Neurosci Lett 95 : 359-364.

Liu CJ, Grandes P, Matute C, Cuenod M, Streit P (1989) Glutamate-like immunoreactivity revealed in rat olfactory bulb, hippocampus and cerebellum by monoclonal antibody and sensitive staining method. Histochemistry 90 : 427-445.

Lowenstein PR, Somogyi P (to be published) Synaptic organization of cortico-cortical connections from the primary visual cortex to the posteromedial lateral suprasylvain area (PLMS). J Comp Neurol.

Luddens H, Pritchett DB, Kohler M, Killisch I, Keinanen K, Monyer H, Sprengel R, Seeburg PH (1990) Cerebellar GABA-A receptor selective for a behavioural alcohol antagonist. Nature 346 : 648-651.

Madl JE, Beitz AJ, Johnson RL, Larson AA (1987) Monoclonal antibodies specific for fixative-modified aspartate : immunocytochemical localization in the rat CNS. J Neurosci 7 : 2639-2650.

Malherbe P, Sigel E, Baur R, Persohn E, Richards JG, Mohler H (1990) Functional characteristics and sites of gene expression of the a1, b1, g2-isoform of the rat GABA receptor J Neurosci 10 : 2330-2337.

Muller CM, Singer W (1989) Acetylcholine-induced inhibition in the cat visual cortex is mediated by a GABAergic mechanism. Brain Res 487 : 335-342.

Mulligan, KA, Törk, I (1988) Serotoninergic innervation of the cat cerebral cortex. J Comp Neurol 270 : 86-110.

Ottersen OP (1989) Quantitative electron microscopic immunocytochemistry of neuroactive amino acids. Anat Embryol 180 : 1-15.

Ottersen OP, Storm-Mathisen J (1984) Neurons containing or accumulating transmitter amino acids. In : Björklund A, Hökfelt T, Kuhar MJ (eds) Classical Transmitters and Transmitter Receptors in the CNS Part II. Elsevier Science Publishers BV, p 141.

Ottersen OP, Storm-Mathisen J (1985) Different neuronal localization of aspartate-like and glutamate-like immunoreactivities in the hippocampus of rat, guinea-pig and Senegalese babcon (Papio Papio), with a note on the distribution of gaminobutyrate. Neuroscience 16 : 589-606.

Ribak CE (1978) Aspinous and sparsely-spinous stellate neurons in the visual cortex of rats contain glutamic acid decarboxylase. J Neurocytol 7 : 461-478.

Sato H, Hata Y, Masui H, Tsumoto T (1987) A functional role of cholinergic innervation to neurons in the cat visual cortex. J Neurophysiol 58 : 765-780.

Schofield PR, Shivers BD, Seeburg PH (1989) The role of receptor subtype diversity in the CNS. Trends Neurosci 13 : 8-11.

Sherman SM (1985) Functional organization of the W-, X-, and Y-cell pathways in the cat : a review and hypothesis. In : Sprague JM, Ejpstein AN (eds) Progress in Psychobiology and Physiological Psychology. Academic Press, New York, p 233.

Shivers BD, Killisch I, Sprengel R, Sontheimer H, Kohler M, Schofield PR, Seeburg PH (1989) Two novel GABA-A receptor subunits exist in distinct neuronal subpopulations. Neuron 3 : 327-337.

Sillito AM, Kemp JA (1983) Cholinergic modulation of the functional organization of the cat visual cortex. Brain Res 289 : 143-155.

Soltesz I, Roberts JDB, Takagi H, Richards JG, Mohler H, Somogyi P (1990) Synaptic and non-synaptic localization of benzodiazepine/GABA-A receptor/Cl-channel complex using monoclonal antibodies in the dorsal lateral geniculate nucleus of the cat. Eur J Neurosci 2 : 414-429.

Somogyi P (1988) Immunocytochemical demonstration of GABA in physiologically characterized, HRP-filled neurons and in their postsynaptic targets. In : van Leeuwen FW, Buijs RM, Pool CW, Pach O (eds) Molecular Neuroanatomy. Techniques in the Behavioral and Neural Sciences. Elsevier, Amsterdam, p 339.

Somogyi P (1989) Synaptic organisation of GABAergic neurons and GABA-A receptors in the lateral geniculate nucleus and visual cortex. In : Lam DKT, Gilbert CD (eds) Neural Mechanisms of Visual Perception, Retina Research Foundation Symposium. Portfolio Publishing Company, Texas, p 35.

Somogyi P (1990) Synaptic connections of neurones identified by Golgi impregnation : characterization by immunocytochemical enzyme histochemical, and degeneration methods. J Electron Micros Tech 15 : 332-351.

Somogyi P, Eshhar N, Teichberg VI, Roberts JDB (1990) Cellular and subcellular localization of a putative kainate receptor using a monoclonal antibody in the chick and fish cerebellar cortex. Neuroscience 35 : 9-30.

Somogyi P, Freund TF (1989) Immunocytochemistry and synaptic relationships of physiologically characterized, HRP-filled neurons. In : Heimer and Zaborsky (eds) Neuroanatomical Tract-Tracing Methods II : Recent Progress. Plenum Publishing Corporation, New York, p 239.

Somogyi P, Freund TF, Hodgson AJ, Somogyi J, Beroukas D, Chubb IW (1985) Identified axo-axonic cells are immunoreactive for GABA in the hippocampus and visual cortex of the cat. Brain Res 332 : 143-149.

Somogyi P, Halasy K, Somogyi J, Storm-Mathisen J, Ottersen OP (1986) Quantification of immunogold labelling reveals enrichment of glutamate in mossy and parallel fibre terminals in cat cerebellum. Neuroscience 19 : 1045-1050.

Somogyi P, Kisvarday ZF, Martin KAC, Whitteridge D (1983) Synaptic connections of morphologically identified and physiologically characterized large basket cells in the striate cortex of cat. Neuroscience 10 : 261-294.

Somogyi P, Takagi H, Richards JG, Mohler H (1989) Subcellular localization of benzodiazepine/GABA-A receptors in the cerebellum of rat, cat, and monkey using monoclonal antibodies. J Neurosci 9 : 2197-2209.

Storm-Mathisen J, Ottersen OP (1986) Antibodies against amino acid neurotransmitters. In : Panula P, Paivarinta H, Soinila S (eds) Neurohistrochemistry : Modern Methods and Applications. Alan R. Liss Inc, New York, p 107.

Törk, I, Hornung, JP, Somogyi, P (1988) Serotonergic innervation of GABAergic neurons in the cerebral cortex. Neurosci Lett [suppl] 30 : S131.

Vanderwolf, CH (1988) Cerebral activity and behavior : control by central cholinergic and serotonergic systems. In : Smythies JR, Bradley RJ (eds) International Review of Neurobiology. Academic Press, San Diego, p 225.

Vitorica J, Park D, Chin G, de Blas AL (1988) Monoclonal antibodies and conventional antisera to the GABA-A receptor/benzodiazepine receptor/Cl-channel complex. J Neurosci 8 : 615-622.

# REGULATED EXPRESSION OF TRANSCRIPTION FACTOR AND NEUROTRANSMITTER GENES IN NEURAL POPULATIONS : STUDIES USING *IN SITU* HYBRIDIZATION WITH RADIOLABELLED OLIGONUCLEOTIDE PROBES

G.R. Uhl, D.A. Appleby and T.N. Nishimori*

Laboratory of Molecular Neurobiology
NIDA/ARC and
Depts. of Neurology and Neuroscience, Johns Hopkins School of Medicine
P.O. Box 5180, Baltimore, Maryland 21224

Quantitated analyses of messenger RNAs (mRNAs) expressed in neurons reveal numerous instances in which changes in mRNA levels mirror changes in the functional activity of the expressing neurons (Uhl and Nishimori, 1990). The brain expresses more different genes than any other organ in the body. Much of this complexity may relate to the expression of different sets of genes by different neuronal populations. Studies of both the nature and the mechanisms for function-related changes in gene expression are thus important. To understand the detailed neurobiology of the these changes, they must be examined in individual neurons. *In situ* hybridization is the method of choice for such studies (Uhl, 1986).

Changes in neural function altering neuronal gene expression can occur : a) trans-synaptically, when a neural input acts via receptors in the membrane of the recipient neuron ; b) hormonally or pharmacologically, when membrane receptor function is altered by substances diffusing from a distance ; c) and by hormones that can diffuse across cell membranes and combine with intra-neuronal, diffusible receptors. Here, we focus on trans-synaptic gene regulation.

Trans-synaptic gene regulation in neurons is thought to proceed by steps localized at the membrane, in the cytoplasm, and in the nucleus (Uhl and Nishimori, 1990)). Agonist occupancy of neurotransmitter receptors takes place at the cell membrane, but can change several cytoplasmic functions. Changes in levels and distribution of calcium, phosphatidyl inositol and arachidonic acid metabolites and cyclic nucleotides can each be observed. These modifications, in turn, can alter the post-translational processing of specific cytoplasmic

* Dept. of Oral Anatomy, Hiroshima University College of Dentistry, Hiroshima, Japan

NATO ASI Series, Vol. H 58
Neurocytochemical Methods
Edited by A. Calas and D. Eugène

proteins through activation of enzymes including the protein kinase family. These enzymes can phosphorylate proteins including transcription factors. Transcription factors are DNA-binding proteins. Further, their affinities for DNA binding and their efficacy in regulating transcription of their target genes can be altered by this phosphorylation (Ptashne, 1988; Yamamoto et al., 1990). This pathway can thus allow information from events taking place at the neural membrane to be transduced into changes in the expression of neuronal genes, including those related to neurotransmission.

Important portions of these putative pathways for trans-synaptic gene regulation can now be studied in neurons using *in situ* hydridization. Examination of genes whose expression is changed when neurotransmitter genes change their expression can help to identify candidate genes involved in this intracellular signalling. Determining which transcription factors are up-regulated in neurons where a neurotransmitter gene that is a potential target of these transcription factors is up-regulated can thus provide clues important for elucidating the biochemical and dynamic pathways of neuronal gene regulation *in vivo*.

One model system for examining mechanisms of trans-synaptic neurotransmitter gene regulation *in vivo* is provided by the dynamically-changing expression of a neuropeptide neurotransmitter gene, preproenkephalin, in neurons of the dorsal horn (Iadarola et al., 1988 ; Nishimori et al., 1988 ; Nishimori et al., 1989a,b ; Noguchi et al., 1989 ; Uhl and Nishimori ; 1990). The medullary dorsal horn has been especially well-studied in this regard (Uhl and Nishimori, 1990). Low intensity primary afferent stimulation leads to rapid up-regulation of the number of neurons that express preproenkephalin in lamina I and II of the trigeminal nucleus caudalis. The effect is noted in animals sacrificed immediately at the end of one hour's electrical stimulation of afferents to these zones. Interestingly, this rapid up-regulation is blocked if small caliber afferents are co-stimulated by high-intensity stimuli. This rapid trans-synaptic regulation also displays a specific cellular basis: it acts largely by recruiting a new, previously nonexpressing population of neurons to the group that previously expressed (Uhl and Nishimori, 1990). When *in situ* hybridization autoradiographic grain densities are analyzed, this phenomenon is reflected in an increased fraction of laminae I and II neurons displaying grain densities above background levels. Conversely, mean grain densities per positive neuron are similar to control values. These cellular changes thus provide a benchmark against which any potential candidate transcription factor gene change can be measured (Nishimori et al., 1989a ; Uhl and Nishimori, 1990). If the factor leads to the preproenkephalin up-regulation, it should be present in an additional 10-15% of all laminae I and II neurons in animals sacrificed at the end of low-intensity primary afferent stimulation.

As a first step toward identifying the genes potentially involved in this trans-synaptic up-regulation of preproenkephalin, the known factors binding to putative regulatory sequences lying in the 5' flanking region of the preproenkephalin gene were surveyed. A "cassette" region in this area contains sequences that could recognize protein products of several transcription factor gene families. These include the AP-I and ATF factors (Jun, Fos and ATF family members), CRE factors (CREB and CRE-BP families), AP-2 factors, AP-4 factors, NF-1 factors, and Zif-268 factors (Comb et al., 1986 ; Comb et al., 1988 ; Van Nguyen et al., 1990 ; Comb, personal communications).

We initially focused on the Jun factors since previous immunohistochemical work had shown that another protein to which Jun may bind to enhance its ability to activate transcription, cFos, was activated in dorsal horn neurons by primary afferent stimulation (Hunt et al., 1987 ; Uhl et al., 1990). Further, the AP-1 site to which these hetero-dimeric factors can bind mediated powerful activation of expression of several genes. Three Jun genes, Jun A (cJun), Jun B and Jun D, have been cloned to date (Nekabeppu et al., 1988 ; Ryder et al., 1988). We identified segments of each of the mRNA products of each of these genes that show little homology to the other Jun mRNAs to serve as targets for hybridization. We constructed radiolabeled oligonucleotide cDNA hybridization probes directed against these mRNA segments that were similar in length, G & C content, and specific activity. Hybridization of brain sections with probes directed against each mRNA resulted in autoradiograms displaying robust stimulus-related up-regulation of Jun B hybridization, more modest up-regulation of Jun A (cJun) hybridization, and no sizable change in low levels of Jun D mRNA (Uhl et al., 1990). Levels and stimulus-responsiveness of other transcription factors that could also bind to the preproenkephalin promoter have also been examined in preliminary experiments. These results support modest roles, at best, for AP-2, NF1-L and NF1-red in this regulation, while the CREB/ATF family appears better positioned for possible involvement (Uhl and Appleby in preparation).

These studies can provide hints of the transcription factor genes that are good candidates for potential involvement in one example of *in vivo* up-regulation occurring in these neurons. The quantitative features of this up-regulation are also of interest. Jun B and preproenkephalin are both up-regulated in approximately 12-15% of laminae I and II neurons by this stimulation (Uhl et al., 1990). This similarity is consistent with a possible mechanistic association between these two rapid gene-regulatory events. Such temporal and cellular association cannot, does not, prove causality. This sort of data does at least suggest which factors and cellular pathways are potentially involved in *in vivo* regulation.

How can this correlative evidence be strengthened ? If a transcription factor is both necessary and sufficient for up-regulation, then the gene segment recognizing the factor will confer up-regulation on a reporter gene (such as chloramphenicol acetyl transferase, CAT) in transgenic animals. We have recently been successful in constructing such transgenic animals, and are using them to study this regulation. (O'Hara et al., in preparation). These results, too, have limitations in interpretation. For example, studies in cell culture systems suggest that several factors working in concert may be necessary for optimal gene expression and regulation. We are approaching this problem by constructing promoter mutants altered in just the bases recognizing one specific transcription factor. This approach is also limited, however, in promoter "cassette" regions where factor binding sites may each recognize several related factors, where sites are densely spaced, and where they may even overlap. Both DNA footprinting and single-base substitution experiments suggest that the binding sites for several factors overlap to a substantial extent in regions of the preproenkephalin promoter (Comb et al., 1988 ; Van Nguyen et al., 1990). The difficulties in producing and characterizing the multiple lines of transgenic animals required for these experiments make this work tedious, however.

A second means for possibly elucidating which of the "candidate" transcription factors found in *in situ* hybridization experiments could involve use of oligonucleotides that could selectively compete for binding of a cellular transcription factor, and thus eliminate binding to the promoter region on specific genes. This tactic has been used successfully in studies of gene expression in Aplysia neurons injected with the oligonucleotides (Dash et al., 1990). Problems with access may limit its usefulness with neurons that cannot be individually injected, however.

As noted above, the activities of several transcription factors can be altered by post-translational modification, especially phosphorylation (Yamamoto et al., 1988). With the availability of antibodies that could specifically recognize the phosphorylated and the dephosphorylated forms of these factors, their differential regulation could also be studied *in vivo*. The recent availability of antibodies allowing study of cytoplasmic second messengers and of dyes elucidating dynamics in free intracellular calcium (detailed elsewhere in this volume) offer the possibility that the cytoplasmic messengers that can trigger transcription factor activation can be directly studied in the appropriate neurons. Since the second messenger systems that can help regulate many transcription factor genes *in vitro* are known, these results would also help to define more- and less-likely pathways of *in vivo* gene regulation.

While trans-synaptic regulation of gene expression is now widely documented in the nervous system, its mechanisms remain largely speculative in virtually each case. Using *in situ* hybridization techniques to elucidate which members of a constellation of possible transcription factors could be involved *in vivo* is one first step toward understanding such mechanisms. Employing the method in transgenic animals can provide a second line of evidence for or against involvement of a specific factor. Elucidation of such gene regulatory mechanisms can thus represent a first step in being able to alter gene expression in the brain in a targeted fashion.

## ACKNOWLEDGEMENTS

We are grateful to Drs. Michael Comb and Michael Kuhar for throughtful discussions, John Dwyer and John Park for help with data collection and analysis, and Ms. Carol

Sneeringer for help with the manuscript. This work was supported by the intramural program of the National Institute of Drug Abuse, the NINDS, the NIMH, the McKnight Foundation, the Sloan Foundation, the American Parkinson's Disease Association, and the Juvenile Diabetes Association.

## REFERENCES

Comb M, Birngerg NC, Seasholtz A, Herbert E and Goodman HM (1986) Cyclic AMP- and phorbol ester-inducible DNA element. Nature 323 : 353-356.

Comb M, Mermod N, Hyman SE, Pearlberg J, Ross ME and Goodman HM (1988) Proteins bound at adjacent DNA elements act synergistically to regulate human proenkephalin cAMP inducible transcription. EMBO 7 : 3793-3805.

Dash P, Hochner B and Kandel ER (1990) Injection of the cAMP-responsive element into the nucleus of Aplysia sensory neurons blocks long-term facilitation. Nature 345 : 718-721.

Hunt SP, Pini A and Evan G (1987) Induction of c-fos-like protein in spinal cord neurons following sensory stimulation. Nature 328 : 632-634.

Iadarola MJ, Douglass J, Civelli O and Naranjo R (1988) Differential activation of spinal cord dynorphin and enkephalin neurons during hyperalgesia : evidence using cDNA hybridization. Brain Res 455 : 205-212.

Nakabeppu Y, Ryder K and Nathans D (1988) DNA binding activities of three murine Jun proteins : stimulation by Fos. Cell 55 : 907-915.

Nishimori T, Buzzi MG, Moskowitz MA and Uhl GR (1989a) Differential effects of small- and large-caliber primary afferents on nucleus caudalis preproenkephalin expression. J Comp Neurol (in press).

Nishimori T, Buzzi MG, Moskowitz MA and Uhl GR (1989b) Preproenkephalin mRNA expression in nucleus caudalis neurons is enhanced by trigeminal stimulation. Mol Brain Res 6 : 203-210.

Nishimori T, Moskowitz MA and Uhl GR (1988) Opioid peptide gene expression in rat trigeminal nucleus caudalis neurons : normal distribution and effects of trigeminal deafferentiation. J Comp Neurol 274 : 142-150.

Noguchi K, Morita Y, Kiyama H, Sato M, Ono K and Tohyama M (1989) Preproenkephalin gene expression in the rat spinal cord after noxious stimuli. Mol Brain Res 5 : 227-234.

Ptashne M (1988) How eukaryotic transcriptional activators work. Nature 335 : 683-689.

Ryder K, Lau LF and Nathan D (1988) A gene activated by growth factors is related to the oncogene v-jun. Proc Natl Acad Sci USA 85 : 1487-1491.

Uhl GR (ed.) (1986) *In situ* Hybridization in Brain. Plenum Press, New York.

Uhl GR, Appleby D, Nishimori T, Buzzi G and Moskowitz M (1990) Jun B, cJun, Jun D and cFos mRNAs in nucleus caudalis neurons : rapid selective enhancement by afferent stimulation. Mol Brain Res (submitted).

Uhl GR and Nishimori T (1990) Neuropeptide gene expression and neural activity : assessing a working hypothesis in nucleus caudalis and dorsal horn neurons expressing preproenkephalin and preprodynorphin. Cell Mol Neurobiol 10 : 73-98.

Van Nguyen T, Kobierski L, Comb M and Hyman SE (1990) The effect of depolarization on expression of the human proenkephalin gene is synergistic with cAMP and dependent upon a cAMP-inducible enhancer. J Neurosci 10 : 2825-2833.

Yamamoto KK, Gonzalez GA, Biggs III WH and Montminy MR (1988) Phosphorylation-induced binding and transcriptional efficacy of nuclear factor CREB. Nature 334 : 494-498.

# THE USE OF MARKERS TO STUDY THE REGULATION OF GLIAL CELLS IN THE PNS AND CNS

Rhona Mirsky and Kristján R. Jessen
Department of Anatomy and Developmental Biology
University College London
Gower Street
London WC1E 6BT
England

The main glial cells of the PNS and CNS can be identified both in situ and in culture by the use of combinations of different antibodies which act as markers for the different cell types. The particular combination of antibodies chosen depends very much on the developmental stage and anatomical area of the nervous system chosen, on whether the cells are being studied *in situ* or in culture, and on the question that is being asked. In studies of glia from the CNS antibodies have been particularly useful in identifying different glial types in dissociated cell cultures derived from various brain regions. These cultures normally contain a mixture of cell types, including neurons, unless special efforts are made to separate them, and although with practice some of the cell types can be identified on the basis of morphology, in many cases cells lose their typical *in vivo* shape, flatten and become more amorphous when grown on the flat surface of a glass coverslip or tissue culture dish. Thus a prime use of markers has been for identification (Mirsky, 1980). Astrocytes have most often been identified by the expression of glial fibrillary acidic protein (GFAP), although other markers such as glutamine synthetase are occasionally used. All astrocytes in culture appear to express high levels of GFAP. Many astrocytes *in situ*, particularly the fibrous astrocytes of white matter can also be labelled with antibodies to GFAP in tissue sections. In the cortex, however, many grey matter astrocytes are poorly labelled with GFAP antibodies. After lesion, GFAP expression is extremely strong in reactive astrocytes in tissue sections.

NATO ASI Series, Vol. H 58
Neurocytochemical Methods
Edited by A. Calas and D. Eugène

Oligodendrocytes have most commonly been identified in culture on the basis of expression of the glycolipid galactocerebroside but antibodies to other molecules such as myelin basic protein, proteolipid protein, CNP'ase, sulfatide and 04 antigen have also been used. Since some of these molecules appear earlier than others during oligodendrocyte differentiation in vivo there is some disagreement between different laboratories as to which marker should be used to define the transition to an oligodendrocyte from a progenitor or blast cell. Some groups, notably Raff and his associates (for review see Richardson et al., 1990) define an oligodendrocyte as a galactocerebroside positive cell, whereas other groups, notably Skoff and his associates (Skoff, 1990) define an earlier cell as an oligodendrocyte because it is sulfatide positive. All of these markers can, under appropriate conditions, be used on cells in situ to identify oligodendrocytes.

A progenitor cell, which gives rise in culture to both oligodendrocytes ($GC^+$) and an astrocytic ($GFAP^+$) cell (type II astrocyte) has been defined on the basis of studies of single cell cultures and bulk dissociated cell cultures from optic nerve by Raff and his associates (see Richardson et al. 1990 for review). Analysis in this CNS culture system has been facilitated by the fact that the optic nerve does not contain neuronal cell bodies and thus the cultures contain no neurons. The progenitor cell, which has a bipolar morphology, can be labelled with the A2B5 antibody, with an anti-$GD_3$ ganglioside antibody and with tetanus toxin, and can be induced to differentiate into an A2B5 negative, galactocerebroside-positive oligodendrocyte in serum-free medium or to an A2B5-positive type II astrocyte in serum containing medium. Although it is not certain that the type II astrocyte, a process-bearing stellate cell in vitro, exists in vivo, this approach, using markers combined with appropriate culture analysis, has resulted in the identification of a progenitor cell for the oligodendrocyte lineage and enabled its development, regulation and differentiation to be studied.

We have been more interested in studying the development and differentiation of glial cells in the PNS, with a particular emphasis on the development and differentiation of Schwann cells in peripheral nerves. These cells, found in intimate association with axons, are one of the main glial cell types in the PNS. Other major categories of peripheral glia include the satellite cells that surround the cell bodies of neurons in sensory, sympathetic and parasympathetic ganglia, and enteric glia, the glial cells of the gut nervous system which resemble astrocytes both morphologically and in their relationship to neurons. All these cells derive from the neural crest and share molecules in common.

The Schwann cells in mature nerves are found as two major morphological variants, myelin-forming Schwann cells that form sheaths round the larger axons and non-myelin-forming Schwann cells, that enclose several smaller axons in individual troughs in the Schwann cell surface membrane. The molecular phenotype of these cells in peripheral nerves has been studied using antibodies in combination with a number of different tissue preparations. These include dissociated cell cultures, tissue sections, teased nerve preparations and single cell suspensions which can be labelled and dried down on to a microscope slide. Since each of these methods has its own strengths and weaknesses it has proved most fruitful to use them in combination in our studies. In particular the use of teased nerve preparations has enabled us to distinguish the molecular phenotype of Schwann cells in myelinated and unmyelinated fibres much more easily than in tissue sections where unmyelinated fibres are not readily separated by eye from components of the endoneurial compartment of the nerve. Once an antigen has been localised to a particular fibre type, it is, however, necessary to show that the molecule is associated with the Schwann cell since the antibody also has access to the axons of the fibres. Dissociation of Schwann cells from axons can be achieved using enzyme cocktails (collagenase or collagenase and trypsin) and the cells can then be immunolabelled immediately in suspension or after a few hours in culture to determine whether a particular molecule is associated with the Schwann cell or not. In this process it is necessary to have a marker which distinguishes Schwann cells from all other cells in the preparation since morphological distinction is lost, particularly in suspensions of cells. We have found that the calcium binding protein S100 is an excellent pan-Schwann cell marker both in suspension stained and cultured Schwann cells. Using antibodies to S100 in combination with other antigens in double immunolabelling studies it has been possible to show that in situ myelin-forming Schwann cells express molecules not expressed by non-myelin-forming cells. These include the myelin proteins $P_o$, MAG, $P_2$ and myelin basic protein. Conversely non-myelin-forming Schwann cells in situ express molecules which are not associated in substantial amounts with myelin-forming Schwann cells. These include GFAP, N-CAM, L1, A5E3 protein (130kD), Ran-2 protein (140kD) and NGF receptors. In addition both variants of Schwann cell express a number of glial associated molecules in common. These include S100, CNP'ase and the glycolipids galactocerebroside and 04 antigen (probably sulfatide) (for review see Mirsky and Jessen 1990).

The use of Schwann cell cultures from rat sciatic nerve has revealed that when a Schwann cell is removed from axonal contact it undergoes very rapid changes in molecular phenotype which are a reversal of differentiative changes that occur under axonal influence during the development of myelinated and unmyelinated fibres in vivo. Thus myelin-forming Schwann cells lose the myelin protein $P_o$, myelin basic protein and $P_2$ and reexpress GFAP, the cell adhesion molecules N-CAM and L1, the cell surface protein A5E3 (130kD) and the NGF receptor (Jessen et al. 1987; Jessen et al. 1990) while both myelin-forming and non-myelin-forming Schwann cells lose galactocerebroside and 04 antigen (probably sulfatide) (Mirsky et al. 1980; Jessen et al. 1985, 1987; Mirsky et al. 1990). Cells in culture and in vivo express S100. The changes observed are similar to those seen in Schwann cells after nerve transection in vivo and the phenotype of a Schwann cell in culture resembles that found in a Schwann cell in situ on embryo day 16-17 sciatic nerve. Thus the phenotype is $S100^+$, $N\text{-}CAM^+$, $L1^+$, $A5E3^+$, $NGFR^+$ although unlike the E16-17 Schwann cell, the cultured cell is also $GFAP^+$.

The changes in Schwann cell phenotype that occur in culture provide a useful starting point for studying the intracellular signals in axons that influence Schwann cell differentiation, and in particular induction of the myelin phenotype. By adding exogenous agents which activate or mimic second messenger systems inside the cells to cultures from 5 day old rat sciatic nerve we have been able to determine that elevation of intracellular cAMP levels in the presence of growth factors (e.g. serum) induces the Schwann cells to express the 04 antigen, normally present on all Schwann cells at E19 in vivo, but cells with a myelin phenotype are not found. If DNA synthesis is halted (e.g. by withdrawal of growth factors) elevation of intracellular cAMP induces a myelin related phenotype in many Schwann cells in the cultures. This includes induction of $P_0$ and down-regulation of GFAP, N-CAM, A5E3 and GFAP, all events which occur during differentiation in vivo (Jessen et al. 1990). Thus we are able to mimic a series of developmental changes that occur in vivo by manipulating the culture conditions, leading to the hypothesis that early in development axons may cause elevation of cAMP in Schwann cells and that later in development withdrawal from the cell cycle causes the elevated cAMP within the Schwann cells to initiate the formation of the myelin sheath round the larger axons (Morgan et al. 1990).

In conclusion, it should be clear that the use of antibodies both in situ and in vitro to study the major categories of glia within the nervous system has been extremely useful not only for cell identification but also for studying fundamental problems. In the future the ability to combine these methods with in situ hybridization for mRNA should extend its possibilities still further.

## Materials and methods

### Tissue preparation

Cell suspensions from sciatic nerves or cervical sympathetic trunk were prepared from Sprague-Dawley rats. Embryos, 16 to 19 days old, were removed from pregnant rats killed by ether anaesthesia followed by bleeding. Newborn, 5 day and 10 day rats were killed by decapitation and 28- to 32-day or older rats were killed by either anaesthesia followed by bleeding. The sciatic nerves were excised and the epineurial sheath was removed. The tissue was placed in an enzyme mixture contain 2mg $ml^{-1}$ collagenase (Lorne Diagnostics), 1.2mg $ml^{-1}$ hyaluronidase (Sigma Chemical Co.) and 0.3mg $ml^{-1}$ trypsin inhibitor from chicken egg white (Sigma Chemical Co.) in Dulbeccos Modified Eagle's Medium (DMEM) without calcium or magnesium, and was then chopped finely. The tissue was incubated at 37°C and 5% $CO_2$ for 2 3/4h (30-day or older rats), 1h(10-day rats), 45 min (5 day rats, or 30 min (17-to 18-day embryos and newborn rats.) An equal volume of Minimum Eagle's Medium with 0.02M HEPES buffer (MEM-H) containing 10% calf serum was added and the cells gently dissociated by trituration through a plastic pipette tip. The cells were centrifuged for 10 min at 500g and resuspended in either MEM-H with 10% calf serum and antibody if they were to be suspension stained, or DMEM with 10% foetal calf serum (FCS) if they were to be placed in culture.

**Cell culture**

Cell suspensions of Schwann cells, prepared as described above in DMEM with 10% FCS, were plated in 50µl droplets onto glass coverslips, 13mm in diameter and coated with poly-L-lysise in a 24-well multiwell plate and kept at 37°C in an incubator gassed with 5% $CO_2$. After 3h the cultures were topped up with 400µl DMEM with 10% FCS. In some experiments cytosine arabinoside ($10^{-5}$M) was added after 16 hours for 2 - 3 days to kill dividing cells. After removal of the cytosine arabinoside Schwann cell cultures were 95% pure and in most studies were used within 5 days (for detailed methods see Mirsky et al. 1980; Jessen et al. 1987 and Morgan et al. 1990).

Cell cultures of optic nerve are described by Richardson et al. (1990).

**Denervation**

Adult Sprague-Dawley rats, weighing 90-110 gm, were anaesthetized and the left sciatic nerve either cut or crushed 2-4 mm below the sciatic notch for 8 min using artery clips. In other experiments, the left cervical sympathetic trunk was either cut and the proximal stump rerouted into the sternomastoid muscle, or crushed 5-7 mm below the superior cervical ganglion. The site of the crush was marked with carbon powder. After surgery, rats were kept for periods of time ranging from 4 d to 2 months before the nerves were removed for dissociation and analysis. The distal stump of the cut or crushed nerve was measured and an equivalent length of the unoperated nerve on the right side was used as a control.

**Teased nerve preparations**

Mixed nerves (myelinated and unmyelinated) such as sciatic nerve, and predominantly unmyelinated nerves such as cervical sympathetic trunk, were excised from rats of various ages and partially teased into nerve bundles and individual fibres with fine syringe needles in a small amount of phosphate-buffer saline (PBS) on microscope slides. Samples were allowed to dry before immunofluorescent staining, and in some cases were prefixed with 95% ethanol/5% acetic acid or with 4% paraformaldehyde in PBS for 10 min prior to labelling.

**Frozen sections**

Dorsal root ganglia, superior cervical ganglia, sciatic nerves, cervical sympathetic trunk, cerebellum and proximal colon were removed from Sprague-Dawley rats of various ages. Frozen sections of 3 - 7μm were cut and thawed on to dry or polylysine-coated microscope slides. In some experiments sections were briefly rinsed with 0.15% Triton in phosphate buffered saline prior to application of antibodies.

**Immunolabelling**

Immunofluorescence was the method used in most of our studies although peroxidase based methods have also been used for visualisation in some cases. In general frozen tissue sections, teased nerves, cultured cells or cell suspensions were incubated for 30 min at room temperature or overnight at 4° with the appropriate antibodies. Pre-fixation prior to immunolabelling was sometimes employed depending on whether the antibody recognized an antigen in an extracellular or intracellular location, with the exception of frozen sections which were not normally prefixed. Antibody binding was usually visualised either with fluorescein or rhodamine conjugated to goat anti-mouse Ig or goat anti-rabbit Ig. In some studies a biotinylated anti-mouse Ig or anti-rabbit Ig second layer was used, followed by streptavidin-fluorescein or streptavidin-Texas Red. All these antibodies were incubated with the preparation for 30 min at room temperature. In experiments where two antibodies from different species were used the fluorescent rabbit Ig antibodies were absorbed with mouse Ig and the fluorescent mouse Ig antibodies were absorbed with rabbit Ig to remove cross-reactivity and antibodies were normally applied sequentially. After post-fixing for 5 min in 2% paraformaldehyde in PBS all samples were mounted in Citifluor anti-fade mounting medium and examined with either X40 or X63 objectives on a Zeiss fluorescence microscope.

**Antibodies**

Many different antibodies have been used in our studies of glial cells. Among the most important of these are antibodies to the intermediate filament proteins GFAP and vimentin (both polyclonal and monoclonal), to the cell adhesion molecules N-CAM (Gennarini et al. 1984) and L1 (polyclonal)(Rathjen and Schachner, 1984), to the cell surface protein A5E3 (monoclonal) (Jessen et al. 1990) and to Ran-2 (monoclonal) (Mirsky and Jessen 1984), to the NGF receptor, using either the 217c monoclonal antibody (Peng and de Vellis 1982; Ferrari et al. 1990) or the 192 IgG antibody (Taniuchi and Johnson, 1985), the glycolipids O4 (monoclonal) (Sommer and Schachner, 1981) and galactocerebroside, (monoclonal) (Ranscht et al. 1982) and the calcium binding protein S100. In the CNS studies described here monoclonal antibody A2B5

(Eisenbarth et al. 1979) and monoclonal antibody LB1 which recognizes ganglioside $GD_3$ (Curtis et al. 1988) have also been used.

## References


Curtis R, Cohen J, Fok-Seang J, Hanley MR, Gregson NA, Reynolds R and Wilkin GP (1988) Development of macroglial cells in rat cerebellum. Use of antibodies to follow early in vivo development and migration of oligodendrocytes. J Neurocytol 17:43-54

Eisenbarth GS, Walsh FS and Nirenberg M (1979) Monoclonal antibody to a plasma membrane antigen of neurons. Proc Natl Acad Sci USA 76:4913-4917

Ferrari G, Fabris M, Polato P, Skaper SD, Skaper MD, Fiori MG and Yan Q (1990) Rat NGF receptor is recognised by the tumor associated antigen monoclonal antibody 217c. Exp Neurol in press

Gennarini G, Rougon G, DeAgostini-Bazin H, Hirn M and Goridis C (1984) Studies on the transmembrane disposition of the neural cell adhesion molecule N-CAM. A monoclonal antibody recognizing a cytoplasmic domain and evidence for the presence of phosphoserine residues. Eur J Biochem 142:57-64

Jessen KR, Morgan L, Brammer M and Mirsky R (1985) Galactocerebroside is expressed by non-myelin-forming Schwann cells in situ. J Cell Biol 101:1135-2243

Jessen KR, Morgan L and Mirsky R (1987a) Axonal signals regulate the differentiation of non-myelin-forming Schwann cells: an immunohistochemical study of galactocerebroside in transected and regenerating nerves. J Neurosci 7:3362-3369

Jessen KR, Morgan L, Mirsky R (1987b) Myelinated, but not unmyelinated axons reversibly down-regulate N-CAM in Schwann cells. J Neurocytol 16:689-700

Jessen KR, Morgan L, Stewart HJS and Mirsky R (1990) Three markers of adult non-myelin-forming Schwann cells, 217c (Ran-1), A5E3 and GFAP: development and regulation by neuron-Schwann cell interactions. Development 109:91-103

Mirsky R (1980) Cell-type-specific markers in nervous system cultures. TINS, 3: 190-192

Mirsky R, Dubois C, Morgan L and Jessen KR (1990) Prenatal Schwann cell development:appearance of 04 differentiation antigen in rat embryo sciatic nerve and its regulation by axon-Schwann cell signals. Dev 109:105116

Mirsky R and Jessen KR (1990) Schwann cell development and the regulation of myelination. Seminars in Neurosci 2, in press

Mirsky R, Winter J, Abney ER, Pruss RM, Gavrilovic J and Raff MC (1980) Myelin-specific proteins and glycolipids in rat Schwann cells and oligodendrocytes in culture. J Cell Biol 84:483-494

Morgan L, Jessen KR and Mirsky R (1990) The effects of cyclic AMP on differentiation of cultured Schwann cells: progression from an early phenotype (04+) to a myelin phenotype ($P_O^+$, $GFAP^-$, $N\text{-}CAM^-$, NGF $receptor^-$) depends on growth inhibition. J Cell Biol in press

Peng WW, Bressler JP, Tiffany-Castiglioni and de Vellis J (1982) Development of a monoclonal antibody against a tumor-associated antigen. Sci 215:1102-1104

Ranscht B., Clapshaw PA, Price J, Nobel M and Seifert W (1982) Development of oligodendrocytes and Schwann cells studied with a monoclonal antibody against galactocerbroside. Proc Natl Acad Sci USA 79:2709-2713

Rathjen F and Schachner M (1984) Immunocytological and biochemical characterization of a new neuronal surface component (L1 antigen) which is involved in cell adhesion. EMBO Journal 3, 1-10

Richardson WD, Raff MC and Noble M (1990) The oligodendrocyte-type 2 astrocyte lineage. Seminars in Neurosci 2, in press

Skoff RP (1990) Gliogenesis in rat optic nerve: astrocytes are generated in a single wave before oligodendrocytes. Dev Biol 139:149-168

Sommer I and Schachner M (1981) Monoclonal antibodies (01 to 04) to oligodendrocyte cell surfaces:an immunocytological study in the central nervous system. Dev Biol 83:311-327

Taniuchi M and Johnson EM Jr (1985) Characterization of the binding properties and retrograde axonal transport of a monoclonal antibody directed against the rat nerve growth factor receptor. J Cell Biol 101:110-1106

# REPORT ON THE DISCUSSION OF THE SECOND SESSION: DETECTION OF SPECIFIC MOLECULES, AND THE SECOND PLENARY LECTURE: MOLECULAR NEUROANATOMY OF SYNAPSES, CELLS AND SYSTEMS IN THE BRAIN

Margaret R. Matthews
Department of Human Anatomy
University of Oxford
South Parks Road
Oxford OX1 3QX
United Kingdom

## Discussion of the paper by Dr O.Bosler

This centred on questions of relative sensitivity and specificity of the various approaches described, with particular reference to problems of establishing co-localization of transmitters which like the monoamines may be subject to uptake after release. Dr Bosler reported that in his experience the use of Fab fragments rather than the whole antibody slightly reduced the sensitivity of immunocytochemical detection, as compared with the conventional PAP method, but that other workers claimed to achieve a greater sensitivity by their use. The applications and relative value of uptake radioautography and immunocytochemistry for the monoamines were considered. It was emphasized that labelling by uptake of radiolabelled transmitter, or uptake radioautography, gave a clearly defined localization in nerve terminals as well as in cell bodies, and gave a clearer definition of nerve terminals than did radioimmunocytochemistry, making them easier to find, but that uptake radioautography could not discriminate between neurones which synthesized the specific transmitter and those which did not synthesize it but had an uptake system for it. The latter are possibly receptive for the transmitter in question. The significance of the finding that there are neurones which apparently co-localize dopamine and serotonin was specifically considered. For example, neurones in the substantia nigra have shown apparent double labelling for

NATO ASI Series, Vol. H 58
Neurocytochemical Methods
Edited by A. Calas and D. Eugène

dopamine and for serotonin in their cell bodies but not in their terminal fields; there, in particular, double labelling was never seen in the same nerve terminal; therefore the apparent double labelling in the cell bodies was tentatively attributed to non-specific attachment of antibody (Steinbusch, 1984). In the dorsomedial nucleus of the hypothalamus, Dr Bosler has found neurones which in their cell bodies accumulate serotonin but do not take up tritiated dopamine, i.e. they have an uptake system which is selective for the one but not the other. Could this be used as an argument that, if there is tritiated dopamine labelling of serotonin-immunoreactive nerve terminals, this must indicate specific rather than non-specific uptake? It was conceded that, even if so, the uptake might signify receptivity with an associated uptake system, i.e. uptake without production, rather than true co-existence of the two transmitters. To resolve this question would require dual immunocytochemical labelling, rather than the combination of immunocytochemistry for one transmitter and uptake radioautography for the other. This would moreover entail the demonstration of the presence of the appropriate transmitter-synthesizing enzymes, respectively, for each transmitter, since the mere demonstration of the co-localization of endogenous monoamine transmitters would still not exclude the possibility of uptake without synthesis, from local sources such as other monoaminergic nerve endings in the neighbourhood.

This presentation placed a salutary emphasis upon the necessity for double-checking conclusions about co-existence by using as many alternative approaches as possible, and revealed the prospect that it would soon be, or in some cases already was, possible to localise simultaneously not just two but even three or more antigens at the ultrastructural level, by appropriate combinations of an expanding range of techniques. Throughout, there was an insistence upon the importance of rigorous control of the conditions of processing, particularly where there was need for discrimination between two forms of end-product of enzymic activity, or for the complete blocking of one set of antigenic sites before the initiation of the second stage of certain double immunocytochemical protocols. Justifiable optimism was thus tempered with experimental rigour and with an emphasis upon caution in the interpretation of findings.

Discussion of the plenary lecture by Dr P. Somogyi

Topics of co-localization and of experimental rigour were raised also in relation to the lecture of Dr Somogyi. To the question whether there was physiological evidence that a nerve ending which showed co-localization of choline acetyltransferase (ChAT) and GABA could function both as a cholinergic and a GABA-ergic ending, he replied that there was not such evidence for the visual cortex but that such a phenomenon had been described in the retina. Intracellular recording in the visual cortex has shown that activation of the cholinergic system leads in most neurones to a fast, $GABA_A$ mediated inhibition followed by a slow muscarinic facilitation, but the inhibition has been attributed to fast cholinergic activation of GABA-ergic interneurones. The inhibitory phase is thus thought to be indirectly mediated and the facilitation directly mediated. It is not known whether there may also be release of GABA from the cholinergic nerve terminals, which could contribute to the inhibition directly. In the retina, however, there is evidence for the presence of GABA, and glutamic acid decarboxylase, in all cholinergic amacrine cells. Here, it appears that the two transmitters are released on different time scales and possibly by different mechanisms: acetylcholine, by calcium-dependent exocytotic release and GABA, it has been suggested, in non-vesicular fashion by reverse transport.

Asked whether his material indicated a different subcellular location of GABA and ChAT within the same terminal, Dr Somogyi replied that in his opinion that would be too much to ask of the present techniques. He pointed out, moreover, that peroxidase immunocytochemistry can give both light and heavy end-product in, for example, nerve terminals in a specimen but that this does not necessarily indicate differences in the antigen content. It might, however, affect the probability of demonstrating co-localization in the same terminals. For example, he had found in the visual cortex that 10% of ChAT-immunoreactive terminals, demonstrated by a pre-embedding immunoperoxidase technique, were also immunoreactive for GABA, detected by post-embedding colloidal gold. Those doubly labelled terminals were however only very lightly immunoreactive for ChAT. The failure to demonstrate GABA immunoreactivity in terminals showing a heavier peroxidase reaction need not imply the absence of GABA from these terminals: it might mean merely that the antigenic sites for the second antibody detection system were masked by peroxidase end-product. The figure of 10% must therefore be regarded as a minimum, not necessarily indicating the full extent of co-localization.

In a plea for greater circumspection in the interpretation of results, Dr Somogyi cited two areas in which he felt unjustified conclusions were being reached. The first concerned the demonstration of immunoreactivity in cell bodies for amino-acids which were used metabolically as well as acting as transmitters. If non-quantitative methods were used, such as the immunoperoxidase technique, which had been extensively used in studies on cell bodies in the brain, then no valid conclusions could be drawn about the transmitter status of those neurones. If, on the other hand, a quantifiable technique such as immunogold were used, then a transmitter function might be concluded if it were possible to demonstrate that a neurochemically-different cell population, or nerve terminal population, in the same material showed either no immunoreactivity or, for example in the case of glutamate, a lower level of immunoreactivity. In nerve terminals it might thus be possible to recognise 'enrichment' in the amino-acid, as was found for glutamatergic terminals in comparison with GABAergic terminals, where indeed the metabolic pool of glutamate might for good reason be smaller than the average. In any case, however, this would be only a part of the evidence, and the eventual conclusion about the identity of a transmitter should be addressed by physiological and pharmacological methods. In further discussion of putative amino-acid transmitters, such future possibilities were envisaged as the ability to demonstrate calcium-dependent release in situ, and to detect the transmitter at its moment of release.

Dr Somogyi's second criticism related to the subcellular localization of receptors. He had been fortunate in being able to use antibodies which gave membrane-associated immunoreactivity, selectively located to specific membranes; but antibodies for receptors which had equally clearly been shown to be membrane constituents, for example cholinergic receptors, were being reported as giving a homogeneous labelling of the cytoplasm of neurones. This could not be accepted as a full description of the localization of such receptors; and it was moreover attracting the disbelief of biochemist colleagues and inspiring in them a mistrust of all attempts at receptor localization. In some cases it was unlikely to be the antibodies which were at fault, since these had been shown, in non-immunohistochemical tests, to be epitope-specific, but the technology which was inappropriate; and the uncritical publication of such observations was to be avoided. Hence, indeed, the value of meetings like the present workshop, which could and should function to safeguard the standards of the discipline. It was nevertheless recognised that receptor molecules might truly be located intracytoplasmically in an antigenically recognisable form, perhaps in association with membranes of the endoplasmic reticulum, during their manufacture or after internalization. Work which was quoted as exemplary was that of A. Beaudet (Hamel & Beaudet, 1987) in localizing opioid receptors by an autoradiographic technique, who showed in a recent quantitative analysis and with controls that over 75% of specific binding sites were at the plasma membrane and about 20%, intracellular. In this study the tissue was pre-fixed and incubated as Vibratome slices without additional permeabilization, and these factors will have governed access of radioligand to the cytoplasm. Dr Somogyi reported his own finding in respect of the GABA receptor, studied by pre-embedding immunocytochemistry in tissue slices after freezing and thawing to aid penetration of antibody : a substantial proportion was intracellular, associated with intracellular membranes. Evidence was found for internalization of receptor, by way of vesicles and multivesicular bodies, but not for receptor lying free in the cytoplasm.

It was asked to what extent uptake of a transmitter and the presence of a receptor for that transmitter might be correlated. That this holds true for the monoamines was illustrated by O. Bosler, who reported that the thalamic paraventricular nucleus contains some neurons which can take up [$^{3}$H]adrenaline but lack its biosynthetic enzyme, PNMT, and also exhibit a high density of adrenergic receptors; these are thus adrenoceptive, but not adrenergic. H.Steinbusch confirmed this as applying to all the catecholamines and to serotonin, and emphasized that in all these cases, at sufficiently low concentrations, the uptake system involved was not non-selective but was specific for each transmitter. G.R. Uhl pointed out that uptake would depend on the type of receptor involved: there are different molecular classes of receptors, and those which also transport the agonist molecule tend to be of the class which has a single transmembrane region, like the NGF or LDL receptors, as opposed to those with 5 or 7 transmembrane regions or the ligand-gated or G-protein-linked families, which are held not to internalize as a general rule. He noted that the transporters which had been cloned had 12 or so hydrophobic regions, were sodium-dependent and formed a different, distinct group from the various classes of receptors mentioned.

In response to a question whether there was neglect of necessary precautions Dr Somogyi raised the problem of antibody controls. Wherever possible rigorous criteria should be applied to antigen identity as well as to localization, following the procedures laid down for example by Swaab (1982) and by Pool & Buijs (1988). Where antibodies were obtained from other laboratories, however, this might not always be possible: for example, in the case of a polyclonal antibody the pre-immune serum was not always available. In such cases, attachment of antibody to tissue sections that could not be explained on the basis of attachment by recognition sites, which was shown by perhaps 50% of antibodies, was perhaps the most frequent source of error in interpretation. In his view it had not yet been adequately resolved how an antibody might adhere to a section, in a tissue compartment-selective way, without involving the recognition site.

Discussion of G.R. Uhl's paper

Dr Uhl was asked how the boundary was defined between conditions of high and of low stringency in hybridization. He explained that for nucleic acids in solution proportional relationships had been established between the extent of hybridization and the length of the segment, the salt concentration, the concentration of formamide and the temperature, each of which could be 'set' to establish a desired level of stringency, which would depend on the requirements of a particular experiment. Once a pattern of hybridization had been established, other tests could be applied to determine whether it was specific or non-specific. On the other hand, it could not be assumed that exactly the same kinetics as determined for RNA in solution would apply to RNA immobilized in tissue sections, where for example protein might interfere, making the _in vivo_ 'melting curve' quite different. Here additional factors might have to be manipulated.

On the relative sensitivities of RNA versus DNA probes, Dr Uhl replied that many factors were involved. An RNA-RNA hybrid is more stable than a DNA-DNA hybrid, giving an advantage at the level of hybridization _per se_. Some of the means whereby DNA probes are labelled, as compared with RNA probes, may give a higher specific activity per number of nucleotides, but by making a relatively longer complementary RNA probe it is possible to obtain higher specific activities per molecule. One disadvantage of this is that the longer the hybridization probe, the greater is the potential for cross-hybridization, especially where a gene is a member of a 'family', a situation that is now known to apply to virtually all genes. Another disadvantage is that the larger the molecule, the harder it is for it to penetrate into tissues. A strategy that has been used to deal with this, for complementary RNA probes, is to hydrolyse the probe. This however yields a mixture of different hybridizing segments, differing in length and in specific activity, which potentially complicates quantitative analysis and alters specificity. In general, for a system where RNA is completely mobilized and in an open configuration, as in 'northern' blot analysis, complementary RNA probes offer an advantage (and are used as a standard technique in Dr Uhl's laboratory). In a system involving tissue sections, DNA probes have relative advantages, for example in terms of greater stability, since RNA is more biodegradable by, potentially, residual ribonucleases in tissue sections or as contaminants on glassware, etc. The choice of probe thus depends on the setting in which it will be used. Even for messages having a low frequency, such as the transcription factors on which he had been working, Dr Uhl found that DNA oligonucleotide probes, which could be multiply internally labelled with $^{35}S$ to give specific activities as high as 2 to 4 Ci per millimole, could give satisfactory results in tissue sections, and that complementary RNA probes had not given any real advantage in terms of sensitivity to outweigh the disadvantages of, for example, higher background from the relatively greater non-specific stickiness of RNAs.

To the question whether hybridization studies to date had permitted the prediction from the amount of a given mRNA in a cell whether the protein for which it encoded had a high or a low rate of turnover, or whether no simple relationship had been revealed, Dr Uhl replied that several different stages are encompassed in the concept of gene expression, and that the content or concentration of mRNA in a cell at a given time, which has been termed 'ribosomal loading', is only one index among many possible translational or post-translational regulatory processes, the latter including proteolytic cleavage, phosphorylation or glycosylation, which can serve to control 'gene expression' in its wider sense. In the case of various neuropeptides the stages of post-translational processing, and sometimes of translation, have been at least partly determined; and in the adrenal gland, for example, many neuropeptide precursors are demonstrably present as higher molecular weight forms. In the brain this does not seem to occur, in general, so that here this form of post-translational processing does not seem to be rate-limiting, at least to the extent that precursors to the final products are accumulated. Neuropeptides are not actively phosphorylated; and the glycosylation of pro-peptide hormones tends, with some exceptions, to involve areas which are not important for their biological activity. That still leaves open the question of the extent to which 'ribosomal loading' determines regulation of neuropeptides. Neuropeptides form in a sense the simplest possible system. Some systems are known in Prokaryotes in which translational regulation is in active, day-to-day occurrence, mostly in relation to availability of nutrients; but in Eukaryotes the number of examples that have been worked out is relatively limited, though others are currently being explored. Dr Uhl's inclination was to say that translational regulation does not occur until it is proven otherwise: it remains an open question, unresolved in most systems and difficult to approach in tissue, _in vivo_, and thus has tended to be avoided. This is clearly an area for future study.

On the question of the regulation of transmitter-synthesizing enzymes, which might show very stable levels, as opposed to neuropeptides, it was pointed out that for the enzymes synthesizing monoaminergic transmitters much of the regulation may be post-translational, such as phosphorylation, which in the case of tyrosine hydroxylase strongly influences the activity of the enzyme, and that this may occur even in the nerve terminal. Post-translational regulation, which need not be confined to the nucleus, could be of great importance for some synthetic enzymes. It remains to be determined how much of the change in activity of a transmitter-synthesizing enzyme like tyrosine hydroxylase is produced through regulation of transcription and how much is due to post-transcriptional modification. Both processes certainly occur, but their relative contributions are uncertain.

Any regulation occurring in the nerve terminal must be post-translational, and it may well be that some of the regulation occurring in cell bodies is also post-translational. This also merits future investigation.

Neuropeptides are, in a sense, a more tractable model system than transmitter-synthesizing enzymes. On the other hand, it has been difficult to gain recognition for the concept that changes in neuropeptide gene expression might be coupled to neural activity; and the idea that such changes could occur with a time-course of minutes has only recently been accepted as thinkable, although there has been evidence to that effect for some time. The time-course of the changes in c-FOS expression found in spinal trigeminal neurones after nerve stimulation was discussed. Like pro-enkephalin mRNA, FOS was found to increase early following primary afferent stimulation, and might thus be regarded as a candidate for involvement in the regulation of enkephalin expression; but their relationships with stimulus intensity differed in a way which threw doubt on this possibility: pro-enkephalin mRNA increased very rapidly following stimulation of low-threshold primary afferents, whereas FOS expression increased strongly only if there was co-stimulation of high-threshold afferents, and in the latter case the early increase of enkephalin expression did not occur. It would be of value to determine which of the transcription factors that have so far been identified as binding to a limited part of the pro-enkephalin promoter, or indeed whether others which might bind at greater distances, are actually involved in regulation of the gene, and in what circumstances. To be able to say that the expression of FOS and enkephalin is up-regulated in some of the same cells at appropriate times is not the same as to say that regulation of pro-enkephalin actually operates through a mechanism involving FOS, and such a conclusion has not so far been possible in any circuit in the brain in vivo, when followed through from the externally applied stimulus to the changes in neurotransmitter synthesis.

Next was considered the topic of turnover of RNA. The level of RNA in a cell must be the resultant of both transcription and degradation. Dr Uhl had shown regulation of transcription in the case of both FOS and JUN. How much was known about degradation? Was this involved in the regulation of peptide mRNA in neurones? Dr Uhl replied that processes of degradation and turnover were proving exceptionally difficult to study in most systems in vivo, and that while model systems yielded useful data they could not necessarily replicate the particular features of in vivo regulation. Some principles had, however, been established: first, that RNA half-lives do change and that these changes are in general correlated with changes in transcription rates. A marked increase of transcription seems to be accompanied by a longer half-life, with a resultant equivalent of far more message. Second, there have been revealed some mechanisms which can regulate the half-life of RNA, although these may be fairly idiosyncratic. The most notable of these is the demonstration, for example in relation to dehydration-induced changes in vasopressin mRNA, of the involvement of zinc in regulation of the extent of poly-adenylation, that is the length of the poly-A tail which protects RNA from degradation. This is a major potential means for regulation, but on the evidence so far obtained from other systems it appears unlikely to be a general mechanism. Much remains to be discovered concerning the selective up- and down-regulation of gene expression in its widest sense, at both transcriptional and post-transcriptional levels, and the regulation of the lifetime of individual mRNAs in vivo. As to whether apparent changes in the number of cells which expressed a specific gene were truly on-off changes in transcription or changes in level, 'gated' by limitations in the sensitivity of detection, Dr Uhl replied that this was problematic, but that distribution histograms of the level of signal detected per neurone autoradiographically by use of radiolabelled probes revealed a continuum, rather than a simple on-off pattern of expression. There is now increasing evidence that many cells may have just a very few copies of almost all RNAs, and this has given rise to a concept of 'leakiness' in gene expression. It is thus possible that levels of label which are currently regarded as indistinguishable from background may represent hybridization to these very scanty mRNAs. Regional and temporal differences in expression are certainly easier to demonstrate when they are large. Dr Uhl suggested that the best way to analyse the raw data is therefore by frequency distribution histogram, and by asking whether this reveals a monotonic normal population which may be shifting to the right or left. One example of this was shown by Dr Uhl, for his data on pro-enkephalin expression; another was the regulation of neuropeptide expression which has been observed in the striatum on blocking nigral dopaminergic inputs or cortical excitatory amino-acid inputs, by drug-induced or surgical lesions: in both cases there was a shift to the right or left in the level of expression shown in the population by the frequency distribution histogram, but the fraction of neurones showing label remained the same. In the case of pro-enkephalin regulation, however, clear examples had been seen of regulation in the one case by 'turning on' more cells and in the other case by 'turning up' expression in the same number of cells. In transgenic animals, Dr Uhl's group had found that by using particular pieces of the pro-enkephalin promotor region in the constructs which were inserted into the genome it seemed to be possible to dissociate these two kinds of regulation, and this gave some confidence that these might be potentially two distinct biological processes, using different constellations of transcription factors operating on the same gene.

In his presentation and in the ensuing discussion, Dr Uhl thus emphasized that, in addition to improving the localization of specific DNA or RNA sequences, increasing resolution in techniques of _in situ_ hybridization is now permitting the quantitative study of relative levels of gene expression, and of changes in these levels, and that the mechanisms of such changes are beginning to be explorable, for example through the study of correlated changes in the expression of genes for transcription factors: this is likely to provide a fertile field for future research.

Discussion of R.Mirsky's paper

Following her description of the successes obtained through immunohistochemical labelling and double-labelling of astrocytes and Schwann cells in culture and in teased preparations, Dr Mirsky was asked for her views on the possibilities for the use of antibody markers in cell sorting. She replied that this approach had been used by others in sorting progenitor cells preparatory to experiments on their transformation. Various markers had proved very useful for monitoring the purity of a cell preparation, and this was an important step since mixed primary cultures contained many types of cells. The cell sorter was potentially one means of purifying such cultures.

On being asked to comment on why Type 2 astrocytes had not been found in vivo, Dr Mirsky pointed out that it might be that they were present in very small numbers, unlike their precursors and derivatives, and that this would render them difficult to demonstrate. A simple alternative explanation would be that the phenotype was changing under the conditions of culture. The question was still an open one, which M. Raff and his group were endeavouring to solve.

Several questions centred on the differential distribution of glial fibrillary acidic protein (GFAP) in astrocytes and in supporting cells of the peripheral nervous system, and on possible reasons for the observed differences. In answer Dr Mirsky reported that all the supporting cells (glial cells) of the enteric plexuses are very strongly GFAP immunoreactive, more so than Schwann cells, at least in the rat. In the sympathetic ganglia most or all of the satellite cells are GFAP positive, and the Schwann cells of the associated unmyelinated nerve fibres are strongly positive. For this reason she uses the cervical sympathetic trunk as a good model, since it has few myelinated nerve fibres (of which the Schwann cells are GFAP negative) and the two types of fibres can be separated by teasing. In the dorsal root ganglion of the rat most of the satellite cells are GFAP negative, but some ganglion cells are surrounded by GFAP positive satellite cells. The reason for this is so far obscure; the GFAP cells are not, as was at first thought, consistently associated with the larger neurones. In some species most or all of the satellite cells of the dorsal root ganglion are GFAP positive. It was remarked that astrocytes in different regions in vivo could show strong variations in intensity of immunolabelling for GFAP, as could also occur in culture, and it was asked what mechanisms might be responsible for such differences. Dr Mirsky replied that the mechanism was unknown, but that the form of an astrocyte might be correlated with the amount of GFAP which it contained, since electron microscopy showed considerable differences in the concentration of glial filaments between, for example, fibrous and protoplasmic astrocytes.

Finally were considered some of the other antigenic markers for peripheral Schwann cells. Asked whether there were differences in immunoreactivity of Schwann cells surrounding motor axons, Dr Mirsky said that she knew of no absolute differences but that there were indications of a greater immunoreactivity *in situ* in adult nerve for the epitope L2/HNK-1 in the myelin sheaths of motor than of sensory axons. As to the possible function of N-CAM in these cells, this was so far unknown, though it had been suggested that it might act as an adhesion molecule between neurones and supporting cells. Other tests had however tended to indicate that another cell adhesion molecule expressed by these cells, L1, was the more important in that respect. In addition, it had been shown that during development the sialic acid in the peripheral nervous system was highly regulated (Landmesser et al. 1988, Rutishauser & Jessell, 1988). It would be an interesting topic for a future study, to explore the role of the Schwann cells in this, to see whether there is a change in the level of sialylation of N-CAM at the stage early in development when the Schwann cells begin to penetrate and subdivide the large bundles of young axons in the early nerves.

The work described by Dr Mirsky thus lays a substantial groundwork, and opens up numerous possibilities and challenges for the future, in the elucidation of differential features of glial and other supporting cells of the nervous system, not only in the adult but also during development and in the dynamic processes of repair. This part of the discussion, pointing the way to gaps and potential advances in the understanding of these vitally important but poorly understood cells, formed a forward-looking conclusion to a lively session.

## References

Hamel E, Beaudet A (1987) Opioid receptors in rat neostriatum: radioautographic distribution at the electron microscopic level. Brain Res 401:239-27.

Landmesser L, Dahm L, Schultz K, Rutishauser V (1988) Distinct roles for adhesion molecules during innervation of embryonic chick muscle. Devel Biol 130:645-670.

Pool CW, Buijs RM (1988) Antigen identity in immunocytochemistry. In: Van Leeuwen, FW, Buijs RM, Pool CW & Pach O (eds) Molecular Neuroanatomy. Techniques in the Behavioral and Neural Sciences, Vol. 3. Elsevier, Amsterdam, pp 233-266.

Rutishauser V, Jessell TM (1988) Cell adhesion molecules in vertebrate neural development. Physiol Rev 68:819-857.

Steinbusch HWM (1984) Serotonin-immunoreactive neurons and their projections in the CNS. In: Bjorklund A, Hokfelt T, Kuhar NJ (eds) Classical Transmitters and Transmitter Receptors in the CNS part II. Handbook of Chemical Neuroanatomy, Vol 3, Elsevier, Amsterdam.

Swaab DF (1982) Comments on the validity of immunocytochemical methods. In: Chan-Palay V, Palay SL (eds) Cytochemical Methods in Neuroanatomy. Neurology and Neurobiology, Vol 1, Alan R Liss, New York.

# APPLICATIONS TO THE STUDY OF ACETYLCHOLINE RELEASE OF THE CHOLINE OXIDASE ACETYLCHOLINE ASSAY

N. Morel and M. Israel

Departement de Neurochimie
lab. Neurobiologie Cellulaire et Moleculaire
C.N.R.S.
91198 Gif sur Yvette, France

## Introduction

Numerous chemical or enzymatic reactions have been coupled to light emitting systems to assay biologically important compounds (De Luca, 1978; Van Dyke, 1985), as alternatives to methods relying on radioactive labels. A chemiluminescent method to measure acetylcholine (ACh) has been developped by Israel and Lesbats (1981, 1985, 1987). Oxidation of luminol by H2O2 in the presence of peroxidase produces a blue light emission. Several instruments, luminometers, are now commercially available which allow to measure light emission with great precision and sensitivity (see Van Dyke, 1985, for a comparison of different luminometers). The luminescent reaction is coupled to two enzymatic reactions (Figure 1). First, ACh is hydrolysed by acetylcholinesterase, leading to the production of choline. Then choline is oxidized by choline oxidase, a bacterial enzyme, with formation of betain and H2O2. The choline oxidase chemiluminescent assay allows the measure of picomole amounts of ACh in biological extracts, in the nanomolar range of ACh concentration.

NATO ASI Series, Vol. H 58
Neurocytochemical Methods
Edited by A. Calas and D. Eugène

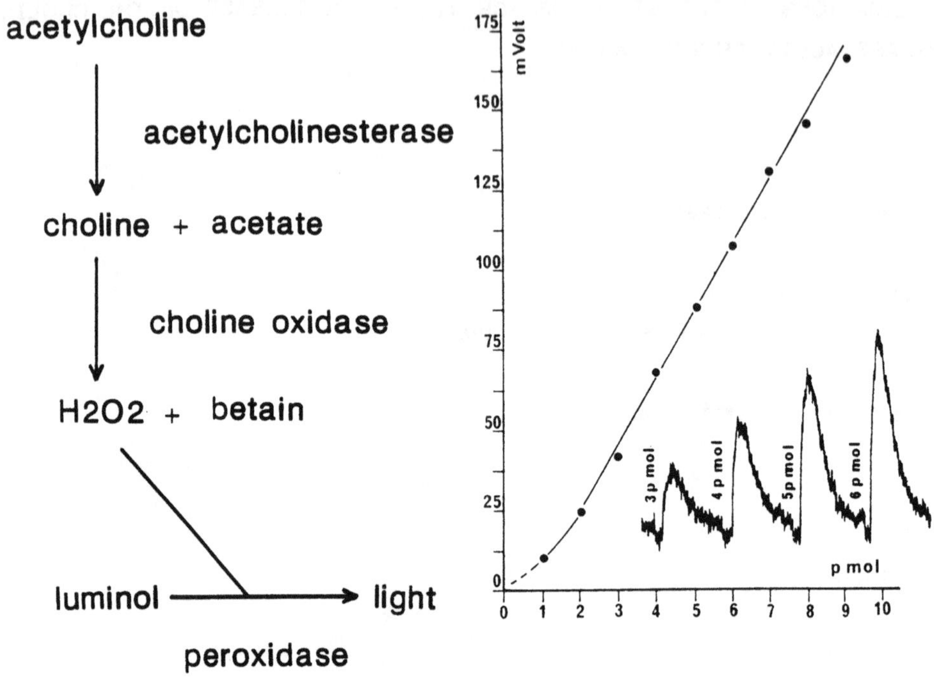


Figure 1: Choline oxidase chemiluminescent ACh assay.

Principle and curve showing the light emission for increasing amounts of ACh in the pmole range.

Assay of ACh in choline containing samples

It is possible to distinguish choline from ACh, without having to separate the two compounds, using the choline oxidase method (Figure 2). In the absence of acetylcholinesterase in the reaction medium, the addition of choline (15 pmol) gives a light emission while the addition of ACh gives no response. Upon injection of acetylcholinesterase, ACh is hydrolysed and is detected by the corresponding light emission. The figure also shows that a fresh choline injection again triggers a light response and that injection of ACh plus choline gives the expected summation.

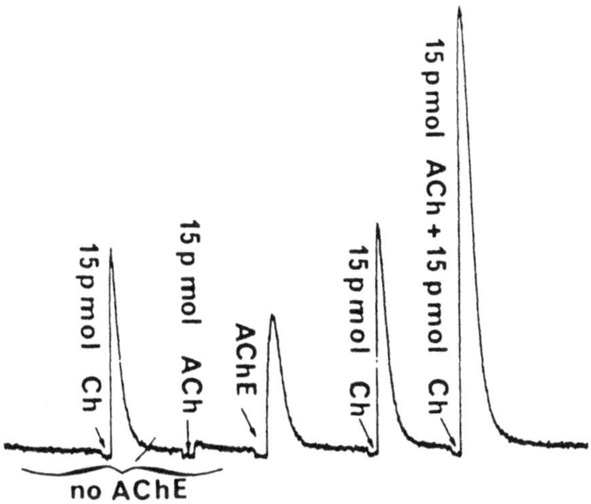


Figure 2: Distinction between choline and ACh.

Practically, to assay ACh in choline containing samples, it is just necessary to mix the sample with the reaction mixture without acetylcholinesterase, wait untill the end of the initial light emission, and then add acetylcholinesterase. The light emitted is now proportional to the concentration of ACh. The same sort of protocol is used if the biological sample contains peroxides or oxidant substances.

Several tissue extracts, especially in mammals, contain interfering reducing substances which block the chemiluminescent reaction. This is demonstrated by the decrease or suppression of the light response to the injection of an ACh standard in the reaction mixture in the presence of the tissue extract. It is essential, in this case, to oxidize these samples before the assay, using small amounts of H2O2 or sodium metaperiodate, amounts which are adjusted to restore the expected light response to the ACh internal standard (see Israel and Lesbats, 1987).

## Continuous monitoring of ACh release from Torpedo electric organ synaptosomes

The essential breakthrough resulting from the choline oxidase-chemiluminescent method was to record, in line and in real time, the release of ACh from nerve terminals in suspension by a biochemical procedure.

Isolated nerve terminals (synaptosomes) prepared from Torpedo electric organ as previously described (Israel et al, 1976; Morel et al, 1977) have two major advantages for physiological studies. First, due to the homogeneity of Torpedo electric organ innervation, the synaptosomal fraction contains only cholinergic nerve endings and, since the fraction is highly purified, its ACh content is very high (around 300 pmol ACh per mg protein). Second, physiological preservation of the isolated nerve terminals appears to be excellent, most probably because physiological osmolarity and salt composition are maintained throughout the isolation procedure (Morel et al, 1977).

An aliquot of the synaptosomal suspension is constantly mixed with the reaction mixture in a physiological medium. ACh release is triggered by a calcium entry into the synaptosomes, obtained either by KCl depolarization in the presence of calcium (Figure 3) via presynaptic voltage dependent calcium channels, or directly using a calcium ionophore A23187 or a presynaptic neurotoxin (Figure 4) extracted from a sea annelid Glycera convoluta (Manaranche et al, 1980; Morel et al, 1983).

After KCl depolarization of synaptosomes (Figure 3A), the light emission resulting from ACh release lasts about 2 minutes, reaching its maximum in about 10 seconds. It is calibrated by injection in the same test tube of a known amount of ACh. The rising phase of the release curve is less steep than for the ACh standard, showing that the release rate is the limiting factor, which was also the case with all the releasing agents tested. Two parameters of synaptosomal ACh release can be determined from such light emission recordings. First, the total amount of ACh which is released is measured from the surface area under the curve as compared with that of the ACh standard. Second, an index of the rate of ACh release is estimated from the mean slope of the rising phase of the light emission. It can be seen from Figure 3B that, for increasing KCl depolarization, both parameters increase (Meunier, 1984). ACh release stops in spite of the maintained depolarization, when only a small proportion of the synaptosomal ACh content has been released (about 10 % for depolarization by 50 mM KCl). This was attributed to the desensitization of the presynaptic voltage dependent calcium channels (Birman and Meunier, 1985).

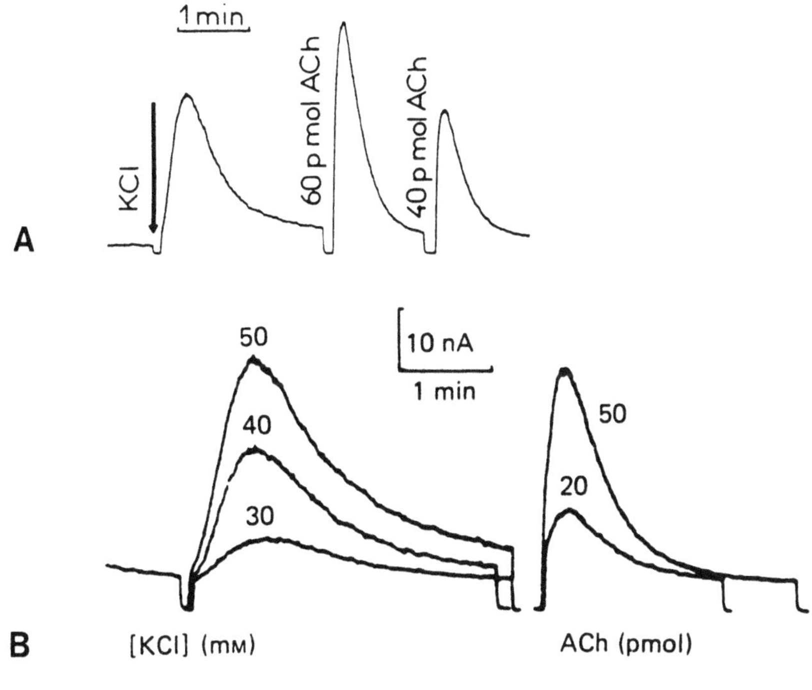


Figure 3: ACh release from synaptosomes depolarized by KCl.

By short circuiting the natural calcium channels, it is possible to elicit the release of larger amounts of ACh. Figure 4 presents superimposed ACh release recordings obtained from synaptosomes stimulated by increasing concentration of the presynaptic neurotoxin of Glycera convoluta. Both the total amounts of released ACh and the rates of ACh release were determined and plotted as a function of the neurotoxin concentration in the inset.

Whereas the rate of release increases as the neurotoxin concentration does, a constant amount of ACh is released (Morel et al, 1983). In this case, ACh release stops when the synaptosomal releasable ACh pool is exhausted.

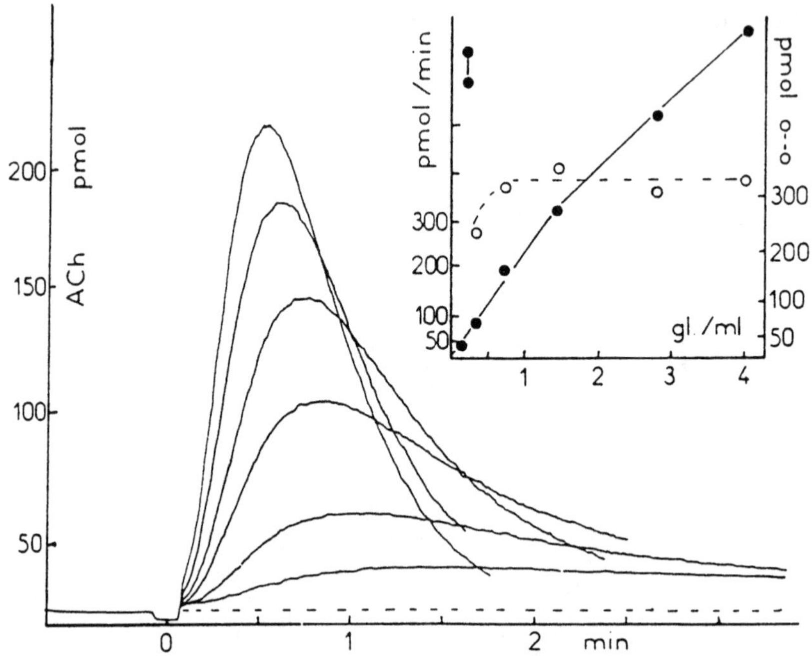


Figure 4: Synaptosomal ACh release triggered by Glycera convoluta neurotoxin.

Superimposed recordings obtained with increasing neurotoxin concentrations. In inset: slope of the rising phase, surface of the light response.

Evaluation of ACh compartments in synaptosomes

ACh is distributed in two different compartments in cholinergic nerve endings: a vesicular pool corresponding to ACh entrapped in synaptic vesicles and a free cytoplasmic pool where ACh synthesis takes place. When

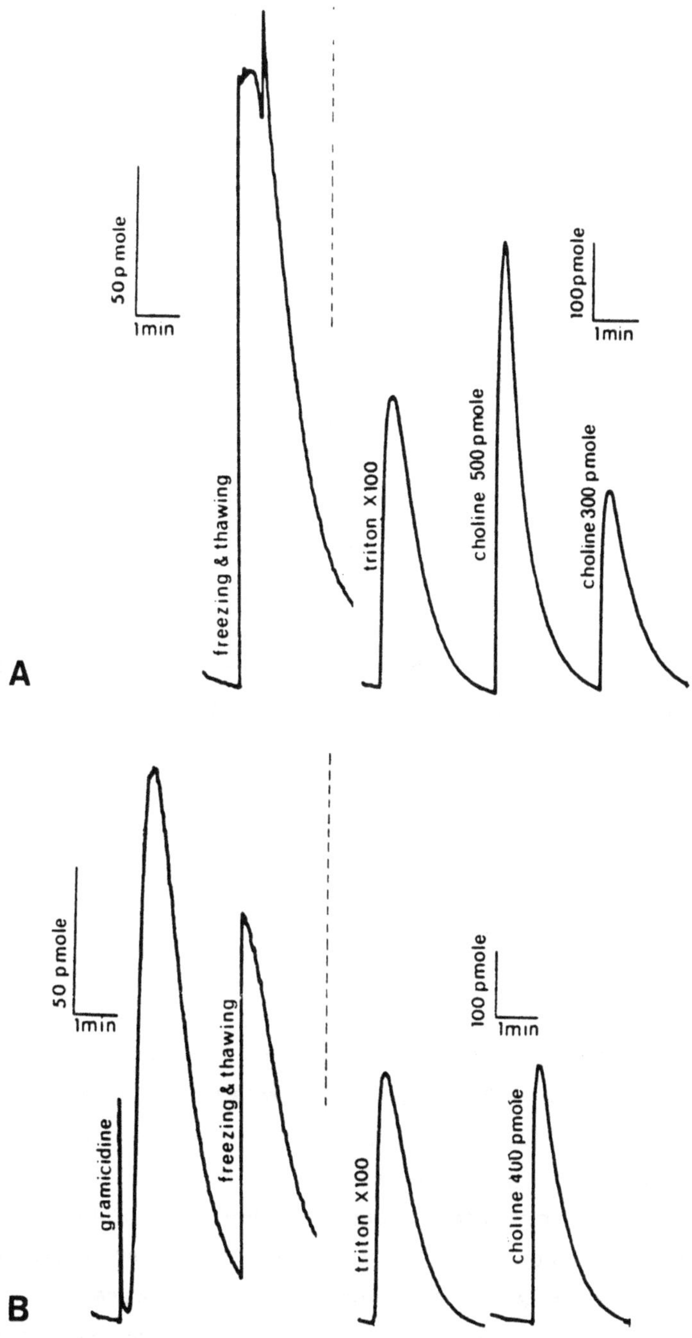


Figure 5: Determination of ACh compartments in synaptosomes, in resting conditions (A) or after stimulation by Gramicidin D (B).

synaptosomes are frozen and thawed in the reaction mixture, the first light emission recorded corresponds most probably to the cytoplasmic ACh compartment which leaks out during the thawing process. If the preparation is frozen and thawed again several times, no significant amounts of ACh are detected. This is in agreement with the observation that vesicular ACh remains occluded even after freezing and thawing the tissue, synaptic vesicles full of ACh being purified from frozen electric organ. In order to evaluate the vesicular ACh compartment (60-70 % of the total), it is necessary to add a detergent (Triton X100) to permeabilize the vesicular membrane. Figure 5A illustrates the measure of cytoplasmic and vesicular ACh compartments on synaptosomes in resting conditions. Same determinations are performed on synaptosomes stimulated by the cation ionophore Gramicidine, when ACh release was over (Figure 5B). The cytoplasmic ACh pool is markedly decreased whereas the vesicular compartment is not modified. The sum of the released transmitter plus the remaining cytoplasmic ACh pool equals the initial soluble ACh store. This result was observed in all conditions of ACh release tested so far (Israel and Lesbats, 1981) and suggests that the cytoplasmic ACh compartment is the source of the released transmitter.

## ACh release from proteoliposomes

The observation that ACh was released from the cytoplasmic neurotransmitter pool, taken with the demonstration, after ultrarapid freezing, of a redistribution of intramembrane particles in the membrane of stimulated nerve endings (Israel et al, 1981; Muller et al, 1987), led to the hypothesis that the presynaptic plasma membrane contains a protein, that was named mediatophore, able to translocate ACh when activated by calcium. To test this hypothesis, reconstitution experiments were performed (Israel et al, 1984, 1986; Birman et al, 1986). The plasma membrane of Torpedo synaptosomes was purified (Morel et al, 1982). Membrane fragments, or proteins extracted from this membrane, were mixed by synthetic lecithin. The mixture was resuspended in an ACh containing internal milieu and sonicated. Proteoliposomes were formed which have incorporated presynaptic membrane proteins in their membrane (as shown by the presence of intramembrane particles after freeze fracture) and have entrapped ACh. The external solution was replaced by gel filtration. ACh release was triggered by creating a calcium influx into the liposomes by the successive additions of a calcium ionophore (A23187) and of calcium (Figure 6), and was monitored as was done for synaptosomal suspensions.

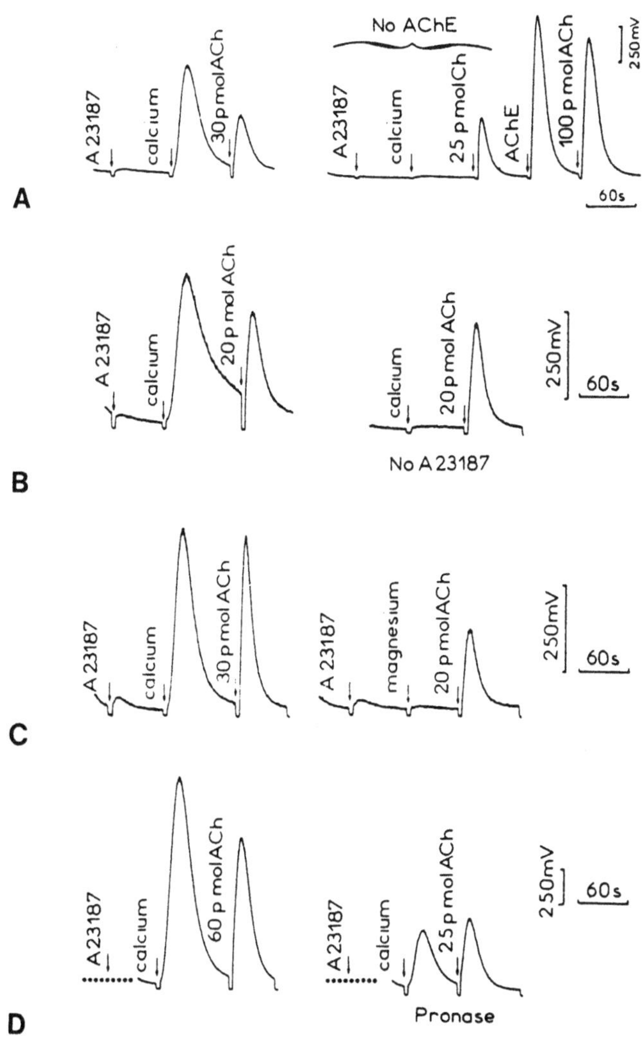


Figure 6: ACh release from proteoliposomes prepared from presynaptic membrane proteins.

A light emission was recorded in response to a calcium influx into the proteoliposomes (Figure 6A). It is due to the release of ACh since it was absent when acetylcholinesterase is omitted from the medium, in spite of the fact that choline was detected by the reaction mixture (Figure 6A). Upon introduction of the enzyme, an immediate light response was monitored, with a very steep rising phase, showing that ACh was already released at that time. A calcium influx into the proteoliposomes is required to trigger the release of transmitter since the addition of calcium on the outside, in the absence of the ionophore, is unefficient (Figure 6B). Magnesium, which is also transported by the ionophore A23187, is unable to replace calcium (Figure 6C). ACh release is markedly reduced when membranes treated by the non specific protease pronase are used to prepare the proteoliposomes (Figure 6D), showing that presynaptic membrane proteins are required.

Mediatophore was first identified by its activity, measured as the rate of calcium dependent ACh release from proteoliposomes prepared from mediatophore containing protein fractions. This functional test was used to follow the purification of the protein mediatophore.

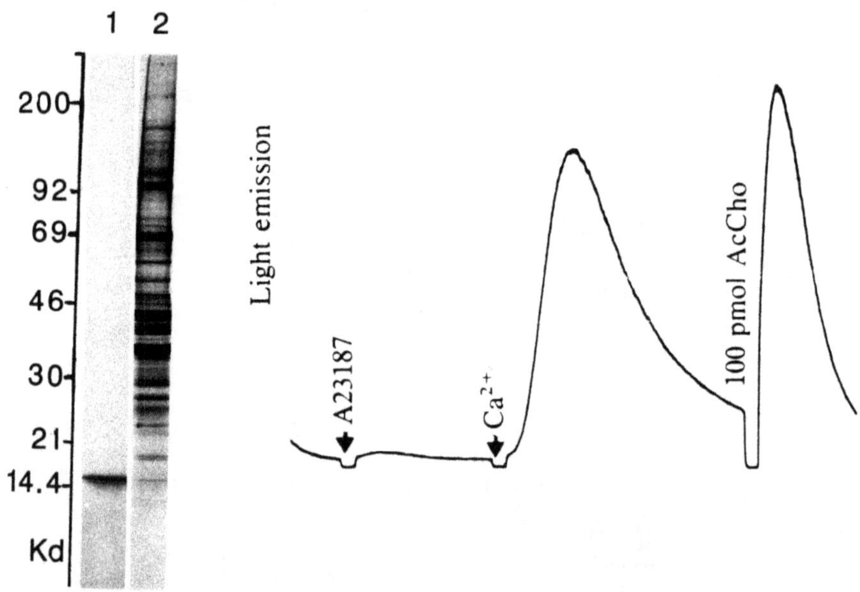


Figure 7: ACh release from proteoliposomes prepared with purified mediatophore.

1: 15 kDa proteolipid subunit of purified mediatophore.

2: synaptosomal plasma membrane.

A highly purified mediatophore preparation was obtained by a two step procedure (Israel et al, 1986): mediatophore was first extracted from presynaptic plasma membranes at alkaline pH and then it was solubilized in a chloroform-methanol (v/v) mixture. Mediatophore has a molecular mass of about 200 kDa and appears to be built up by the association of 15 kDa proteolipid subunits (Figure 7A). The complete aminoacid sequence of the mediatophore proteolipid has been established and shown to have a high homology to that of the bovine chromaffin or yeast vacuolar ATPase proteolipids (Birman et al, 1990). Proteoliposomes prepared from the purified mediatophore and synthetic lecithin are able to release ACh in a calcium dependent way (Figure 7B). A number of functional properties of synaptosomal ACh release were still conserved by the purified reconstituted protein: calcium dependent desensitization (Israel et al, 1987) or same pharmacological properties (Morot Gaudry-Talarmain, 1989).

The mediatophore purification procedure was applied on membranes from rat brain and mediatophore activity was found associated with synaptosomal membranes, more abundant in cholinergic areas (striatum, cerebral cortex) as compared to cerebellum (Israel et al, 1988). Mediatophore activity was also present and concentrated in innervated portions (as compared to aneural zones) of rat sternocleido mastoid muscles (Israel et al, 1990).

Reconstitution experiments have shown that a protein, the mediatophore, able to translocate ACh when activated by calcium, is indeed present in the membrane of cholinergic nerve terminals. How the mediatophore participates to neurotransmitter release in physiological conditions remains to be elucidated. Two working hypothesis can be put forward. Either mediatophore could act as a transmembrane channel allowing the direct flow of cytoplasmic ACh. Or it could be some sort of docking protein for a subpopulation of active synaptic vesicles (Zimmermann and Whittaker, 1977) with channel properties, thus allowing the release of the transmitter that these organelles have concentrated from the cytoplasmic newly synthetized compartment.

## Conclusions

The choline oxidase chemiluminescent method developped by Israel and Lesbats to assay ACh offers many advantages, some of which were reviewed here (for a more complete and technical description, see Israel and Lesbats, 1987). It is possible to make easily more than 50 tests in an afternoon, using rather unexpensive reagents. Moreover this is the only biochemical method which allows to see ACh release in real time and, therefore, to adapt the experimental conditions while the physiological experiment is in progress. Several modifications of the choline oxidase method are possible. For instance, it has been adapted to measure acetylcholinesterase activity using its natural substrate ACh (Birman, 1985) or to detect the release of this enzyme from brain slices (Llinas and Greenfield, 1987).

## References


Birman S. (1985) Determination of acetylcholinesterase activity by a new chemiluminescence assay with the natural substrate. Biochem.J., 225, 825-828.

Birman S., Israel M., Lesbats B. and Morel N. (1986) Solubilization and partial purification of a presynaptic membrane protein ensuring calcium dependent acetylcholine release from proteoliposomes. J.Neurochem., 47, 433-444.

Birman S. and Meunier F.M. (1985) Inactivation of acetylcholine release from Torpedo synaptosomes in response to prolonged depolarizations. J.Physiol.(London) , 368, 293-307.

Birman S., Meunier F.M., Lesbats B., Le Caer J.P., Rossier J. and Israel M. (1990) A 15 kDa proteolipid found in mediatophore preparations from Torpedo electric organ presents high sequence homology with the bovine chromaffin granule protonophore. FEBS Letters, 261, 303-306.

De Luca A., ed. (1978) Bioluminescence and chemiluminescence. Methods in Enzymology, 57, Academic Press, New York.

Israel M. and Lesbats B. (1981a) Chemiluminescent determination of acetylcholine, and continuous detection of its release from Torpedo electric organ synapses and synaptosomes. Neurochem. Int., 3, 81-90.

Israel M. and Lesbats B. (1981b) Continuous determination by a chemiluminescent method of acetylcholine release and compartmentation in Torpedo electric organ synaptosomes. J.Neurochem., 37, 1476-1483.

Israel M. and Lesbats B. (1985) Chemiluminescent determination of acetylcholine and continuous detection of its release from tissues. in Bioluminescence and chemiluminescence: Instruments and Applications. Van Dyke K. ed. C.R.C., Boca Raton, Florida.

Israel M. and Lesbats B. (1987) The use of bioluminescence techniques in neurobiology, with emphasis on the cholinergic system. in Neurochemistry, a practical approach. Turner A.J. and Bachelard H.S. eds, IRL Press, 113-135.

Israel M., Lesbats B., Morel N., Manaranche R., Gulik-Krzywicki T. and Dedieu J.C. (1984) Reconstitution of a functional synaptosomal membrane possessing the protein constituents involved in acetylcholine translocation. Proc. Natl Acad. Sci. USA, 81, 277-282.

Israel M., Lesbats B., Morel N., Manaranche R. and Le Gal la Salle G. (1988) Is the acetylcholine releasing protein mediatophore present in rat brain? FEBS Letters, 233, 421-426.

Israel M., Lesbats B., Sbia M. and Morel N. (1990) Acetylcholine protein protein mediatophore at rat neuromuscular synapses. J.Neurochem., 55, 1758-1762.

Israel M., Manaranche R., Mastour-Frachon P. and Morel N. (1976) Isolation of pure cholinergic nerve endings from the electric organ of Torpedo marmorata. Biochem.J., 160, 43-55.

Israel M., Manaranche R., Morel N., Dedieu J.C., Gulik-Krzywicki T. and Lesbats B. (1981) Redistribution of intramembrane particles related to acetylcholine release by cholinergic synaptosomes. J.Ultrastructure Res., 75, 162-178.

Israel M., Meunier F.M., Morel N. and Lesbats B. (1987) Calcium-induced desensitization of acetylcholine release from synaptosomes or proteoliposomes equipped with mediatophore, a presynaptic membrane protein. J.Neurochem., 49, 975-982.

Israel M., Morel N., Lesbats B., Birman S. and Manaranche R. (1986) Purification of a presynaptic membrane protein that mediates a calcium dependent translocation of acetylcholine. Proc.Natl Acad. Sci. USA, 83, 9226-9230.

Llinas R.R. and Greenfield S.A. (1987) On-line visualization of dendritic release of acetylcholinesterase from mammalian substantia nigra neurons. Proc.Natl.Acad.Sci.USA, 84, 3047-3050.

Manaranche R., Thieffry M. and Israel M. (1980) Effect of Glycera convoluta venom on the spontaneous quantal release of transmitter. J.cell Biol., 85, 446-458.

Meunier F.M. (1984) Relationship between presynaptic membrane potential and acetylcholine release in synaptosomes from Torpedo electric organ. J.Physiol.(London) 354, 121-137.

Morel N., Israel M., Manaranche R. and Mastour-Frachon P. (1977) Isolation of pure cholinergic nerve endings from Torpedo electric organ. J.cell Biol., 75, 43-55.

Morel N., Manaranche R., Israel M. and Gulik-Krzywicki T. (1982) Isolation of a presynaptic plasma membrane fraction from Torpedo cholinergic synaptosomes: evidence for a specific protein. J.cell Biol., 93, 349-356.

Morel N., Thieffry M. and Manaranche R. (1983) Binding of a Glycera convoluta neurotoxin to cholinergic nerve terminal plasma membranes. J.cell Biol., 97, 1737-1744.

Morot Gaudry-Talarmain Y., Diebler M.F., Robba M., Lancelot J.C., Lesbats B. and Israel M. (1989) Effect of cetiedil analogs on acetylcholine and choline fluxes in synaptosomes and vesicles. Eur.J.Pharmacol., 166, 427-433.

Muller D., Garcia-Segura L.M., Parducz A. and Dunant Y. (1987) Brief occurence of a population of presynaptic intramembrane particles coincides with transmission of a nerve impulse. Proc.Natl.Acad.Sci.USA, 84, 590-594.

Van Dyke K., ed. (1985)
Bioluminescence and chemiluminescence: Instruments and Applications. C.R.C. Press, Bota Raton, Florida.
Zimmermann H. and Whittaker V.P. (1977) Morphological and biochemical heterogeneity of cholinergic synaptic vesicles. Nature, 267, 633-635.

# MORPHOMETRIC EVALUATION OF POPULATIONS OF NEURONAL PROFILES IN THE CENTRAL NERVOUS SYSTEM

Michele Zoli and Luigi F. Agnati
Institute of Human Physiology
University of Modena
Via Campi 287
41100, Modena
Italy

The present paper will review some developments in the quantitation of populations of neuronal profiles (cell bodies, dendrites and nerve terminals). Some general tools will be introduced as a background for specific neurobiological applications. The theoretical bases for each procedure will be illustrated.

## MATERIALS AND METHODS

### Histology

While for morphometric evaluations the main goal of the staining is to get the best possible signal to noise ratio, for microdensitometric evaluations the relationship between stained substance content and staining intensity the type of quantification which can be obtained. No reference to such a relationship is necessary for semiquantification, which simply put the measured data in rank order. Relative quantification necessitates the demonstration of the function which links staining intensity to stained substance content. Absolute quantification necessitates the demonstration of the correspondence between absolute amounts of stained substance and the staining intensities, and can be obtained by using standards with known stained substance content processed together with the biological preparations. In the case of immunocytochemistry, both peroxidase-antiperoxidase and avidin-biotin procedures have been proven effective for relative and absolute quantification provided that diaminobenzidine and $H_2O_2$ are used in optimal concentrations (Benno et al. 1982, Gross and Rothfeld 1985, Zoli et al. 1990b) and the antibody concentration is far from saturation, in order to avoid the Bigbee effect (Bigbee et al. 1977).

NATO ASI Series, Vol. H 58
Neurocytochemical Methods
Edited by A. Calas and D. Eugène

Computer-assisted image analysis

Morphometric and microdensitometric analyses were carried out by means of IBAS I/II (Zeiss Kontron, Münich, FRG) image analyzer. We tested the system response to an array of standard optical density values. To this end we compared measurements obtained by means of the photomultiplier (MPC 64, Zeiss Kontron, Münich, FRG), which gives absolute transmittance values, with the measurements obtained by means of the TV camera. A fairly good linear relationship exists between grey tone values and absolute transmittance values within grey tone values 50 and 200, which can be selected as the working range (Zoli et al. 1990b). Of particular relevance for the present procedures are the "discrimination function", the "close function" and the "boolean operation AND". The "discrimination function" consists of the selection of a chosen range of grey tones. Thus, it is feasible, by means of this function, to separate the specific profiles (darker grey tones) from the non-specific ones (lighter grey tones) or the discrimination of a particular subpopulation of specific profiles, such as the cell bodies which are usually darker than the neuropil. The "close function" consists, at each step, of a "dilatation" of 2 pixels of the original object followed by an "erosion" of 2 pixels of the newly formed objects. The "boolean operation AND" is based on the superimposition of two discriminated images, and gives a new image containing only the pixels which are positive in both the original images.

## RESULTS

It may be useful to define two terms: neurochemical area and cluster. The first term indicates a neuroanatomical field where a certain type of neurochemically identified profiles have been visualized. The second term indicates a set of profiles within the neurochemical area which can be distinguished from the remaining population on the basis of its lower interprofile distance.

### 1. Identification and characterization of neurochemical areas

1.1 Acquisition of profiles and sampling. The acquisition of the positive objects present in the image can be obtained by automatically discriminating the positive profiles from the background according to microdensitometric or morphometric criteria. In the case of microdensitometric discrimination, the profiles are selected on the basis of their optical density (specific optical density) which is higher than the optical density of the background (non-specific optical density). Objective criteria for the choice of the discrimination threshold have been previously proposed (see Agnati et al. 1988b, Zoli et al. 1990b). For example, the discrimination threshold can be selected as the mean grey value - n (usually 3 to 5) standard deviations of the grey tone present in an area where only non-specific staining is detectable. In the case of morphometric discrimination the profiles are selected on the basis of one or more morphometric features, such as shape, size etc. While microdensitometric discrimination is a common feature of image analyzers, only a few devices (such as GOP system, Context Vision AB, Linkoeping, Sweden) are capable of a morphometric discrimination.

When one is dealing with a distribution of isolated profiles (fig. 1A) it is possible to adopt an exhaustive procedure which takes into account the entire population of profiles present in the field. When the number of objects is very large, or when there is a coalescence of the original objects (fig. 1B), it is usually convenient to make a sampling of the profiles present in the field. In the case of a semiautomatic device, this can be, e.g., obtained by superimposing on the original image an opaque overlay bearing "holes" of suitable dimensions; in the case of an automatic device, the sampling can be performed by means of the AND operation which compares the original image with an image bearing an array of dots of appropriate size and at appropriate distance (see fig. 1C). The size of the holes (dots) and the interhole (interdot) distance are chosen in relation to the size and the density of the profiles. In the semiautomatic procedure only the holes in which profiles are present and cover at least 50% of the hole surface are considered positive holes. In the case of the automatic procedure a size threshold on the resulting image can be put to select only the dots above a certain size. At any rate the final aim of the sampling procedure, i.e. to obtain a spatial distribution of discrete objects, is achieved (fig. 1D).

1.2. Identification of neurochemical areas. Two methods have been proposed to objectively identify neurochemical areas: the method of boundary construction based on density maps and the polygon method. The former method is more suited to a numerical characterization of neurochemical areas, whereas the latter is more suited to a geometrical characterization. It should be noted that the polygon method gives satisfactory results only when the number of objects in the neurochemical area is relatively large (usually more than 100), whereas the density map method has not such a limitation.

1.2.1 The method of boundary construction based on density maps. By subdividing the plane in unitary squares (whose side is usually 5 times the mean profile diameter in the case of cell bodies and 5 times the interhole/interdot distance in the case of nerve terminals and dendrites, since they are usually analyzed by the sampling procedure) a density map of the object distribution can be obtained (fig. 2A,B).

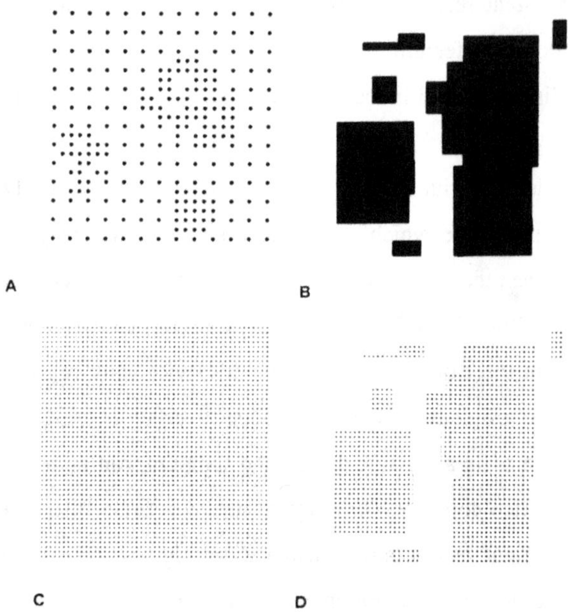


Fig. 1. Acquisition of profiles and sampling.
Panel A. Discrete distribution of profiles. On this distribution the exhaustive procedure can be performed. The diameter of the profiles is 8 pixels.
Panel B. Continuous distribution of profiles. On this distribution the sampling procedure can be performed.
Panel C. Sampling grid used to sample the continuous distribution reported in panel B. The dot area is 12.6 pixel$^2$ and the interdot distance is 6 pixels.
Panel D. Sampling procedure performed by using the boolean operator AND on panels B and C. By this procedure a discrete distribution can be obtained.
The side of the sampled field is 512 pixels.

The square with maximal density is identified. The neurochemical area is defined as the area made by all the touching squares with a number of profiles higher than 5% of the maximal profile density per square. Squares containing less than 5% of the maximal profile density per square are included in the neurochemical area only if they are completely surrounded by positive squares.

1.2.2 The polygon method. To determine in an objective way the boundaries of a neurochemical area the following procedure can be used.

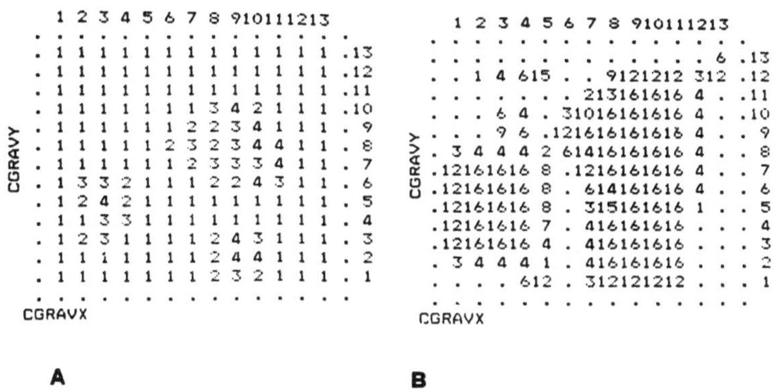


Fig. 2. Objective identification of neurochemical areas: the method of boundary construction based on density maps. In panels A and B are reported the density maps of figs. 1A and 1B, respectively. The size of the elementary square is 40 pixels, i.e., 5 times the diameter of the profiles of fig. 1A. The neurochemical area is defined as the area made by all the touching squares with a number of profiles higher than 5% of the maximal profile density per square. Thus, in the present examples the neurochemical areas are constituted by all non-empty squares.

First, suitable neuroanatomical landmarks are chosen to allow the superposition of Cartesian axes. As shown in fig. 3, a field is selected which covers the entire group of profiles as they appear to the observer. Each profile is reduced to a single point. The scatter diagram of these points together with the gravity center (GC) is plotted in the Cartesian plane. The plane is scanned by rotation of a polar axis around the GC of the group by steps of $\alpha$ degrees. The angle $\alpha$ is inversely related to the number (n) of points scanned (the $\alpha$ angle becomes progressively lower when the profile density encountered by the polar axis becomes progressively higher), so that at each step an equal number of objects is scanned. Thus, when the n-value is fixed by the operator (e.g., n = 20) the $\alpha$ value of each scanning step is determined by the density of profiles in that region of the field. After each step n points are present in the sector ($a_i$) of the Cartesian plane delimited by the starting and final positions of the polar axis. From each sector an isosceles triangle is derived, whose base is located at such a distance from the GC that it contains 95% of the n points present in the sector. This base is taken as the boundary of the neurochemical area. It should be noticed that the sizes of the bases are usually different from each other. Therefore an irregular polygon is obtained. From the polygon, the area, the perimeter and the shape factor for the neurochemical area can be calculated. The regression line (main axis) for the scatter diagram of the points within the polygon is calculated by means of the least squares method. Its slope is a useful parameter for characterizing the group of profiles.

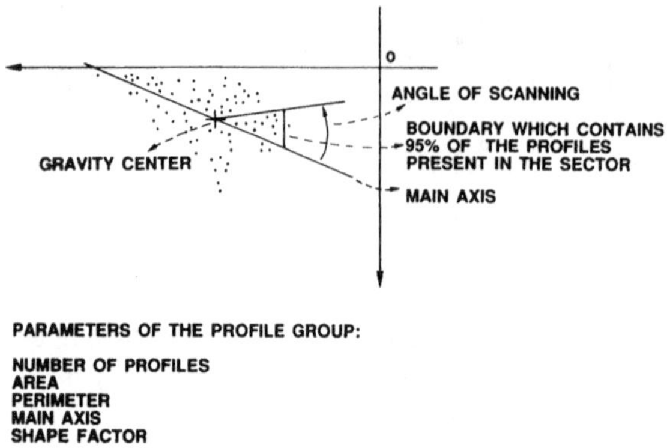


Fig. 3. Objective identification of neurochemical areas: the polygon method. Schematic representation of the procedure developed to two-dimensionally delimit a group of profiles. For further details, see text.

1.3 Quantitative approaches to the characterization of neurochemical areas. From the density map it is possible to calculate a profile density histogram, where classes characterized by increasing numbers of profiles per square are given on the X-axis and the respective frequencies are reported on the Y-axis. The numerical parameters of the histogram (the arithmetic mean, the standard deviation, the skewness and the kurtosis) summarize the features of the density map (for further details see, e.g., Snedecor and Cochran 1967). Also an index of the evenness of profile distribution can be obtained by the Gini's index (see Agnati et al. 1982, 1984, Zoli et al. 1990a), which evaluates the evenness of profile distribution among the positive squares of a density distribution. The geometrical parameters allow a characterization of the geometry of the neurochemical area and of its topological relationships with respect to some neuroanatomical landmarks. Geometrical parameters to be considered are the area, the perimeter, the shape factor, the GC and the main axis slope. Also a geometrical evaluation of the evenness of distribution can be obtained, i.e., the index of uniformity (see Zoli et al. 1990a,b). This procedure is based on the comparison between the interobject distances in the distribution under study and in a theoretical distribution where the same number of objects are uniformly distributed. In conclusion, a two-dimensional (2D) representation of a group of profiles present in a certain field can be achieved. In this plot the GC and the main axis are reported. A table is usually given with some selected numerical and geometrical parameters. The above mentioned approaches can also be extended to provide a three-dimensional (3D) representation of the neurochemical area under study (Zoli et al. 1986, 1990b).

2. Identification of subregions inside a neurochemical area on the basis of different profile spatial and/or optical density.

2.1. The overall procedure. Subpopulations of profiles inside a neurochemical area can be detected on the basis of their higher spatial density (morphometric approach), or on the basis of their higher optical density (microdensitometric approach). In the former case we are dealing with the analysis of clusters of profiles (e.g., cluster of cells within a region, or subgroups within a cell group), whereas in the latter case we are dealing with the analysis of patch of profiles vs a matrix of the same kind of profiles. Thus, in principle, it is necessary to state morphometric and microdensitometric criteria. Sometimes specific subpopulations are characterized by both their higher spatial and optical densities. Moreover, profile spatial density increases (such as in the case of small profiles as, e.g., nerve terminals) tend to appear also as profile optical density increases in view of the progressive overlapping of profiles in the image. Therefore, section thickness is very important in relation to mean profile diameter and spatial density. Finally, one should also consider that the magnification used for the analysis changes the resolution level, i.e., the capability of discrimination between two close-by objects. This can lead to a conversion of areas with high spatial density of profiles into areas with high optical density profiles.

In the case of the microdensitometric approach to patch detection (high optical density profiles and/or high spatial density of profiles for which the relationship among mean profile diameter, profile spatial density, section thickness and magnification of the analysis leads to profile overlap), we proceed by selecting, via visual inspection, an area where the patch is clearly apparent. Then, the area is interactively delimited and its grey tone histogram is plotted. The median value (or sometimes the 75th percentile value) is selected as the lower limit to perform a discrimination in order to visualize the higher intensity profiles. On this image it is possible to proceed to the application of a quantitative analysis of the patches, namely to consider the spatial distribution, optical density, and size of the patches within the entire field under study.

In the case of the morphometric approach to cluster detection (high spatial density of profiles for which the relationship among mean profile diameter, profile spatial density, section thickness and magnification of the analysis does not lead to profile overlap) a method has been proposed based on the fully automatic device IBAS I/II (Agnati 1988a). After discrimination of the specific profiles the cluster identification is based on the close function. By means of this procedure objects lying close together are merged to form larger individual objects. Profiles merge together according to the distance between them, their size and shape and finally the number of times that the step "close" is repeated. From this image all the scattered profiles are erased. By applying the AND function on the discriminated image and the image obtained using the close function plus the erasing of scattered profiles, the profiles present in the cluster can be identified and measured.

Another morphometric approach is based on the construction of density maps. The identification of clusters or patches within density maps rests on arbitrary but objective criteria for the selection of suitable spatial density thresholds below which the squares are considered as empty squares and above which each square maintains its number of profiles (spatial density number) (see Zoli et al. 1990b).

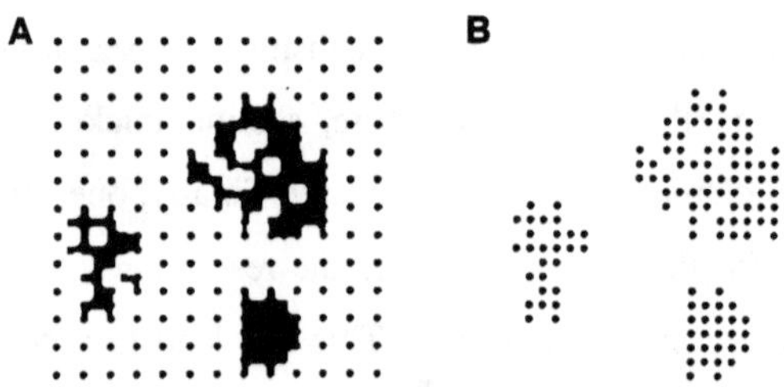


Fig. 4. Morphometric approach to cluster detection.
Panel A. Application of the close function on fig.1A. In the example the close function has been repeated 4 times to reach the merging of profiles which are located around 1 profile diameter from each other. With this procedure three clusters (one large and two small) have been identified.
Panel B. The profiles within the clusters are identified and can be counted (69 profiles in the large cluster and 27 profiles in each small cluster). This figure is obtained by performing the boolean operation AND. For further details, see text.

3. Comparison between profile distributions. Topological relationships between two (or more) populations of profiles.

3.1. Procedures based on field areas (simple overlap): The degree of the overlap with respect to the overall area covered by the two stainings can be obtained by means of the index of the overlap (IO):

$IO = [A_O / (A_{S1} + A_{S2} - A_O)] \times 100$, where $A_O$ = area of overlap, $A_{S1}$ and $A_{S2}$ = field areas of the stainings S1 and S2. The IO can take any value between 0 (absence of overlap) and 100 (complete overlap).Another index (index of inclusion, II) can be obtained from the percent ratio

$(A_O / A_{S1}) \times 100$, assuming $A_{S2} \geq A_{S1}$, since this index evaluates the amount of the smaller field area contained in the larger one. Indexes similar to those developed for the 2D analysis can be used for 3D analysis. To this aim, the volume of overlap ($V_O$) is calculated according to the following approximate formula, which considers the solid under study as made by a set of trunks of cone:

$V_O = [(\Sigma \{[(r_{O,i}^2 + r_{O,i+1}^2 + r_{O,i} \times r_{O,i+1}) \pi d] / 3\}$

The sum ranges from 1 to (n - 1), where n is the number of rostrocaudal levels. $r_{O,i}$ is the radius of the area of overlap for the $i^{th}$ level. The interlevel distance along the rostrocaudal axis is equal to d.

3.2. Procedures based on density maps (codistribution). These procedures are based on the comparison between two or more density maps. It is possible to correlate square by square the number of profiles present in the corresponding elementary squares of adjacent sections,

- considering only the squares for which both distributions are positive
- considering only the squares positive for one distribution
- considering all the squares present in a particular anatomical area.

The correlation coefficient can give a quantitative evaluation of the degree and significance of the similarity between two, or more, density maps (see Zoli et al. 1986, 1990b). In figs. 5 and 6, as an example of the application of some of the methods described above, the analysis of the codistribution between galanin and growth hormone releasing factor immunoreactvities in the median eminence is reported.

## DISCUSSION

In the present paper various image analysis techniques have been proposed to quantitate populations of neuronal profiles. The methods presented are based on functions, such as the discrimination and the close functions, and the boolean operator AND, which are common features of most automatic image analyzers. Moreover, some of these methods can be also implemented in semiautomatic devices, and, in certain cases, they can be performed even manually.

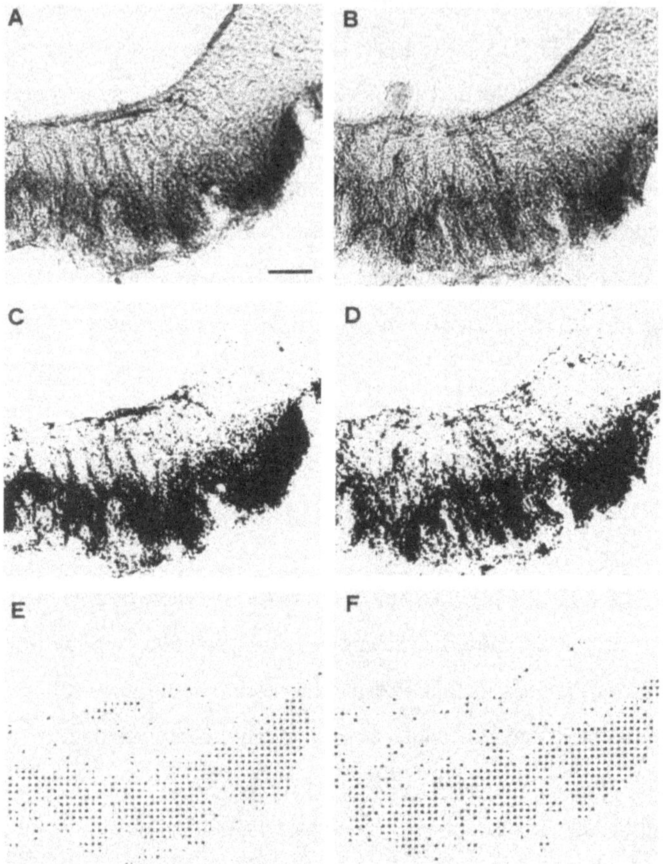


Fig.5. Panels A and B: microphotographs of growth hormone releasing factor (GRF, panel A) and galanin (GAL, panel B) immunoreactivities in adjacent coronal sections of the median eminence. Peroxidase-antiperoxidase method was used. Antibody dilution = 1: 1000. Section thickness = 40 μm. Bregma level -2.8 mm according to Paxinos and Watson atlas (1982). Calibration bar = 50 μm.
Panels C and D: interactive discrimination of the images reported in panels A and B. The field areas were 26730 $\mu m^2$ for GRF (panel A) and 27450 $\mu m^2$ for GAL (panel B) immunoreactivities. Using the boolean operator AND on these two image the overalp area can be obtained. In the example, the overlap area was 16850 $\mu m^2$, giving an index of overlap of 45% and an index of inclusion of 63%. For further details, see text.
Panels E and F. Images resulting from the application of the AND function to a grid (intercenter distance between adjacent dots = 8 μm, dot diameter = 4.8 μm) and the discriminated images reported in fig.5C (GRF imunoreactivity = panel E) and D (GAL immunoreactivity = panel F). By means of this procedure a discrete distribution, coding the density of profiles present in the images of figs. 5C and D can be obtained.

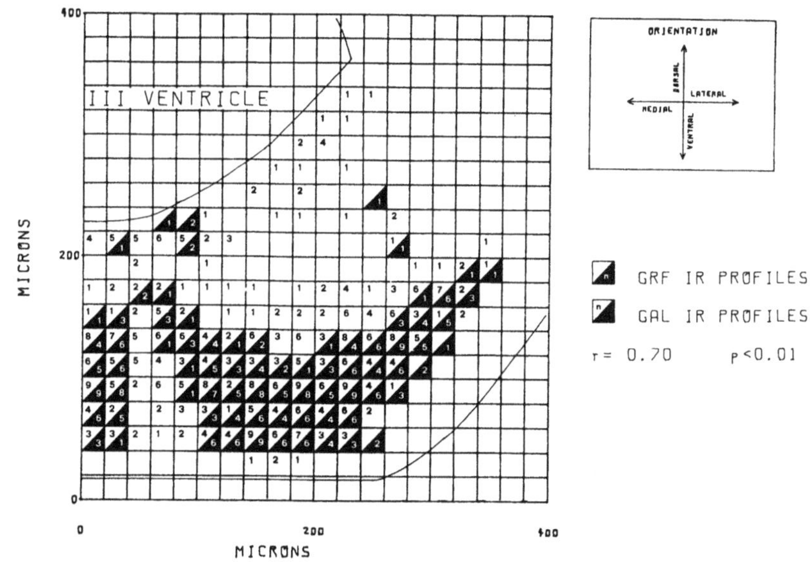


Fig.6. Density distribution of GRF (white numbers in black triangles) and GAL (black numbers in white triangles) immunoreactivities in the median eminence. The images reported in figs. 5E and F were subdivided into unitary squares of 400 $\mu m^2$. The number of profiles present in each unitary square is given. The correlation between the two density distributions was performed by means of Pearson's procedure (only squares positive for both stainings were considered).

The quantitative evaluation of the spatial distribution of a certain profile population and of its relationships with the distribution of other profile populations is a general objective of the analysis of histological preparations. The present image analysis techniques can be applied to standard problems of chemical neuroanatomy. However, many of these methods can be easily adapted to the analysis of histological and histochemical preparations from any other tissue.

Computer-assisted morphometry and microdensitometry allow one to describe in quantitative terms neural structures visualized by chemical anatomical techniques (Agnati et al. 1977, 1984, Benno et al. 1982, Sterberger and Sternberger 1986, Mize et al. 1988). Therefore, the full potentiality of chemical neuroanatomy can be exploited only thanks to image analysis. Moreover, computer assisted image analysis is particularly suited to the study of brain plasticity. These methods have been employed by our group in the evaluation of plastic phenomena induced by physiological and pathological events in the brain (see e.g., Agnati et al. 1983, 1988b, Zoli et al. 1988, 1990a).

It should also be stressed that quantitative approaches are leading to the development of a new type of neuroanatomy. This discipline is based on unambiguous definitions of terms, which until now have been only broadly and subjectively defined, such as cell group, patch of terminals, layer,... All these terms can, in principle, be defined exactly.

ACKNOWLEDGEMENTS

This paper has been supported by grants from MPI and CNR.

REFERENCES

Agnati LF, Fuxe K, Hökfelt T, Goldstein M, Jeffcoate SL (1977) A method to measure the distribution pattern of specific nerve terminals in sampled regions. Studies on tyrosine hydroxylase, LHRH, TRH and GIH immunofluorescence. J Histochem Cytochem 25:1222-1236

Agnati LF, Fuxe K, Zini I, Calza L, Benfenati F, Zoli M, Hökfelt T, Goldstein M (1982) A new approach to quantitate the density and antigen contents of high densities of transmitter-identified terminals: immunocytochemical studies on different types of tyrosine hydroxylase immunoreactive nerve terminals in nucleus caudatus putamen of the rat. Neurosci Lett 32:253-258

Agnati LF, Fuxe K, Calza L, Benfenati F, Cavicchioli L, Toffano G, Goldstein M (1983) Gangliosides increase the survival of lesioned nigral dopamine neurons and favour the recovery of dopaminergic synaptic function in striatum of rats by collateral sprouting. Acta physiol Scand 119:347-363

Agnati LF, Fuxe K, Benfenati F, Zini I, Zoli M, Fabbri L, Härfstrand A (1984) Computer assisted morphometry and microdensitometry of transmitter identified neurons with special reference to the mesostriatal dopamine pathway. I. Methodological aspects. Acta Physiol Scand Suppl 532:5-32

Agnati LF, Fuxe K, Zoli M, Härfstrand A, Kalia M, Janson AM, Zini I, Benfenati F, Grimaldi R, Goldstein M, Gustafsson J-A (1988a) Morphometric studies on epinephrine- and neuropeptide Y (NPY)-costoring neurons in the rat brain. In: Epinephrine in the central nervous system. JM Stolk, D U'Prichard, and K Fuxe eds, Oxford University Press, New York, pp 32-45

Agnati LF, Fuxe K, Zoli M, Zini I, Härfstrand A, Toffano G, Goldstein M (1988b) Morphometrical and microdensitometrical studies on phenylethanolamine-N-methyltransferase- and neuropeptide Y-immunoreactive neurons in the rostral medulla oblongata of the adult and old male rat. Neuroscience 26:461-478

Benno RH, Tucker LW, Joh TH, Reis DJ (1982) Quantitative immunocytochemistry of tyrosine hydroxylase in rat brain. I. Development of a computer assisted method using the peroxidase antiperoxidase technique. Brain Res 246:225-236

Bigbee JW, Kosek JC, Eng LF (1977) Effects of primary antiserum dilution on staining of antigen-rich tissues with the peroxidase antiperoxidase technique. J Histochem Cytochem 25:443-447

Gross DS, Rothfeld JM (1985) Quantitative immunocytochemestry of hypothalamic and pituitary hormones: validation of an automated, computerized image analysis system. J Histochem Cytochem 33:11-20

Mize RR, Holdefer RN, Nabors LB (1988) Quantitative immunocytochemistry using an image analyzer. I. Hardware, image processing, and data analysis. J Neurosci Meth 26:1-23

Paxinos G, Watson C (1982) The rat brain in stereotaxic coordinates. Academic Press, New York

Sternberger LA, Sterberger NH (1986) The unlabeled antibody method: comparison of peroxidase-antiperoxidase with avidin-biotin complex by a new method of quantification. J Histochem Cytochem 34:509-605

Zoli M, Fuxe K, Agnati LF, Härfstrand A, Terenius L, Toni R, Goldstein M (1986) Computer assisted morphometry of transmitter-identified neurons: new openings for the understanding of peptide-monoamine interactions in the mediobasal hypothalamus. In: Neurohistochemistry. Modern methods and applications. P Panula, H Paivarinta, S Soinila eds Alan R Liss, New York, pp 137-172

Zoli M, Agnati LF, Fuxe K, Zini I, Merlo Pich E, Grimaldi R, Härfstrand A, Goldstein M, Wikström AC, Gustafsson J-A (1988) Morphometrical and microdensitometrical studies on phenylethanolamine-N-methyltransferase- and neuropeptide Y-immunoreactive nerve terminals and glucocorticoid receptor-immunoreactive nerve cell nuclei in the paraventricular hypothalamic nucleus in adult and old male rats. Neuroscience 26:479-492

Zoli M, Merlo Pich E, Cimino M, Lombardelli G, Peruzzi G, Fuxe K, Agnati LF, Cattabeni F (1990a) Morphometrical and microdensitometrical studies on peptide- and tyrosine hydroxylase-like immunoreactivities in the forebrain of rats prenatally exposed to methylazoxymethanol acetate. Brain Res 51:45-61

Zoli M, Zini I, Agnati LF, Guidolin D, Ferraguti F, Fuxe K (1990b) aspects of neural plasticity in the central nervous system. I. Computer-assisted image analysis methods. Neurochem Int 16:383-418.

# POSITRON EMISSION TOMOGRAPHY : FUNCTIONAL NEUROANATOMY *IN VIVO*

J.C. Baron
INSERM U. 320 and CYCERON,
Boulevard H. Becquerel, BP 5027,
F-14021 Caen Cedex. France

Positron emission tomography (PET) is a well-established technique that allows to map in the living animal or human the distribution of radiopharmaceuticals in brain tissue, following the intravenous or inhalational administration of a molecule labeled with a positron emitting radionuclide. The main advantages of PET relative to other functional imaging methods are : 1) the availability of a wide array of compounds naturally labeled with $^{15}O$, $^{11}C$, $^{13}N$ or $^{18}F$ ; 2) the absolute quantitation, on a pixel-by-pixel basis, of the radioactivity concentration present in the tissue ; and 3) the excellent time and spatial-resolutions of the data obtained, allowing, with state-of-the-art equipment, to determine the time course of radioactivity in brain structures that are only a few millimeters in size (Phelps and Mazziotta, 1985). Relative to *in vitro* autoradiography, the distinct advantage of PET is that it provides the opportunity to carry out sequential measurements in the same subject ; in experimental animals, a pre-lesion (control) study can be performed allowing to use each subject as its own control.

## I - Generalities

### 1 - Principles of PET

The reader is referred to recent reviews on this topic (Links, 1990). Briefly, the principe of PET is to detect the pair of high energy (511 KeV) photons emitted at 180° from one another as a result of annihilation of one positron and one electron. This process of annihilation occurs following a brief course of the positron in biological tissues (1 to 2 mm on average, depending on the energy of the positron). Scintillation detectors (NaI, $BaF_2$, CeF, etc...) are arranged in a ring fashion around the subject ; the photomultiplier tubes are

NATO ASI Series, Vol. H 58
Neurocytochemical Methods
Edited by A. Calas and D. Eugène

connected in facing pairs and possess an electronic clock such that one event is registered only if two photons are detected, one in each of the facing detectors, within a very short time window (<1 nsec). Each event is registered together with its time of appearance, which will allow the computer to reconstruct tomographic images of the distribution of radioactivity (actually of annihilations), directly corrected for decay and according to selected time frames. Because several detector rings (or a large crystal block) can be fitted into a PET camera, up to 15 contiguous images are generally obtained, easily covering the entire brain. The attenuation of the 511 KeV photons in the tissues can be directly measured in each plane of interest for each PET study using an external source containing a positron-emitting compound ("transmission" data). Thanks to adequate fantoms containing a solution of a positron-emitting radiopharmaceutical with known radioactive concentration, it is possible to calibrate the brain images obtained which are displayed on the video screen according to a quantitative scale in units of nCi/pixel (voxel). The statistics in these images depend on the total number of events recorded in each pixel, which in turn depends on the amount of radioactivity administered, the integrated acquisition time of the image, time of acquisition relative to physical decay of the radionuclide, the statistics in the "transmission" images, the 3D-size of the pixel (which is a function of the spatial resolution of the PET camera and of the reconstruction filter used), and, last but not least, the biological behaviour of the compound (e.g., entry through the blood-brain-barrier, drug metabolism, clearance rate, specific binding). The optimal performances of state-of-the-art PET cameras are ≅ 1 sec temporal resolution, and ≅ 5 mm spatial resolution (full width at half-maximum) in each of ≅ 15 slices. It must be emphasized that although this spatial resolution allows to distinguish two hot spots distant from 5 mm, the absolute quantitation of radioactivity concentration remains very sensitive to the "partial volume effect". Hence, smearing of radioactivity from one structure to another results in "blurring" of structure borders and, hence, inaccurate quantitation unless the structure has a size at least twice the spatial resolution (Links, 1990). This effect also occurs in the Z axis, and is such that a given structure must encompass the whole axial resolution if quantitation is to be accurate.

## 2 - Radiochemistry

The practical aspects of positron radiochemistry and cyclotron science will not be dealt with here (see Dannals *et al.*, 1990). The short half-life of most positron-emitters is such that a cyclotron (preferably a small one fully dedicated for PET research) is generally needed on site in order to synthetize the labeled compounds when needed. If this is not possible, special delivery systems from neighbouring cyclotrons, or by means of rapid earth- or air-delivery procedures, is necessary.

Regarding radiochemistry for PET, four major points need be emphasized : 1) the "natural" labeling allows to synthetize radiopharmaceuticals whose biochemical, physical, and pharmacological properties are strictly unaltered ; 2) the synthetic procedures allow to obtain chemically pure radiopharmaceuticals ; 3) they can be prepared with very high specific radioactivity (SRA), e.g. in the order of 0.5-3 Ci/μmol for $^{11}C$- or $^{18}F$-labeled compounds, allowing to avoid untoward pharmacological or receptor saturation effects in PET studies (usually, 5-20 mCi at most are administered for a PET study, generally corresponding to about 0.5 to 5 μg of drug) ; and 4) thanks to the short half-life of positron emitters, not only are the radiation effects limited, but it becomes possible to administer several labeled compounds at short intervals of time in the same experimental session.

3 - Data processing and analysis

Thanks to the large number of compounds that can be labeled with a positron emitter, PET allows access to a variety of physiological, biochemical and pharmacological parameters, such as brain perfusion, oxygen consumption ($CMRO_2$), glucose utilization (CMRglu) (by means of labeled deoxyglucose), protein synthesis, tissue pH, and specific binding sites. It is outside the scope of the present chapter to review the methodological issues related to the compartmental models used to obtain quantitative measures of these parameters. The reader is referred to recent reviews on these topics of further information (Gjedde and Wong, 1990 ; Baron *et al.*, 1989).

In practice, all PET studies require an adequately determined "input function" if quantitation is desired. This is generally obtained by measuring in the arterial plasma (radial or femoral artery catheterization) the time-course of the unchanged radiopharmaceutical, assuming that this will adequately represent the brain arterial or capillary tracer kinetics. This procedure requires fast initial sampling following intravenous or inhalational administration, centrifugation, and counting of radioactivity in gamma-counters perfectly cross-calibrated to the PET camera. A serious complication occurs when metabolism of the radiotracer occurs during the course of the experiment, as often highly complex procedures have to be designed to separate the unchanged compound from its labeled by-products. These complexities with blood radioactivity handling have led to designing new strategies where some "reference structure" in brain is used to estimate the kinetics of "non-specific" input ; this has been applied to the cerebellum in the case of the Dopamine $D_2$ receptors because it lacks these receptors, and to the occipital cortex in the case of $^{18}$F-6-Fluoro-L-DOPA where the activity of the enzyme aromatic-amino-acid-decarboxylase is negligible.

In practice, knowledge of the radiotracer's behavior both in the brain structure of interest and in the arterial plasma or structure of reference allows, using a specifically designed compartmental model and a number of simplifying assumptions, to quantitate the parameter under investigation in absolute biological or pharmacological units (e.g., cerebral blood flow in units of ml blood per 100 ml of brain per minute, or density and affinity of Dopamine $D_2$ receptors in molar units). In most cases, however, it is also necessary to measure some biochemical or pharmacological reference to quantitate the process (e.g., blood glucose or oxygen content, SRA of the radioligand administered).

In most cases, the parameter under investigation can be quantitated on a pixel-by-pixel basis, allowing to obtain "parametric" PET images directly assessable with the eyeball. However, as is often the case with *in vitro* receptor autoradiography, sometimes only the "specifically bound" radioligand or uptake of radiotracer are available and hence displayed in PET images.

The next step is to obtain measures of the parameter under study in any given brain region of interest. This is an extremely complex issue, as the PET images do not directly show the anatomy, but rather a "functional anatomy" which may not be, and is often not, superimposable on the former. The contours of "regions of interest" (ROIs) are determined by any of the following procedures : 1) visually-guided hand or computer-aided delineation of contours of functional structures, preferably with the help of an anatomical atlas for the same species and the same brain cut ; 2) ROIs "libraries" stored in the computer as standard templates specific for each species and each brain cut ; 3) best, definition of ROIs on morphological imaging like X-Ray C.T. scanning or magnetic resonance imaging (MRI), directly displaying the anatomical structures of the individual animal under study, with subsequent copy of the ROIs on the corresponding PET images. Regardless of the method used, a stereotaxic head-frame, or a specially designed head-holder, is generally necessary for orientation and selection of the brain cuts to image with PET. Finally, the limitations relative to the partial volume effect (see above) are important to acknowledge in the ROIs procedure in order to avoid major errors ; for example, structures below 10 mm in diameter may be subject to considerable underestimation of radioactive concentration unless specific measures are taken (Links, 1990).

Following this introductory paragraph, a few illustrative examples of applications of PET as a tool in functional neuroanatomy will be briefly described.

## II - Examples of applications of PET to study brain function *in vivo*

### 1 - Glucose utilization

In analogy to the beautiful experiments done with $^{14}C$-2-Deoxyglucose ($^{14}CDG$) in rats (McCulloch, 1982), studies with $^{18}F$ Fluoro-2-Deoxy-D-glucose ($^{18}FDG$) and PET have demonstrated changes in measured CMRglu in response to functional challenges in the human brain. For example, Blacklock *et al.* (1987) have induced coma in patients with glial-cell brain tumors (astrocytoma) using barbiturates, as part of the therapeutic work-up. They showed that during coma there is a considerable reduction in CMRglu in the normal parts of the brain, while no changes are seen in the tumor itself. This indicates that CMRglu, as measured by PET with $^{18}FDG$, essentially reflects synaptic activity, in agreement with earlier rat studies with $^{14}CDG$. Conversely, focal activation of CMRglu in association with increased synaptic activity has been well demonstrated with PET. For example, large increments in glucose utilization in the primary and association visual cortex can be easily induced by patterned or complex visual tasks (Phelps *et al.*, 1981). Further studies have demonstrated that these focal increases in CMRglu are matched by proportional increases in cerebral blood flow (CBF), indicating that one can use the much more simple measurement of CBF to image focal brain activation ; hence when using $^{15}O$-water, CBF studies can be done serially in the same subject every 15-20 min thanks to the 2 min half-time of $^{15}O$ (Fox *et al.*, 1988). It has been shown that these focal increases in CMRglu and CBF are accompanied by only minor increases in $CMRO_2$, suggesting that during "activation", most of the increment in glucose use results from enhanced "anaerobic" glycolysis (Fox *et al.*, 1988). The CBF paradigm has been used to successfully image focal brain activation during somesthetic activation, visual-color tasks, single word out put and semantic association, anticipatory anxiety, and lactate induced panick attacks (Fox *et al.*, 1987 ; Lueck *et al.*, 1989 ; Petersen *et al.*, 1988 ; Reiman *et al.*, 1989a, b).

The [18]FDG method has also been applied to the study of brain development in a few close-to-normal children from birth to adulthood (Chugani and Phelps, 1986). These authors have demonstrated the changing pattern of CMRglu with age, with increasing glucose use in association cortex and various motor areas in conjunction with behavioral maturation. In mentally retarded children with normal morphologic brain imaging, PET imaging clearly shows immature patterns of CMRglu (Chugani *et al.*, 1987).

Although the adult pattern of CMRglu is reached at around 1 year of age in normally-behaving children, the absolute metabolic rates in mg glucose per 100 g brain per minute increase in all brain structures from low values at birth to values higher than adults between 4 and 9 years of age ; thereafter, there is a subsequent decline in metabolic rates until around age 15 (Chugani *et al.*, 1987). This late decline presumably reflects the secondary elimination of over-numerous synaptic connections. In a study of adults with early blindness (before age 2 years) Veraart *et al.* (1990) report that the CMRglu in the visual cortex was significantly higher than that found in either adults with late blindness or controls ; their likely interpretation was that deafferentation of the visual cortex in the former subjects occurred before the phase of synaptic selection, preventing the CMRglu to fall to adult values after development.

Functional activation of CMRglu has also been well documented during focal or generalized epileptic activity in patients with uncontrolled seizures (Phelps and Mazziotta, 1985). In several instances, the spread of synaptic hyperactivity to cortical fields neighbouring the epileptogenic regions as well as along neuronal relays in the ipsilateral striatum and thalamus and the contralateral cerebellum, has been well demonstrated. Conversely, focal brain damage almost always induces a depression of CMRglu and $CMRO_2$ in distant, connected neuronal structures as a result of deafferentation/disconnection (Baron, 1989). These remote effects can spread over large parts of the brain, affecting the ipsilateral and contralateral cortical mantle, the ipsilateral striatum and thalamus, and the contralateral cerebellum, while their magnitude is larger than what could be explained solely by anterograde or retrograde degeneration ; they also tend to develop immediately after the lesion has occurred, and to recover according to a monoexponential model, attesting to their purely "functional" nature (Baron, 1989). Subhuman primates lend themselves to such sequential PET studies of the distant metabolic effects of lesions, for instance following stereotaxic electrolysis of the Nucleus Basalis of Meynert (Kiyosawa *et al.*, 1989) or section of the corpus callosum (Yamaguchi *et al.*, 1990). The study of these effects has lately attracted interest because, according to the concept of diaschisis, they may well underlie the clinical expression of lesions as well as recovery from functional deficit, and be accessible to therapeutic manipulation (Feeney and Baron, 1986).

2 - Specific synaptic markers

Thanks to selective, high affinity radioligands labeled with the positron emitters $^{11}C$ or $^{18}F$, it is now possible ot obtain PET images of the distribution of specifically-bound radiopharmaceuticals in the brain of the human and subhuman primate. This strategy has been successfully applied to a number of receptor systems, particularly the Dopamine $D_1$ and $D_2$, the $\mu$ and $\delta$ opiate, the central and peripheral types of benzodiazepine, and the $5HT_2$ serotonergic receptors (Baron, 1989). In addition, it is possible to use specific reuptake blockers, such as $^{11}C$-labeled-Nomifensine which labels the dopamine and norepinephine reuptake sites (Aquilonius *et al.*, 1987) ; similarly, the use of $^{18}F$-labeled 6-Fluoro-L-Dopa, an analog of L-Dopa with affinity for the aromatic-amino-acid decarboxylase, allows to label *in vivo* the presynaptic dopamine terminals (Garnett *et al.*, 1983). By means of sophisticated compartmental models using the tissue and plasma kinetics of the administered radioligand, it is possible to obtain in each brain area the relevant quantitative pharmacological parameters e.g. density and affinity of a receptor or unidirectional radioligand-binding site association rates.

As an illustrative example of application of these paradigms, Parkinson's disease (PD) has attracted great interest (Baron, 1990). PET studies in PD patients using $^{18}F$-Fluro-Dopa or $^{11}C$-Nomifensine have demonstrated a major loss of presynaptic dopamine terminals in the striatum (especially in the posterior putamen while the head of caudate is relatively spared) ; post-synaptically, there is an increased density of $D_2$ receptors in the striatum in the early stages of the disease, with normal or reduced density in treated, chronic-stage PD patients. PET studies have also demonstrated a loss of striatal presynaptic dopamine terminals in normal aging and asymptomatic MPTP-exposed subjects, indicating the method's sensitivity to pre-clinical neuronal loss. Similarly, a decrease in striatal $D_2$ receptor binding has been reported with increasing age in normal subjects (Baron *et al.*, 1986), but it is unclear whether this is due to a loss of post-synaptic neurons or reflects a dysregulated dopamine transmission with aging.

<u>Figure legend:</u> Transaxial tomographic cuts displaying the cerebral metabolic rate of glucose (CMRglu) in seven planes of a normal healthy subject. These images were obtained 50-60 min after intravenous administration of 5mCi of $^{18}$F-labelled 2-Fluoro-2-Deoxy-D-glucose ($^{18}$FDG) in tracer amounts, using a TTVO3-LETI-Time-of-Flight positron camera (lateral resolution = 5.5 mm, slice thickness = 9 mm). The CMRglu was calculated in each picture (Pixel) using the "in vivo autoradiographic" approach, which makes use of the pixel radioactive concentration, the arterial plasma $^{18}$F history from t = 0 to t = 60 min, the standard $^{18}$FDG transport and enzymatic rate constants, and the plasma glucose content, and is expressed according to a pseudo-colour scale in units of mg per 100 ml per min. Highest CMRglu Pixel values are shown in yellow-white, and lowest in dark red-black. The anterior is up and the left is on the right of the image. Gray matter areas are depicted in yellow due to their higher CMRglu than white matter ; hence, cerebellar and cerebral cortices, hippocampal areas, head of caudate, thalamus, and brainstem are clearly distinguishable on these images; average CMRglu values for a few areas are indicated on the lower right corner.

⟶

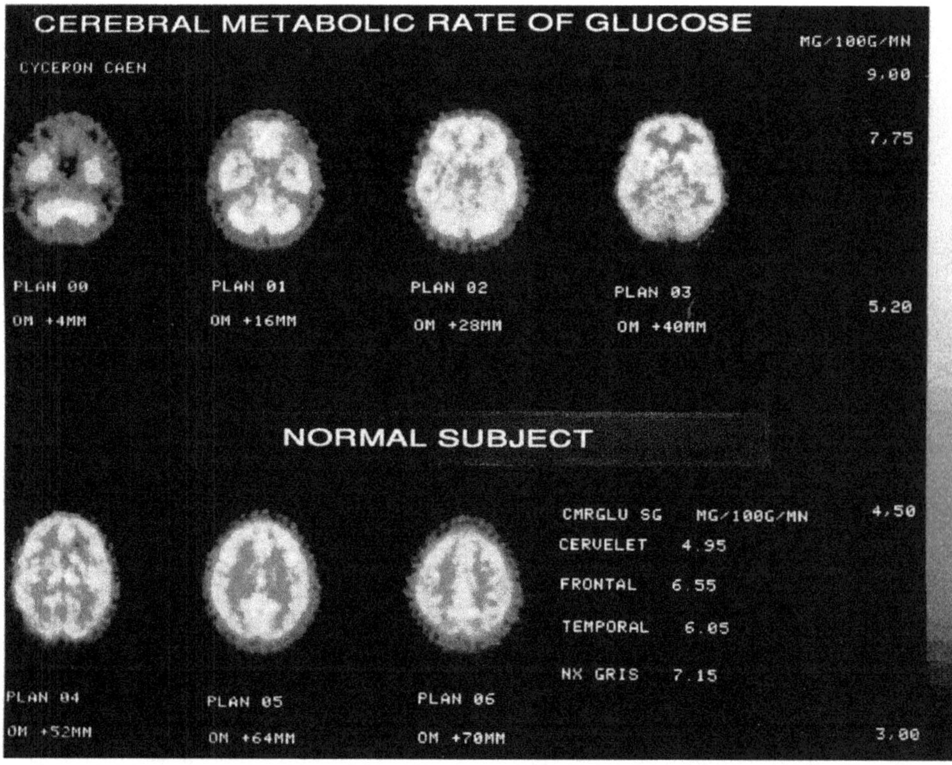
CEREBRAL METABOLIC RATE OF GLUCOSE
MG/100G/MN
CYCERON CAEN
9,00
7,75
PLAN 00
OM +4MM
PLAN 01
OM +16MM
PLAN 02
OM +28MM
PLAN 03
OM +40MM
5,20
NORMAL SUBJECT
CMRGLU SG MG/100G/MN
4,50
CERVELET 4.95
FRONTAL 6.55
TEMPORAL 6.05
NX GRIS 7.15
PLAN 04
OM +52MM
PLAN 05
OM +64MM
PLAN 06
OM +70MM
3,00

REFERENCES

Aquilonius SM, Bergström K, Eckernäs SA, Hartvig P, Leenders KL, Lundquist H, Antoni G, Gree A, Rimland A, Uhlin J, Langström (1987) In vivo evaluation of striatal dopamine reuptake sites using $^{11}$C-nomifensine and positron emission tomography. Acta Neurol Scand 76 : 283-287.

Baron JC (1989) Depression of energy metabolism in distant brain structures : studies with PET in stroke patients. Seminars in Neurology 9 : 281-285.

Baron JC (1990) Maladie de Parkinson et syndromes parkinsoniens dégénératifs : études du système dopaminergique par la tomographie par émission de positon. Cir Metabol Cerv 7 : 73-79.

Baron JC, Frackowiak RSJ, Herholz K, Jones T, Lammertsma AA, Mazoyer MBM, Weinhardt K (1989) Use of positron emission tomography in the investigation of cerebral hemodynamics and energy metabolism in cerebrovascular disease. J Cereb Blood Flow Metabol 9 : 723-742.

Baron JC, Mazière B, Loc'h C, Cambon H, Sgouropoulos P, Bonnet AH, Agid Y (1986) Loss of striatal $^{76}$Br-Bromospiperone binding sites demonstrated by positron tomography in progressive supranuclear palsy. J Cereb Blood Flow Metabol 6 : 131-136.

Blacklock JB, Oldfied EH, Di Chiro G, Tran D, Theodore W, Wright DC, Larson SM (1987) Effect of barbiturate coma on glucose utilization in normal brain versus gliomas. Positron emission tomography studies. J Neurosurg 67 : 71-75.

Chugani HT, Phelps ME (1986) Maturational changes in cerebral function in infants determined by $^{18}$FDG positron emission tomography. Science 231 : 840-843.

Chugani HT, Phelps ME, Mazziotta JC (1987) Positron emission tomography study of human brain functional development. Ann Neurol 22 : 487-497.

Dannals F, Ravert HT, Wilson A (1990) In : Frost, JJ and Wagner, HN (eds). Quantitative Imaging : Neuroreceptors, Neurotransmitters, and Enzymes. Raven Press, New York, p 19.

Feeney D, Baron JC (1986) Diaschisis. Stroke 17 : 817-830.

Fox PT, Burton H, Raichle M (1987) Mapping human somatotensory cortex with positron emission tomography. J Neurosurg 67 : 34-43.

Fox PT, Raichle ME, Mintum MA, Dence C (1988) Nonoxidative glucose consumption during focal physiologic neural activity. Science 241 : 462-464.

Garnett ES, Firnau G, Nahmias C (1983) Dopamine visualized in the basal ganglia of living man. Nature 305 : 137-138.

Gjedde A, Wong DF (1990) In : Frost, JJ and Wagner, HN (eds) Quantitative Imaging : Neuroreceptors, Neurotransmitters, and Enzymes. Raven Press, New York, p 51.

Kiyosawa M, Baron JC, Hamel E, Pappata S, Duverger D, Riche D, Mazoyer B, Naquet R, Mackenzie ET (1989) Time course of effects of unilateral lesions of the Nucleus Basalis of Meynert on glucose utilization of the cerebral cortex : positron tomography in baboons. Brain 112 : 435-455.

Links JM (1990) In : Frost, JJ and Wagner, HN (eds) Quantitative Imaging : Neuroreceptors, Neurotransmitters, and Enzymes. Raven Press, New York, p 37.

Lueck CJ, Zeki S, Friston KJ, Deiber MP, Cope P, Cunningham VJ, Lammerstma AA, Kennard C, Frackowiak RS (1989) The colour centre in the cerebral cortex of man J. Nature 340 : 386.

McCulloch JM (1982) Mapping functional alterations in the CNS with $^{14}C$-deoxyglucose. In : Inversen LL and Snyder SH (eds) Handbook of Psychopharmacology, vol. 15, Plenum Press, New York, 321-340.

Petersen SE, Fox PT, Posenr MI, Mintun M, Raichle ME (1988) Positron emission tomographic studies of the cortical anatomy of single-word processing. Nature 331 : 585-589.

Phelps ME, Mazziotta JC (1985) Positron emission tomography : human brain function and biochemistry. Science 228 : 799-809.

Phelps ME, Mazziotta JC, Kuhl DE (1981) Tomographic mapping of human cerebral metabolism : visual stimulation and deprivation. Neurology 31 : 517-529.

Reiman EM, Fusselman MJ, Fox PT, Raichle ME (1989a) Neuroanatomical correlates of anticipatory anxiety. Science 243 : 1071-1084.

Reiman EM, Raichle ME, Robin E, Mintun MA, Fusselman MJ, Fox PT, Price JL, Hackman KA (1989b). Arch Gen Psychiatry 46 : 493-500.

Veraart C, De Volder AG, Wanet-Defalque MC, Bol A, Michel C, Goffinet AA (1990) Glucose utilization in human visual cortex is abnormally elevated in blindness of early onset but decreased in blindness of late conset. Brain Res 510 : 115-121.

Yamaguchi T, Kunimoto M, Pappata S, Chavoix C, Riche D, Chevalier L, Mazoyer B, Maziere MA, Naquet R, Baron JC (1990) Effects of anterior corpus callosum section on cortical glucose utilization in baboons : a sequential positron emission tomography study. Brain 113 : 937.

# MICROFLUOROMETRIC IMAGING OF TIME AND SPACE DEPENDENT CALCIUM CONCENTRATION CHANGES IN MAMMALIAN NEURONS

Wolfgang Müller and John A. Connor
Roche Institute of Molecular Biology, Roche Research Center
340 Kingsland St., Nutley, NJ 07110, USA

## INTRODUCTION

Excellent signal to noise ratio and resolution has made fluorescence microscopy a favorite tool for molecular probing in the biosciences. Cytoplasmic indicator molecules allow monitoring of intracellular ion concentration changes with high spatial and temporal resolution even in cells not dissociated from their surrounding tissue (tissue slice). General interest has focused especially on intracellular calcium, an ubiquitous second messenger whose many vital functions are well known. Detailed knowledge of the spatial intracellular calcium dynamics appears to be a key for the understanding of the parallel functioning and specificity of the many pathways utilizing $Ca^{2+}$ as a second messenger. Consequently, development of high efficiency fluorescent indicators concentrated to some extent on $Ca^{2+}$ as the target ion. Today, fluorescent indicators are available for a series of ions and molecules including $H^+$ (pH), $Cl^-$ and $Mg^{2+}$ (see Tsien, 1989) and more indicators become available almost monthly. The following treatment is limited to the measurement of calcium, although the basic considerations are generally the same for the measurement of other target ions.

For measurement of $Ca^{2+}$, at the present fura-2 is the most popular dye for a series of reasons (Grynkiewicz et al. 1985; Tsien and Poenie 1986):

1) high fluorescence efficiency
2) good resistance to photodestruction

NATO ASI Series, Vol. H 58
Neurocytochemical Methods
Edited by A. Calas and D. Eugène

3) high sensitivity for physiological intracellular Ca-concentrations
4) easy loading of cells by the membrane permeable acetoxymethylester of fura-2 (Tsien, 1981)
5) a clearly defined isobestic wavelength permitting ratio measurements
6) high selectivity for $Ca^{2+}$ over $Mg^{2+}$.

## BASIC PRINCIPLES

The optical principles involved in fluorometry are briefly summarized below, detailed treatments may be found elsewhere (Pesce et al., 1971; Lakowicz, 1983; Udenfriend, 1969). The fluorescence intensity/unit area of an indicator at a given excitation wavelength $\lambda$ is determined by the following factors:

1) the relative absorbance of light energy for the given wavelength, known as molar extinction coefficient $E\lambda$;
2) the number of indicator molecules, measured by concentration, C, and optical pathlength $\Delta z$;
3) the quantum efficiency, Q, the percentage of absorbed energy released as fluorescence rather than being dissipated;
4) the intensity of the excitation light, $I\lambda$.

The fluorescence intensity is given in first approximation by the following equation:

$$F \propto I\lambda * C * \Delta z * Q * E\lambda \qquad (1)$$

In many experimental situations some of the above cited factors cannot be determined or only roughly estimated, e.g. dye concentration, optical pathlength or excitation intensity. Consequently only relative concentration changes can be estimated which depend, in addition, on the assumption that the unknown factors remain constant throughout the measurements. In the case of dyes which undergo a large spectral shift in their excitation or emission spectrum during binding of their target ion it is possible to eliminate many of the factors one is not interested in by taking measurements at two different wavelengths and forming the ratio of the fluorescence values. Usually the ratio is chosen to increase with an increase in calcium concentration. In this case it is sufficient to know the limiting fluorescence ratios $R_{min}$ and $R_{max}$ of the free and bound indicator and the fluorescence ratio of free and bound indicator $F_{\lambda(f/b)}$ at the second wavelength to calculate the calcium concentration by the following simple formula (Grynkiewicz et al. 1985):

$$[Ca^{2+}] = K_d * F_{\lambda(f/b)} * (R - R_{min}) / (R_{max} - R) \qquad (2)$$

The only uninteresting factor remaining in the ratio, the ratio of excitation intensity (or, in the case of a shift in the emission spectrum, the ratio of emission intensity) is determined implicitly by $R_{min}$ and $R_{max}$ which depend of course on the individual setup used for the measurments. In principle any two wavelengths may be chosen to form the ratio. In practice, there are a number of considerations which should be kept in mind.

First, equation 1 simplifies the actual situation by assuming all terms to be constant throughout the specimen being examined rather than being terms under an integral sign. Therefore equation 1 is applicable only where the terms are indeed constant or else vary in the same way throughout the sample. Of most concern is usually variation in the Z-direction (optical pathway). One must assume that light at both wavelengths is attenuated to the same degree during passage through the sample and additional layers in the optical pathway like coverslip, glial cell layer or slice tissue. Absorption, refraction and scattering are generally wavelength dependent. Therefore, light of one wavelength may be significantly more attenuated than the other while passing through the sample. The ratio of excitation intensities then deviates from the value determined implicitly by $R_{min}$ and $R_{max}$. The deviation usually decreases as the two excitation wavelengths are made close together.

Also, fluorescence can be quenched in the sample differentially, depending on the viscosity of the sample. For fura-2 ratioing at 340 and 380 nm excitation wavelengths, the intracellular fura-2 fluorescence ratio has been found to be reduced by up to 30 % compared to fluorescence ratios in saline (Poenie, 1990). Differential quenching by the viscosity of the sample would be also minimized by the two excitation wavelength close together. Also, measurements at the two wavelengths are usually taken both with the exposure time necessary for resolution of the weaker fluorescence signal. The two wavelength closer to the fluorescence maxima therefore allow higher time resolution. The trade off in choosing two wavelengths close together is a loss in signal. Reference to the fura-2 excitation spectra of Fig. 1A shows that a wavelength pair of 340 and 390 nm will just about maximize the signal for changes in $Ca^{2+}$. Depending on the equipment used and the kind of the sample, usually excitation wavelengths of 380 and 340, 350 or 360 nm have been found as the best compromise.

Intensity variation in the X-Y directions of the microscope field are not a particular problem as long as variations are proportional for the two excitation wavelengths and the camera is reasonably linear. Variation of the single wavelength signals is under this condition eliminated by the ratioing technique.

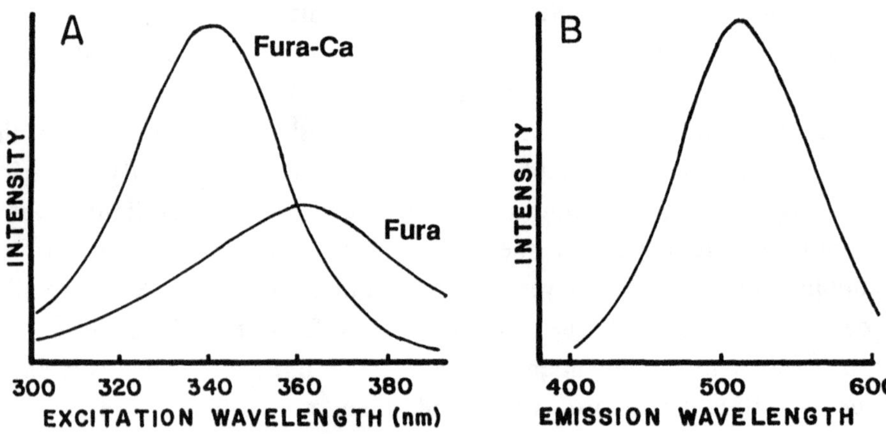


Fig. 1 : Excitation and emission spectra of fura-2. Upon binding $Ca^{2+}$, the fura-2 fluorescence maximum shifts from an excitation wavelength of about 365 nm to about 340 nm. The fluorescence signal undergoes maximal variation with $Ca^{2+}$-binding at excitation wavelengths of 340 and 390 nm.

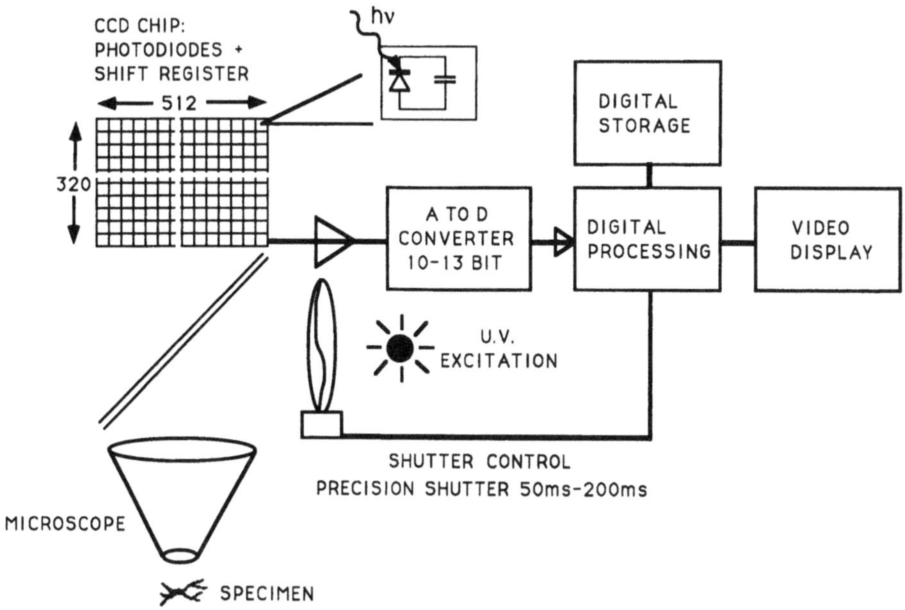


Fig. 2: Schematic representation of the imaging system. The specimen is mapped by the microscope onto a cooled CCD chip representing an array of 320x512 imaging pixels. The values of stored charge per pixel are read out serially, digitized and stored as an array in the computer memory. A personal computer is used in collecting, analyzing and displaying of data. The interference filter for limiting UV excitation to the desired wavelength is located behind the shutter and not shown in the diagram. Filter change can be controlled by the imaging computer.

## EXPERIMENTAL SETUP

The most critical elements of the measurement system are:

- an objective with good transmission of UV at the chosen wavelengths,
- a good quality low light level camera, which represents the most expensive part of the equipment and
- a reliably constant UV source.

The usual experimental arrangement is schematically shown in Fig. 2. For the photometry a charge coupled device (CCD) camera is used. The image of the sample is focused onto the CCD chip functioning like a rectangular array of light sensing diodes coupled with storage capacitors (pixel). Since diode and capacitor are physically the same piece of material, the distraction is made only for instructive purposes. Each diode generates photoelectrons by absorption of the fluorescence photons. The capacitors integrate the photoelectrons during the exposure or capture period and the amount of charge is then read out serially through a high impedance buffer amplifier. Thermal noise interacting with the capture process is sufficiently reduced by cooling the CCD chip to -40 degrees C (by a closed liquid cooling circuit). The time resolution of the system is determined by the sum of exposure times and readout times and is at the present between 250 and 500 ms where large fields are studied.

The readout pixel values are digitized and stored as corresponding arrays on a harddisk. Data are archived on mass storage devices like tapes or WORM (optical) disks.

For online judgement of measurements, the ratio for two corresponding arrays is calculated, transformed into an array of color or grey values and displayed as a ratio map of the field. False color display of the ratio map considerably facilitates the perception of more subtle $Ca^{2+}$ changes. For display of only the interesting parts of the field, i.e. the sample, the image can be masked. The simplest masks are created by selecting a fluorescence threshold for one of the two fluorescence images for display of the ratio map. Structures out of focus may be masked manually with the help of a computer mouse. Images may be corrected for unspecific background or transformed linearly into calcium concentration maps.

Inverted as well as upright microscopes have been used for microfluorometry. Usually inverted microscopes are preferred because they offer more free space above the preparation for electrical recording, stimulation or application of transmitters. Since imaging has to be done from underneath the bath, the structure under study must be close to the chamber bottom because of the short working distance of the microscope objectives. This is no problem with monolayers of dissociated cells but might cause difficulty when imaging cells in tissue slice or organs where cells are impaled from the other end of the slice. Often it is necessary to impale and fill cells with indicator and then flip the slice over prior to optical recording.

In an upright microscope, impalement and imaging is done from the same side of the probe and there is no need for a cover, making focusing on the target cell more simple. Difficulties one has to keep in mind in this situation is varying UV-reflection from the surface of the organ bath and optical distortion by lensing effects of curved fluid surfaces. Varying UV-reflection from the surface can be controlled by measuring background fluorescence of the tissue. With brain slice positioned on a nylon mesh at the interface between perfusate and a humid $O_2/CO_2$ atmosphere, we could reliably exclude the above mentioned problems.

## LOADING CELLS WITH MEMBRANE PERMEABLE INDICATOR

Membrane permeable forms of indicators are highly attractive for their ease of loading. Lipid soluble groups attached by ester bonds to the probe molecule make this very membrane permeable. Inside the cell, the acetoxy-methyl (AM) chains are cleaved by cellular esterases and the indicator is trapped inside the cell. While theoretically extraordinary easy, in practice there can be problems with the deesterification. Either the cells under study have intrinsically low esterase activity or the esterase activity is impaired by experimental conditions. For fura-2, all 4 of the AM groups have to be removed for fura-2 to bind calcium with high affinity, however, fura-2 is trapped in the cells when fewer than four are removed. This difficulty has been inadvertently demonstrated by any number of experimenters who have loaded their cells under improper conditions. Here the cells are fluorescent but the ratio will not change even after treatment with $Ca^{2+}$ ionophore. In less extreme cases a mixture of the two species inside cells will underestimate actual calcium levels. Such an error can be tolerated where qualitative calcium changes are sufficient to make the points desired. A good control for calcium changes measured in cells loaded with fura-2-AM is by comparison to measurements of calcium changes in cells injected with the membrane impermeable, active form of fura-2. Where this is not possible, application of a $Ca^{2+}$ ionophore allows a quick control whether the maximal fluorescence ratio in the cell equals the one obtained in the calibration.

There is a wide range of indicator trapping and holding among mammalian CNS cells of various types and ages. Freshly plated embryonic cells (<3days in vitro, DIV) trap fura-2 or quin-2 very effectively, i.e. 100-300 $\mu$M intracellular indicator are achieved by 30 min incubation in 1-2$\mu$M permeable indicator. In older cultures first trapping in glial cells and fibroblasts becomes poor (4-20 DIV) and beyond 30 DIV trapping often becomes poor in neurons which appear healthy in morphological and electrophysiological terms. At all stages, however, the trapped indicator appears to become completely deesterified during 30-60 min post-load incubation. Acutely isolated neurons from adult animals load and activate the indicator very rapidly (Connor et al. 1988).

Another pecularity of the AM-loading of indicator is that, depending on the cell type and experimental conditions, the indicator might be trapped most effectively not in the cytosol but in cellular subcompartments. Again, a useful control is possible with direct injection of indicator (see Connor and Tseng 1988). Almers and Neher (1985) reported significant differences between signals obtained from mast cells loaded by different methods. In these cells most of the fura-2-AM wound up in internal vesicles and reported none of the $Ca^{2+}$-changes observed in dye-injected mast cells. In imaging, fura-2 compartmentization of this kind can often be identified morphologically in the single wavelength fluorescence image. In cases, where the fura-2 fluorescence from cellular subcompartments can not be resolved in the single wavelength fluorescence images, the subcompartments often show up in the ratio image by their free $Ca^{2+}$-concentration differing from the cytosolic one, either in resting conditions or during transmembraneous $Ca^{2+}$-influx into the cytosol (Connor and Tseng, 1988; Connor, in preparation). Thus, the disadvantage of fura-2 AM

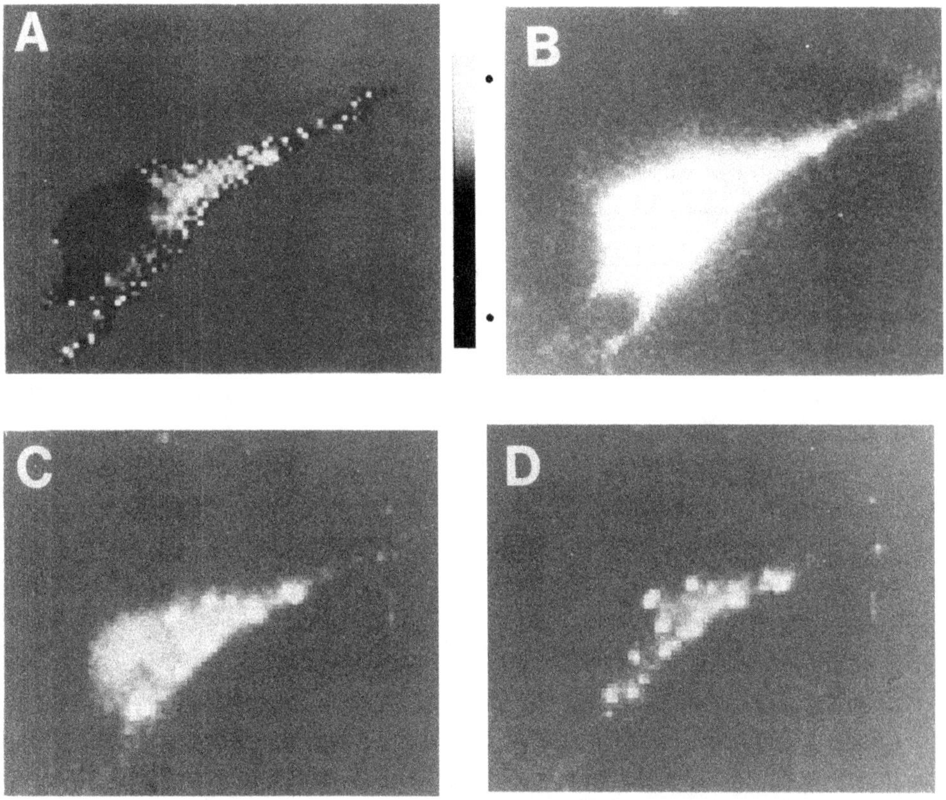


Fig. 3: A: Ratio image of Purkinje neuron loaded with membrane permeant fura-2 AM. Points on the bar correspond to 200 nM (upper) and 50 nM $Ca^{2+}$. $Ca^{2+}$ levels are elevated at around 140 nM in the upper right area while in the nuclear area (upper left portion of the soma) they were around 80 nM. B: 340 nm fluorescence image of the neuron. C and D: Fluorescence pictures of the neuron approximately 10 s and 40 s after rupturing the cell membrane with a microelectrode. (from Connor and Tseng, 1988)

loading of cells for the study of cytosolic $Ca^{2+}$-changes turns into a big advantage where one is interested in the $Ca^{2+}$-changes of cellular subcompartments which are not accesible to injection. The resolution of subcompartments versus sourrounding cytosol depends critically on the depth of focus of the objective and the thickness of the cell studied. Another way is to remove cytosolic fura after AM loading of the cell and its compartments. Fig. 3 shows how, after permeabilizing the cell membrane mechanically with a microelectrode, the fura-2 signal of subcompartments shows up strongly after fura leaked out of the cytosol. In cases where it is desireable to control the membrane potential, whole cell patch recording should wash out cytosolic fura-2 similar to endogenous cytosolic molecules.

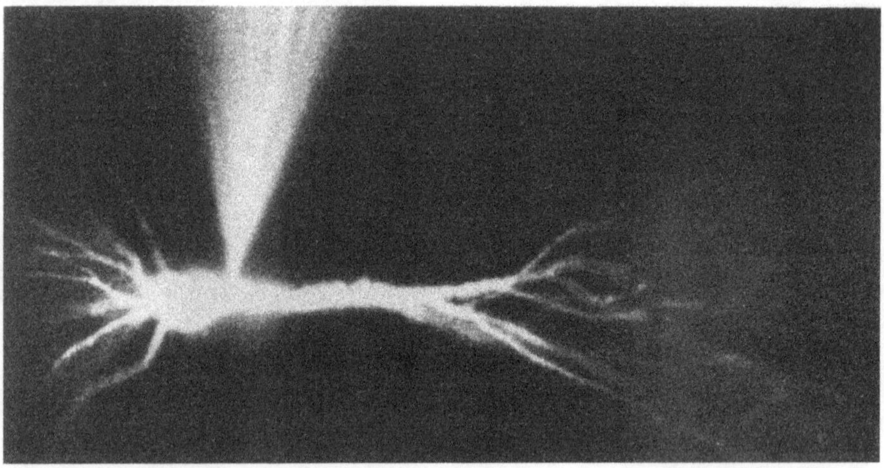

Fig. 4: 380 nm fluorescence image of a CA3 pyramidal neuron in guinea pig hippocampal slice loaded by intracellular injection of membrane impermeable fura-2. The microelectrode, coming from the left, impales the apical dendrite just above the soma. Fura-2 was allowed to equilibrate in the cell for about one hour.

## LOADING CELLS BY INJECTION

Fura-2 injection is not only useful as a control for $Ca^{2+}$-signals measured from AM loading but in non-dissociated tissue preparations it often is the only way to obtain a high resolution fluorescence signal selectively from the cell under study. With the current burst of interest in $Ca^{2+}$-triggered phenomena related to synaptic modification there is a strong need to understand $Ca^{2+}$-dynamics in tiny dendrites, presynaptic terminals and postsynaptic spines. These elements as well as the induction and maintenance of synaptic modification seem to be well preserved in the brain slice preparation. Usable fura-2 concentrations being limited by the $Ca^{2+}$-buffering properties of fura, the fura-2 fluorescence signal from the small volumes of synaptic elements is less strong. With AM loading, fluorescence from surrounding elements then obscures the signal from small synaptic elements to an unacceptable degree.

Fig. 4 shows the 380 nm fluorescence image of a CA3 neuron in guinea pig hippocampal slice loaded by injection. After iontophoretic injection the dye equilibrates within the soma and the dendritic tree

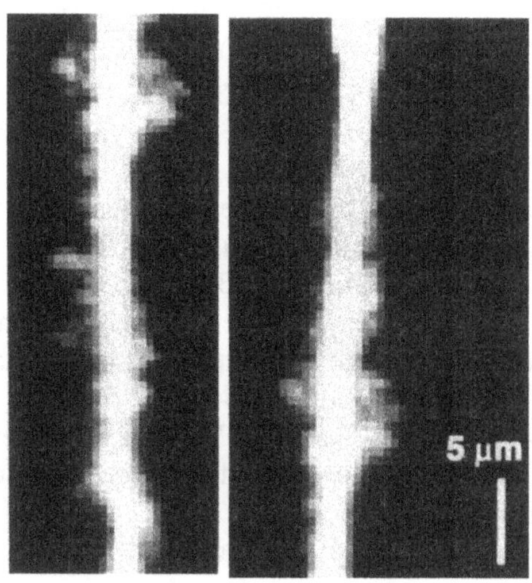


Fig.5: 380 nm fluorescence images of spiny dendrites in the distal stratum radiatum of a CA3 pyramidal neuron in guinea pig hippocampal slice. The cell was loaded with the membrane impermeable form of fura-2 by microinjection into the soma.

within some 60 min. The fura-2 signal then reflects roughly the differences in optical pathlengths between the soma and the various dendrites. After adequate intracellular fura-2 equilibration, the signal to background ratio appears to be good enough to resolve calcium concentration changes in postsynaptic thorny excrescences or spines. Fig. 5 shows 380 nm fluorescence images of fura-2 filled CA3 dendrites in the stratum radiatum covered densely with spines. The spines give obviously sufficient fluorescence signal over background for a critical test of current hypotheses with regard to the role of intra-spine calcium for the induction of synaptic modifications.

ACKNOWLEDGEMENT

We thank D. Tank for making available imaging software and M. Smeyne and R. Sun for technical assistance. W.M. was supported in part by the DFG.

REFERENCES

Almers, W. & Neher, E. (1985) The Ca signal from fura-2 loaded cells depends strongly on the method of dye loading. Febs Letters 192:13-18

Connor, J.A. & Tseng, H.-Y. (1988) Measurement of intracellular $Ca^{2+}$ in cerebellar Purkinje neurons in culture: Resting distributions and responses to glutamate. Brain Res Bull 21:353-361

Connor, J.A., Wadman, W.J., Hockberger, P.E. & Wong R.K.S., (1988) Sustained dendritic gradients of $Ca^{++}$ induced by excitatory amino acids in CA1 hippocampal neurons. Science 240: 649-653

Grynkiewicz, G., Poenie, M. & Tsien, R.Y. (1985) A new generation of Ca indicators with greatly improved fluorescence properties. J Biol Chem 260:3440-3450

Lakowicz, J.R. (1983) Principles of Fluorescence Spectroscopy. Plenum, New York

Pesce, A.J., Rosen, C. & Pasby, T. (1971) Fluorescence Spectroscopy: An introduction for biology and medicine. Marcel Dekker, New York.

Poenie, M. (1990) Alteration of intracellular Fura-2 fluorescence by viscosity: A simple correction. Cell Calcium 11:85-91

Tsien, R.Y. (1989) Fluorescent probes of cell signaling. Ann Rev Neurosci 12:227-253

Tsien, R.Y. (1981) A non-disruptive technique for loading calcium buffers and indicators into cells. Nature 280:527-528

Tsien, R.Y. & Ponie, M. (1986) Fluorescence ratio imaging: A new window into intracellular ionic signalling. Trends Biochem. Sci. 11:450-455

Udenfriend, S. (1969) Fluorescence Assay in Biology and Medicine. Academic Press, New York.

# REPORT ON THE DISCUSSION OF THE THIRD SESSION : QUANTITATIVE ASPECTS

F. Fonnum
Norwegian Defence Research Establishment,
Division for Environmental Toxicology,
P.O. Box 25, N-2007 Kjeller, Norway

Dr Morel reviewed the method developed for studying the release of ACh. The method is based on the specific reactions of two enzymes namely choline oxidase for the assay of choline and acetylcholinesterase for the hydrolysis of acetylcholine. In this way both compounds can be assayed in the same sample. This method will also allow the determination of the ACh during release. KCl was found to release only a small proportion of ACh because the $Ca^{++}$ channels become desentisized. The ionophore gramicidine released a much higher proportion of ACh. In tissue samples substances interferring with the assay may be present. They can be renewed by treatment with oxidising substances.

The discussion focused on the mechanism of ACh release and the possible involvement of synaptic vesicles. Israel and coworkers have isolated a proteolipid named "mediatophore". The mediatophore is 75% homologous of the proton pump which is a constituent of the synaptic vesicles. Further investigation will have to focus very sharply on the possibility that the mediatophore is obtained from very active vesicles which have fused with the plasma membrane.

The last point brought forward in the discussion where the proportion between vesicular and cytoplasmic pools of ACh. This has been estimated in several tissues to be about 1:1. The point was made that the socalled cytoplasmic pool was in part derived from an active vesicle population.

Dr Zoli presented different approaches which may be used to obtain quantitative data from a histochemical preparation. The most urgent questions concerned the reproducibility of the method when comparing different animals and different times. Dr Zoli stressed the importance of preparing all experimental animals in the same standardized way. Usually with this method relative data would be sufficient e.g. when comparing different pharmacological treatments. But absolute values could be obtained provided enough standards were used.

Other people had experienced difficulties in the reproducibility of the method. Even differences in slice thickness of 1-2 μm affected the results. The same level of illumination is a must for high reproducibility. With background subtraction the effect of tissue thickness variation can be reduced. The shadow image is another important factor which must be reset every day.

NATO ASI Series, Vol. H 58
Neurocytochemical Methods
Edited by A. Calas and D. Eugène

The densitometric picture is very good for dynamic and pharmacological studies. The method can also be used for morphometric studies. Groups of heavy stained cells can be identified.

Positron emission tomography was discussed by Baron. The method enables us to measure changes in brain physiology *in vivo*. The method is designed for clinical research although studies of primates is possible. At present very expensive apparatus is under development for studies in cat and rodent. The physical limitation of the method is 2 mm.

The preparation of labelled tools requires fast and excellent chemists. The half life $0^{15}$ is two minutes. With $C^{11}$ the compound must be used as soon as possible to have high specific radioactivity. With $F^{18}$ the half life is two hours, so that $F^{18}$-deoxyglucose must be prepared in the morning and used during the day.

Receptors are assayed at low doses to give high resolution by autoradiography. Some questioned the possible interference with endogenous ligands.

The PET scan showed that lactic acid increased during rapid metabolic changes in the brain. This reminded some participants of muscle physiology where training gave high densities of capillary (high blood flow) whereas sprint gave high lactic acid production. This discussion also brought forward the importance of glial cell for these rapid processes.

The points raised in Dr Muller's talk that attracted much attention was the localization of $Ca^{++}$ in dendrites and the increase in $Ca^{++}$ images after cholinergic activation of the hippocampus slice.

NMDA contains a $Ca^{++}$ channel, but treatment with the agonist had no effect of $Ca^{++}$ in proximal dendrites. With a less efficient technique other authors have claimed to see $Ca^{++}$ in the far dendrite. In general the $Ca^{++}$ image was believed to be derived from extracellular sources. The $Ca^{++}$ binding sites were $Ca^{++}$ binding proteins, sequestration of $Ca^{++}$ etc... More $Ca^{++}$ changes were observed in dendrites than in the soma. It was maintained that there were more $Ca^{++}$ channels in the dendrite than in the soma.

Concerning the $Ca^{++}$ image after cholinergic input, much discussion focused on the specificity of the stimulation. Although the area with the highest density of cholinergic fibers was stimulated, also glutamergic and GABAergic fibers were stimulated. The author maintained that pharmacological tools (carbachol, atropine, eserine) showed that the effects were due to cholinergic input. The cholinergic fibers activate both the inhibitory basket cells and the pyramidal cells. The inhibition from the basket cell lasted, however, only a few seconds whereas the cholinergic input lasted for 60 seconds. Since the image was obtained after 20 seconds, only the muscarinic input survived.

# IMMUNOCYTOCHEMISTRY OF NEUROPEPTIDE RECEPTORS USING ANTI-IDIOTYPIC ANTIBODIES AND ANTIBODIES TO PEPTIDES ENCODED BY COMPLEMENTARY RNA

J-Y. Couraud[1], S. Maillet[1], D. Zouaoui[2], W. Kummer[3], J. Grassi[1], P. Pradelles[1] and M. Conrath[2].

[1] Service de Pharmacologie et d'Immunologie
DRIPP, CEN/Saclay
91191 Gif-sur-Yvette Cedex
France

## 1.Introduction: general comments on the methodologies

During the last two decades, the characterization of neuropeptides and of their cellular receptors has constituted one of the major goals in neuroscience research and has given rise to the development of new technologies, especially in neurocytochemistry. We will briefly present and discuss here two immunological methodologies leading to the production of antibodies against receptors for neuropeptides: the anti-idiotypic and "anti-complementary peptide" methodologies.

### 1.1. The anti-idiotypic methodology

This technique can be considered as the application of the immunological network theory developed by N. Jerne (1974). Basically, it consists of an experimental manipulation of the immune system of an animal leading to the production of a subset of antibodies (abs) whose binding specificities are close to those of any selected endogenous receptor ligand. This methodology has recently been extensively reviewed (Köhler et al., 1989). In brief, the major steps for producing anti-receptor anti-idiotypic antibodies are the following (Fig.1):

a) Anti-ligand abs are obtained by conventional immunizations. Among the various abs generally elicited, some may be directed at the ligand epitope(s) involved in the

[2] Institut des Neurosciences, Dépt. de Cytologie, 7 Quai St-Bernard, 75252 Paris Cédex 05, France

[3] Institut fur Anatomie und Zellbiologie, Universität Heidelberg, Im Neuenheimer Feld 307, D-6900 Heidelberg, F.R.G.

NATO ASI Series, Vol. H 58
Neurocytochemical Methods
Edited by A. Calas and D. Eugène

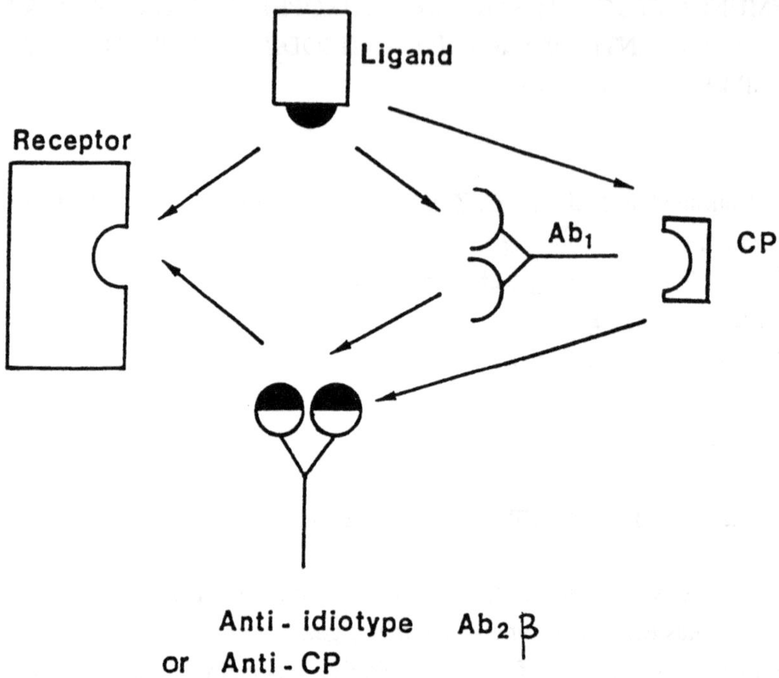


Figure 1: Schematic model for the anti-idiotypic and the anti-complementary peptide immunological approaches to raising anti-receptor antibodies.
In the heterogenous population of anti-idiotypic antibodies, a few (Ab2β, shown here) may display structural and/or functional similarities with the biological ligand because they are complementary to the binding site of anti-ligand antibodies (Ab1). As a consequence, they may cross-react with the receptor. Note that the majority of anti-Id antibodies (Ab2α, not represented here) are not directed at the binding site of Ab1 antibodies and should not, in principle, recognize the receptor. On the other hand, the complementary peptide CP (whose sequence has been deduced from that of the RNA complementary to the RNA encoding for the peptide ligand, see the text) might interact with the ligand and therefore somehow resemble the binding site of an anti-ligand antibody. An anti-CP antibody could thus in turn mimic an anti-idiotypic antibody and recognize the receptor.

interactions with the biological receptor. These idiotypic abs (Ab1) may thus display binding characteristics resembling those of the receptor itself.

b) Immunizations of other animals with these receptor-like abs may then result in the emergence of a second generation of abs (anti-Ab1 or Ab2): the anti-idiotypic abs or anti-idiotypes (anti-Id), a minor proportion of which (Ab2 ) is directed at the binding site of the idiotypic Ab1 antibodies. For this reason, they are thought to mimic (structurally or functionally) the original ligand, of which they are the "internal image". Such anti-Id have been shown in several systems (Strosberg, 1989) to recognize the cellular receptors of endogenous ligands and sometimes to display pharmacological in vitro and in vivo effects. In some instances, the methodology could be simplified since immunization with the ligand has been reported to give rise occasionally not only to anti-ligand antibodies but also to significant amounts of anti-Id (Cleveland et al., 1983), in accordance with Jerne's immune network theory.

Whether performed in one or two steps, this methodology for obtaining original ligands of receptors may afford several advantages: 1) Over conventional anti-receptor abs. Indeed, purified or semi-purified receptors are not needed for immunizations. In addition, and very importantly, anti-Id are expected to recognize the ligand binding site of the receptor and hence to be pharmacologically active, unlike most anti-receptor abs obtained by direct immunization with the receptor. 2) Over classical peptide ligands. A major advantage of anti-Id lies in their great stability, compared with that of most hormone or neurotransmitter derivatives, which are generally small molecules subjected to rapid hydrolysis by endogenous proteases. As will be seen below, the stability of the antibody-receptor interactions will also be determinant for the success of immunocytochemistry at the electron microscope level.

### 1.2. Methodology based on antibodies directed against peptides encoded by complementary RNA

This novel immunological approach is based on a molecular recognition theory proposed by E. Blalock and coworkers (Bost et al., 1987), which may provide a biochemical basis for protein-protein interactions and may in principle allow for the generation of abs of predefined specificity. This theory postulates that complementary nucleic acid sequences encode peptides or proteins (termed complementary peptides or proteins) which interact. This idea arose from the unquestionable observation that codons for hydrophobic amino acids are complemented by codons for hydrophilic amino acids and vice versa, while codons for uncharged amino acids are complemented by those for the same group of uncharged residues. It was then hypothesized that two peptides encoded by complementary nucleic acids, hence whith completely interchanged global hydropathy, would "assume conformations allowing specific binding of the pair". If true, this would mean that, when starting from a peptide ligand of a given receptor, the complementary peptide would assume a conformation somehow mimicking that of the binding site of an anti-ligand antibody (Fig.1). As a consequence, immunization with this receptor-like complementary peptide (CP) would lead to abs recognizing the binding site of the receptor i.e. to antibodies (anti-CP) conceptually identical to anti-Id.

In support of these predictions, specific and rather high-affinity interactions were shown to occur between a dozen or so peptide ligands (including ACTH, γ endorphin, angiotensin II, insulin) and their respective complementary counterparts. Although it seems too early, in our opinion, to generalize this theory (mainly because the reason why a CP should assume a conformation "complementary" to that of the peptide remains poorly understood), it must be emphasized that, as also predicted, anti-CP abs have been demonstrated to bind cellular receptors and were even successfully used to effect their purification (Blalock et al., 1989). At least in one case (Smith et al., 1987), anti-ligand abs and anti-CP abs have been observed to display idiotypic/anti-idiotypic relationships, as represented schematically in Fig.1.

This anti-CP immunological approach for producing anti-receptor abs appears more restrictive than the general anti-idiotypic approach since the former applies only to receptors for peptide ligands whose genes have been already cloned. By contrast, obtaining an anti-CP antibody, i.e. a first-generation antibody, certainly appears easier than raising an "internal image" anti-Id.

We will now describe some practical aspects of the application of these methodologies to the immunocytochemistry of receptors. The neuropeptide with which we have most experience to date is substance P (SP). With two other peptides, NKA (or substance K) and NKB, SP belongs to the neurokinin family. This undecapeptide (Arg-Pro-Lys-Pro-Gln-Gln-Phe-Phe-Gly-Leu-Met) is involved in a large range of biological activities including pain transmission, regulation of blood pressure, and inflammation (Jessell and Woomack, 1985). The cDNAs of the three neurokinins (SP, NKA and NKB), and now of their three respective receptors (NK1, NK2 and NK3), have been cloned very recently (Shigemoto et al., 1990), but the receptors have not yet been purified. This explained our efforts during these last years to produce anti-SP receptor antibodies through the anti-idiotypic approach (Couraud et al., 1985, 1989), and very recently through the anti-complementary peptide approach (manuscript in prep.). Practical details will thus be given essentially by reference to the neurokinin system but appear general enough to be applied for raising and using antibodies to any peptide receptor.

## 2. Antibody generation and characterization

### 2.1. Preparation of antibodies

#### 2.1.1. Anti-idiotypic antibodies

The first step i.e. preparation of anti-peptide abs (Ab1) is, in general, relatively easy.

Preparation of immunogens necessitates the coupling of small peptides to a carrier molecule, but care must be taken that the coupling reaction leaves the ligand epitope involved in receptor interaction intact for immune recognition by idiotypic Ab1 antibodies. In our case, SP was coupled to bovine serum albumin (BSA) through its N-term moiety (since receptor binding occurs through the C-term pentapeptide), using the homobifunctional reagents benzoquinone or 1,5 difluoro-2,4-dinitrobenzene (DFDNB) and conventional techniques (see Couraud et al., 1989). Blanc du Bouscat rabbits or Biozzi strain mice were immunized intradermally at multiple sites with 1mg or 0.2ug of SP-BSA, respectively. Detection of anti-ligand abs is generally carried out through radio-(RIA) or enzymo-immunoassays (EIA): we have developed a very sensitive EIA for SP, using the enzyme acetylcholinesterase (AChE) as marker (Couraud et al., 1987). Within the context of the anti-Id approach, it is important at this stage to compare the specificities of anti-ligand abs with those of the ligand receptor. This is best achieved when a large number of ligand derivatives (agonists, antagonists, fragments, analogs) are available: for each compound, the biological effects (reflecting the binding to physiological receptor) may be compared to the immunoreactivity with the anti-ligand ab (reflecting the interaction with the antibody binding site). Good correlation between these two parameters can be taken as strong indication of a molecular resemblance between the combining sites of the receptor and of the idiotypic ab. Such experiments have allowed us to select for some polyclonal and monoclonal anti-SP abs as SP receptor-like abs (Couraud et al., 1989).

These anti-SP abs have been used in immunocytochemistry studies to reveal SP-like immunoreactivity distribution in nervous and non-nervous tissues, both by us (Couraud et al., 1987) and by others (e.g. Sommer et al., 1985). Practical details concerning the use of these antibodies for optical and electron-microscope observations may be found in these papers. An important point must be kept in mind when using such idiotypic anti-ligand abs, especially polyclonal antisera, in cytochemical experiments: these sera may contain both idiotypic (anti-ligand) and anti-idiotypic (anti-receptor) abs, the latter spontaneously occurring in animals immunized with the ligand, as indicated above. In our case, we checked that anti-SP antisera did not display any significant anti-Id activity, using specific tests (see below).

In order now to obtain anti-Id, polyclonal or monoclonal anti-SP abs were used as immunogens following the two-step method described above. The whole immunoglobulin (Ig) fraction from a pool of rabbit polyclonal anti-SP sera purified by

DEAE-chromatography was used to immunize 5 allotypically matched rabbits, each one receiving 0.5 mg of antibodies emulsified in Freund's complete adjuvant (FCA). Booster injection were given every month and high titers in anti-Id activity were detected after 4 months. On the other hand, we obtained good results by immunizing mice or rats with an equimolecular mixture of two monoclonal anti-SP abs (20 ug for mice, 100 ug for rats) purified by the caprylic acid method (see Couraud et al., 1989) and polymerized with kehole limpet hemocyanin (KLH) using glutaraldehyde. Anti-Id appeared in the sera of both species after two weeks and remained at a high level for at least 2-3 months.

2.1.2. Anti-complementary peptide antibodies

The first step in this approach is to synthesize the complementary peptide (CP). In our case, the complementary RNA sequence was readily obtained from the published sequence of the cloned RNA for SP (Nawa et al., 1983). The CP derived from that RNA sequence read in the 3'-5' direction was: Ser-Gly-Phe-Gly-Val-Val-Lys-Lys-Pro-Asn-Tyr. This peptide, termed PS, was synthesized by Dr. P. Nicolas (Institut J. Monod, Paris) using a solid-phase peptide synthesizer, by t-boc chemistry.

In preliminary experiments, one should test whether or not specific interactions occur between the neuropeptide and its complementary peptide CP. This can be achieved by monitoring the possible binding of $^{3}$H-labeled neuropeptide (widelely available commercially) to microtiter plates coated with the CP. Such preliminary experiments revealed specific binding between $^{3}$H-SP and PS, confirming a recent report (Pascual et al., 1989). This binding was inhibited by an excess of both unlabeled SP and soluble PS. In the light of this result, it is conceivable that antibodies to PS might recognize a SP receptor binding site.

In order to obtain these anti-PS abs, PS was coupled to KLH in the presence of glutaraldehyde. Three rabbits were immunized intradermally at multiple sites with 1mg of PS-KLH emulsified in FCA. Identical booster injections were given at two-week intervals and the appearance of anti-PS in the sera was monitored seven days later, as described below.

2.2. Immunological characterization of antibodies

2.2.1. Anti-idiotypic antibodies

The immunological characterization of the anti-Id abs is very important. Several tests must be developed to reveal the presence of anti-isotypic (in the case of inter-species immunizations), anti-allotypic (inter-strain immunizations) and anti-idiotypic abs and to distinguish further between alpha and beta anti-Id (Fig.1). These tests have been described in detail in Couraud et al (1989). Briefly, two major types of immunological assays are currently used for routine screening of anti-Id. The immunometric (or sandwich) assays will detect both categories of anti-Id. Competitive assays ,which assess competitive binding of the anti-Id and the ligand to the idiotypic Ab1, will be specific for beta anti-Id. These competitive assays are easily performed either by RIA or EIA using the labeled ligand, but must be completed by careful control tests (Couraud et al., 1989) which cannot be detailed here. In order to investigate further the possible mimicry between anti-Id and the ligand (the neuropeptide), additional immunological tests should be carried out: 1) Anti-Id should inhibit the binding of labeled ligand to all anti-ligand idiotypic abs, whatever the origin of the latter abs, 2) Anti-Id should induce the production, possibly in the animal which produces them, of anti-anti-Id (Ab3) with anti-ligand properties.

All these criteria were checked in the case of our polyclonal rabbit anti-SP anti-Id. We have in addition shown by biochemical techniques that these anti-Id specifically bind to membranes enriched in SP receptors (Maillet et al., in press). Furthermore, they are able to exert pharmacological effects in vitro: in rat parotid gland cells, they increase the exocrine release of proteins through an enhancement of membrane phospholipid metabolism, thus behaving as agonists of SP (Couraud et al., 1985, Maillet et al., in press). Taken together, these data strongly suggest that anti-Id abs displaying SP properties will be successfully used for SP receptor cytochemistry.

However, from a practical point of view, insofar as the nature of the molecular interactions between anti-Id and the receptors is not fully understood, we would suggest using anti-Id for cytochemical assays (preferably in a tissue in which the distribution of receptors has been well established with other techniques) as soon as high antibody concentrations are detected in the anti-Id sera and even if no pharmacological data are available. In our opinion, similar advice may also be given to the morphologist when anti-CP abs are to be used. Obviously, appropriate morphological controls for assessing the specificity of immunolabeling (see below) will be of special importance here.

### 2.2.2. Anti-complementary peptide antibodies

The appearance of anti-CP abs in the sera of immunized animals can be detected using at least two general methods. The CP can be radio- or enzymatically labeled in order to monitor its possible binding to the successive bleedings. We have recently prepared a PS-AChE conjugate and developed an enzymo-immunoassay for PS, which will be described elsewhere (in prep). An alternative, less sensitive but perhaps more convenient technique is to detect anti-CP abs in the following way: 1) coat the CP on microtiter plates (10 µg/ml), 2) incubate dilutions of the sera to be assayed, and 3) reveal the presence of bound anti-CP abs by commercially available labeled anti-Ig abs. In our case, this method revealed the production of anti-PS abs in the sera of the immunized rabbits, two months after PS injection.

## 3. Immunocytochemical procedures

We will briefly describe here the experimental conditions successfully applied in SP receptor cytochemistry, using rabbit anti-Id in the rat spinal cord (Conrath et al., 1988a, 1988b) or in the guinea-pig and human trachea (Kummer et al., 1990a, 1990b), and using anti-PS antibodies in the former material (Zouaoui et al., this volume).

### 3.1. Tissue preparation

Adequate fixative conditions appeared very different from one system to another. In the rat spinal cord, the best results were obtained, either with anti-Id or anti-CP abs, with a perfusion via the ascending aorta of rinsing solution followed by 4% paraformaldehyde in 0.1 M phosphate buffer pH 7.4 as fixative. Small blocks were then post-fixed in the same medium and cut into 50 µm vibratome sections. By contrast, in the mammalian trachea, specific labeling with anti-Id could not be obtained following aldehyde fixation, regardless of whether fixation was carried out by perfusion or by immersion of previously unfixed cryostat sections. Specific labeling was only obtained using either unfixed or cold acetone precipitated sections (-20°C, 15 min). It results from these observations that experimental fixative conditions must be adapted to each tissue.

### 3.2. Incubation procedures

Tissue sections were preincubated for 30 min in 0.15 M Sörensen buffer, pH 7.4 or in 0.1M phosphate buffer saline (PBS), supplemented either with 0.25% defatted milk or 1% BSA. The addition of detergent (0.1% Triton or 0.5% Tween) to the preincubation medium increased the immunolabeling by anti-Id or anti-CP abs in the rat spinal cord but totally prevented it in the mammalian trachea.

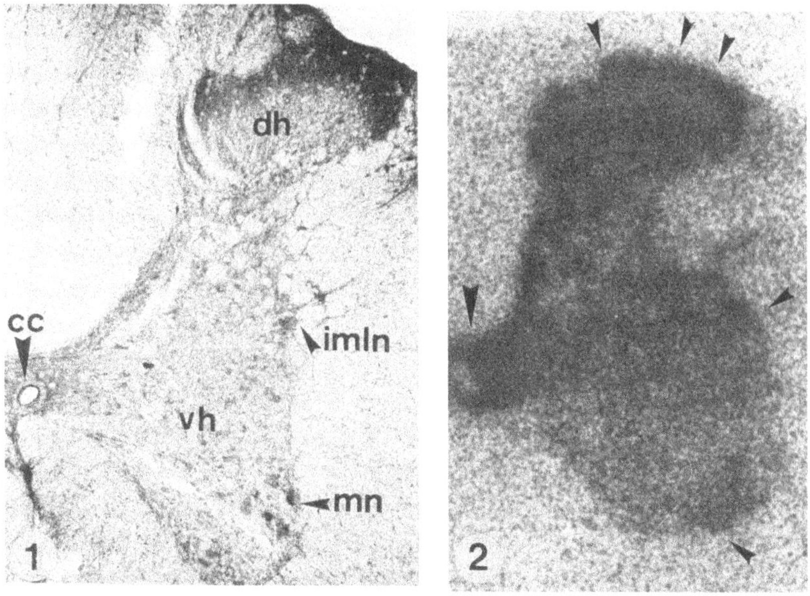


Figure 2: Cytochemical detection of substance P receptor in the rat hemi spinal cord using anti-SP anti-Id antibodies (1) or $^{125}$I-SP (2).
For immunocytochemical experiments (1), bound anti-Id were visualized using HRP-conjugated anti-rabbit Ig goat antibodies. HRP activity was revealed by the diaminobenzidine-$H_2O_2$ technique. Note the intense labeling of superficial laminae (I and II) of dorsal horn (dh) and of large ventral neurons, presumably motoneurons (mn), and the moderate staining of the region adjacent to the central canal (cc) and intermediolateral nucleus (imln). Radioautography experiment (2) was performed on cryostat sections of fresh tissue. The more intensely labeled regions are indicated by arrow heads. They correspond to those which appear immunoreactive using the anti-Id.

Working with anti-Id, we used unfractionated serum (1/50 final dilution) either in rat spinal cord or in trachea, at room temperature in the former tissue, at 4°C in the latter. For fixed sections, incubations lasted for 15 hrs but much shorter incubation times (1 to 2 hrs) were preferred for totally unfixed sections of trachea. The use of the DEAE-purified Ig fraction did not appear to increase the specific labeling by anti-Id. As for incubations of spinal cord sections with anti-CP abs, they were performed for 15 hrs at 4°C, using Ig purified by the caprylic acid method at 4 µg/ml final dilution.

Following extensive washings in buffer, bound primary abs (anti-Id or anti-CP) were visualized according to conventional techniques. In brief, tissue sections were incubated for 1hr at room temperature either in the presence of fluorescein (FITC)-conjugated sheep anti-rabbit Ig (in the case of trachea) or horseradish peroxidase (HRP)-conjugated goat anti-rabbit Ig (in the case of rat spinal cord). HRP activity was revealed by the diaminobenzidine-$H_2O_2$ technique and in some cases, staining was intensified by the method of Liposits et al.(1984). Colloïdal gold-labeled secondary antibodies (anti-rabbit Ig) were also successfully applied in spinal cord sections and staining was intensified by the silver-gold method. The use of colloïdal gold-labeled antibodies greatly reduced the diffusion processes, particularly at membrane sites.

### 3.3. Control experiments

The issue of control procedures is particularly important in receptor immunocytochemistry. Several specifity tests, which appear of general use, have been developed in our systems.

a) Replacement of anti-Id or anti-CP abs by normal rabbit serum (or Ig) or by a preimmune serum should result in no labeling.

b) Preabsorption of the anti-Id with an anti-ligand antibody (preferably but not obligatorily that which served as immunogen for preparing the anti-Id) should prevent any subsequent receptor immunolabeling. Such experiments demonstrate that the anti-Id is bound to the section by an idiotypic determinant mimicking the ligand itself. However, as also shown in our cytochemical study of SP receptor, the ratio between anti-Id and anti-ligand abs must be carefully controlled, since an excess of anti-ligand over anti-Id led to immunolabeling of endogenous peptide, whereas an excess of anti-Id over anti-ligand was inefficient in preventing receptor labeling. For example, in our spinal cord study, polyclonal anti-Id and anti-SP antisera were preincubated together at final dilutions of 1/50 and $1/10^4$, respectively.

A related and important control experiment is to check very carefully by sensitive radio- or enzymo-immunoassays that the anti-Id bleeding to be used in cytochemical experiments does not display any significant anti-ligand activity (resulting either from previously injected Ab1 abs or from auto antibodies -Ab3- developed by the animal in response to anti-Id). Obviously, such contaminations would lead to erroneous images.

c) Immunolabeling of peptide receptor by anti-Id or anti-CP abs should be prevented in the presence of an excess of the ligand or analogs. Such experiments must imperatively be carried out in the presence of a cocktail of protease inhibitors (Kummer et al., 1990a) since degradation of peptide ligands generally occurs very rapidly in the incubation media. Conversely, a preincubation of tissue sections in the presence of anti-Id or anti-CP abs should also inhibit the subsequent binding of radiolabeled neuropeptide.

In our opinion, a negative result in one of these control experiments does not necessarily imply that antibodies are not binding at the receptor site. First, the antibody may exhibit a higher affinity for the receptor than that of the ligand itself. Second, access to the receptor may be quite different for the small ligand and the antibody. Third, the domains on the receptor molecule recognized by the ligand and the anti-Id (or anti-CP) antibody may not overlap extensively. In the latter case, a ligand-protein conjugate (e.g. the immunogen used to prepare the anti-ligand abs) could be preferred to the ligand alone for preventing receptor labeling by the antibody.

Concerning SP receptor cytochemistry in the rat spinal cord, we have shown that immunolabeling with anti-PS was largely decreased in the presence of $10^{-6}$M SP or $10^{-6}$M NKA, but not $10^{-6}$M NKB, suggesting that anti-PS recognized the two closely related NK1 and NK2 receptors, but not NK3. By contrast, the labeling obtained using anti-Id could not be blocked under our experimental conditions, whereas anti-Id did partially reduce the labeling by $^{125}$I-SP in radioautographic experiments (unpublished results). In mammalian trachea, SP readily inhibited anti-Id labeling, providing that phosphoramidon, an inhibitor of neutral endopeptidase, was present in the incubation medium. Taken together, these control experiments point to a labeling of SP receptor both by anti-Id and anti-PS abs.

## 4. Examples: preliminary images of receptors

### 4.1. At the optical microscope level

In the rat spinal cord, we compared the images obtained with $^{125}$I-SP, anti-Id (Couraud et al., in press) and anti-PS abs (Figs.2 and 3). First of all, the procedures involving antibodies were simplified, since autoradiographic, unlike cytochemical, experiments required the use of protease inhibitors (40 µg/ml bacitracin, 5 µg/ml benzamidine and 3mM $MnCl_2$). Second, as seen in Fig.2, labeling was observed in the same regions i.e. dorsal horn, intermediolateral nucleus and motoneurons. However, two facts should be emphasized: 1) the background was lower with antibodies than with iodinated SP. Using anti-PS abs, it was especially reduced, allowing for the detection of immunoreactive sites in layers I, II, VII and X (Fig.3). 2) the resolution of the two techniques was quite different: although in the autoradiogram intense labeling was seen in the region of motoneurons, large immunoreactive identified neurons were clearly observed in that region (and also in laminae II to V), when using anti-SP receptor antibodies (Fig.3).

In the guinea-pig and human trachea, the immunocytochemical approach using anti-Id also proved to be superior to autoradiography discrimining labeled elements, especially in the mucosal layer and the lamina propria (Kummer et al., 1990a).

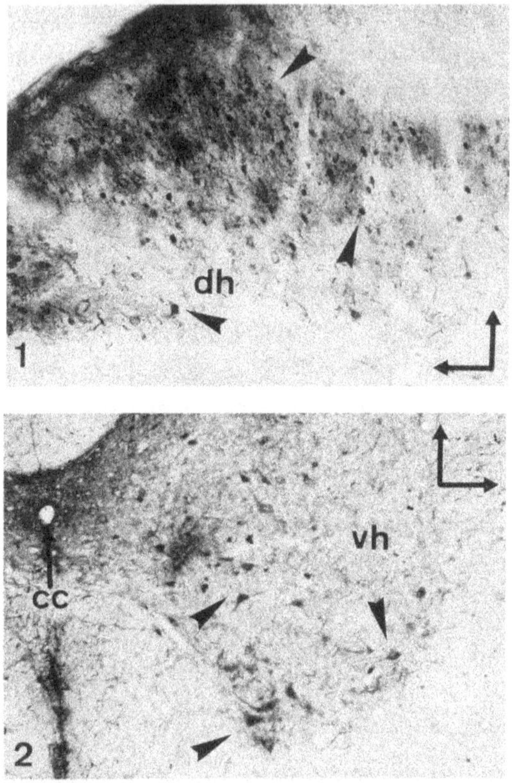


Figure 3: Detection of SP receptor-like immunoreactivity in the rat spinal cord at the O.M. level using anti-CP antibodies.
Experiment was carried out on 50 $\mu$m vibratome sections of paraformaldehyde fixed tissue. Bound anti-CP were visualized using colloïdal gold (1nm)-labeled anti-rabbit Ig antibodies. Staining was intensified by the silver-gold method. (1): dorsal horn (dh). Several cell bodies and fibers appear immunoreactive in the superficial layers, especially in lamina II. (2): ventral horn (vh). Immunoreactive cells (arrow heads) are clearly visible in all layers, around the central canal (cc) and in layer IX, where they could correspond to motoneurons. Immunoreactive fibers are abundant in layer X. Perpendicular arrows indicate the dorsal (vertical arrows) and lateral (horizontal arrows) orientations.

## 4.2. At the electron microscope level

As mentioned above, cytochemical demonstration of peptide receptors at the ultrastructural level is considerably hindered by the instability of most endogenous ligand or of ligand-receptor interactions. Various techniques are being developed to solve these major problems and some are reviewed in this volume. We show here that

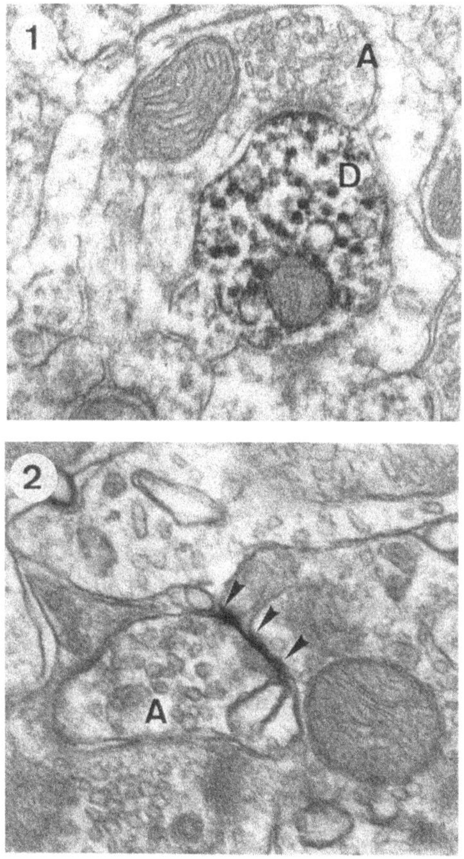


Figure 4: Detection of SP receptor-like immunoreactivity in the rat spinal cord at the E.M. level using anti-CP antibodies.
Bound anti-CP were visualized using HRP-labeled anti-rabbit Ig antibodies (see legend to Fig.2). (1): intracellular labeling in a postsynaptic dendrite (D). The presynaptic axon (A) is devoid of labeling. The enzymatic end-product in D appears associated with neurotubules. (2): membrane labeling over an apposition between an axon (A) and a neurite (probably also an axon). It is difficult here to ascertain whether the apposition corresponds to a synaptic zone between the two elements.

stable and specific anti-receptor antibodies may be very useful tools in this context, since very few images of peptide receptors at the E.M. level (and none at all of SP receptor to our knowledge) have yet been published.

From our first ultrastructural observations using anti-PC abs in spinal cord (Fig.4), two points may be emphasized: 1) SP receptor-like immunoreactivity appeared localized mostly in non-synaptic zones of the membranes, confirming preliminary observations using anti-SP anti-Id abs (Conrath et al., 1988b) and 2) part of the labeling often appeared inside the cytoplasm, either in cell bodies or in dendrites along neurotubules.

The physiological significance of these results is discussed elsewhere (Zouaoui et al., this volume).

## Conclusions

Anti-receptor anti-idiotypic and anti-complementary peptide antibodies appear to be powerful tools for studying biological receptors. The former have been used until now not only for cytochemical (Knigge et al., 1989) but also for biochemical and pharmacological characterization of a limited number of receptors (Strosberg, 1989). The use of anti-receptor antibodies in vivo as therapeutic reagents also seems an important challenge for the near future. As briefly shown here, the use of anti-receptor antibodies in cytochemical studies offers to the experimentator several advantages including 1) the simplicity of experimental protocols, 2) greater resolution at the light microscope level and 3) the possibility of extending the observations to the E.M. level. The latter appears especially promising since such immunological probes, if used with proper controls, might now reveal fundamental information concerning the metabolism (synthesis vs degradation) and the intracytoplasmic pathways for transport and internalisation of cellular receptors. Owing to the possibility of comparing the fine distribution of given ligands, using anti-ligand idiotypic abs, with that of their respective receptors, anti-Id or anti-CP antibodies might also help to clarify outstanding problems in neurotransmission, such as the localisation of receptors in extra-synaptic regions and removed from the releasing terminals.

## Acknowledgments

This work was supported by a financial grant from CEA/IRF. We thank the Fondation Singer-Polignac for financial support to D. Zouaoui.

## References

Blalock JE, Whitaker JN, Benveniste EN, Bost KL (1989) Use of peptides encoded by complementary RNA for generating anti-idiotypic antibodies of predefined specificity. In Antibodies, antigens and molecular mimicry. Methods in Enzymology 178: 49-63 Langone J (ed) Academic Press San Diego New York

Bost KL, Smith LR, Blalock JE (1987) A molecular recognition code: its use for purification of ACTH, endorphin and LHRH receptors. In Anti-idiotypes, receptors and molecular mimicry 35-44 Linthicum S, Farid N (eds) Springer Verlag New York Berlin Heidelberg

Cleveland WL, Wasserman NH, Sarangarajan R, Penn AS, Erlanger BF (1983) Monoclonal antibodies to the acetylcholine receptor by a normally functioning auto-anti-idiotypic mechanism. Nature 305: 56-57

Conrath M, Couraud JY, Pradelles PP ($1988_a$) Anti-idiotypic antibodies as a tool for cytochemical identification of substance P receptors in the central nervous system. J Histochem Cytochem 36: 1397-1401

Conrath M, Couraud JY, Cupo A ($1988_b$) Immunocytochemical study of delta opïoid and substance P receptors in the central nervous system using anti-idiotypic antibodies (abstract). Int. Congress on Histochemical Techniques, Varna (Sept 1988)

Couraud JY, Esher E, Regoli D, Imhoff V, Rossignol B, Pradelles P (1985) Anti-substance P anti-idiotypic antibodies: characterization and biological activities. J Biol Chem 260: 9461-9469

Couraud JY, Frobert Y, Conrath M, Renzi D, Grassi J, Drapeau G, Regoli D, Pradelles P (1987) Monoclonal antibodies to substance P: production, characterization of their fine specificity and use in immunocytochemistry. J Neurochem 49: 1708-1719

Couraud JY, Maillet S, Grassi J, Frobert Y, Pradelles P (1989) Characterization of antisubstance P anti-idiotypic antibodies. In Antibodies, antigens and molecular mimicry. Methods in Enzymology 178: 275-300 Langone J (ed) Academic Press San Diego New York

Couraud JY, Maillet S, Conrath M, Calvino B, Pradelles P (1990) Use of anti-idiotypic antibodies as probes for in vitro and in vivo identification of substance P receptors. Mol Chem Neuropathology, in press

Jerne NK (1974) Towards a network theory of the immune system. Ann Immunol 125C: 373-389

Jessell TM, Womack MD (1985) Substance P and the novel mammalian tachykinins: a diversity of receptors and cellular actions. Trends Neurosci 8 : 43-45.

Knigge KM, Piekut DT, Abood LG, Joseph SA, Michael GJ, Xin L, Berlove DJ (1989) Immunocytochemistry of receptors using anti-idiotypic antibodies. In Antibodies, antigens and molecular mimicry. Methods in Enzymology 178 : 212-221 Langone J (ed) Academic Press San Diego New York.

Köhler H, Kaveri S, Kieber-Emmons T, Morrow WJW, Muller S, Raychaudhuru S (1989) Idiotypic networks and nature of molecular mimicry: an overview. In Antibodies, antigens and molecular mimicry. Methods in Enzymology 178 : 3-35 Langone J (ed) Academic Press San Diego New York.

Kummer W, Fischer A, Preissler U, Couraud JY, Heim C (1990a) Immunohistochemistry of the guinea-pig trachea using an anti-idiotypic antibody recognizing substance P receptors. Histochemistry 93 : 541-546.

Kummer W, Fischer A, Couraud JY, Heim C (1990b) Immunohistochemistry of peptides (substance P and VIP) and peptide receptors in the human trachea (abstract). Autonomic Nervous System Symposium, Oxford (Sept. 1990).

Liposits Z, Setalo G, Flerko B (1984) Application of the silver gold intensified 3-3' diaminobenzidine chromogen to the light and electron microscopic detection of the luteinizing hormone-releasing hormone system in the rat brain. Neuroscience 13: 513-520

Maillet S, Imhoff V, Rossignol B, Grassi J, Pradelles P, Couraud JY (1990) Anti-substance P anti-idiotypic antibodies modulate the secretory process in the rat parotid gland in vitro. Eur J Pharmacol, in press.

Nawa H, Hirose T, Takashima H, Inayama S, Nakanishi S (1983) Nucleotide sequences of cloned cDNAs for two types of bovine brain substance P precursor. Nature 306 : 32-36.

Pascual DW, Blalock JE, Bost KL (1989) Antipeptide antibodies that recognize a lymphocyte substance P receptor. J Immunol 143 : 3697-3702.

Shigemoto R, Yokota Y, Tsuchida K, Nakanishi S (1990) Cloning and expression of a rat neuromedin K receptor cDNA. J Biol Chem 265 : 623-628.

Smith LR, Bost KL, Blalock JE (1987) Generation of idiotypic and anti-idiotypic antibodies by immunization with peptides encoded by complementary RNA: a possible molecular basis for the network theory. J Immunol 138 : 7-9.

Sommer EW, Kazimierczak J, Droz B (1985) Neuronal subpopulations in the dorsal root ganglion of the mouse as characterized by combination of ultrastructural and cytochemical features. Brain Res 346 : 310-326.

Strosberg AD (1989) Interaction of anti-idiotypic antibodies with membrane receptors: practical considerations. In Antibodies, antigens and molecular mimicry. Methods in Enzymology 178 : 179-190 Langone J (ed.) Academic Press San Diego New York.

# NEW APPROACHES FOR RECEPTOR ANALYSIS: HISTOCHEMISTRY OF THE GABA/BZ RECEPTORS IN THE RAT CNS

J-M Séquier* and J-G Richards
Pharmaceutical Research Dept.
F. Hoffmann-La Roche & Co., Ltd
CH-4002 Basel
Switzerland
*Merck Sharp and Dohme Research Laboratories
Neuroscience Research Centre, Terlings Park
Eastwick Road, Harlow, Essex, CM20 2QR

Benzodiazepines (BZ) were fortuitously discovered in the late fifties and are a group of centrally acting drugs which are the most frequently prescribed to treat anxiety, insomnia, epilepsy and muscle tension (Haefely et al., 1981). Initially, research concerning their mechanisms of action was quite unsuccessful until gamma aminobutyric acid (GABA) was recognized to be an important inhibitory neurotransmitter in the central nervous system (Roberts, 1986). The finding that diazepam enhanced GABA-mediated presynaptic inhibition in the spinal cord was decisive in revealing the link between benzodiazepines and GABAergic innervation (Schmidt et al., 1967; Polc et al.,1974). With the use of radiolabelled diazepam in 1977 it was possible to identify high affinity binding sites for benzodiazepines in CNS neuronal membranes (Squires and Braestrup, 1977; Moehler and Okada, 1977). The modulation of GABA binding by benzodiazepines confirmed the link between these drugs and the inhibitory neurotransmitter (Costa et al., 1978). The $GABA_A$/BZ receptor is a ligand gated chloride channel that mediates synaptic inhibition in the mammalian brain (Haefely et al., 1985). Its activation by agonists such as GABA or muscimol generally lead to the hyperpolarisation of the cell membrane by increasing chloride conductance and results in an increase of the threshold for the production of action potentials. This activation can be specifically

NATO ASI Series, Vol. H 58
Neurocytochemical Methods
Edited by A. Calas and D. Eugène

blocked by bicuculline. This receptor is allosterically modulated by a widely range of drugs such as benzodiazepines or barbiturates or by convulsant drugs such as picrotoxin and t-butylbicyclophosphorothionate (TBPS) which bind directly to the chloride channel. Receptor solubilisation and purification has lead to a better characterisation of the $GABA_A$/BZ receptor complex and allowed the generation of antibodies with different specificities (Sigel et al., 1983; Schoch et al., 1985). Primary sequences of potential subunits of the receptor have been obtained by molecular cloning and it was then possible not only to elucidate the possible architecture of the receptor in the membrane but also to express it and to study its biophysical and pharmacological properties (Schofield et al., 1987). The heterogeneity revealed by molecular cloning is greater than previously suspected (Levitan et al., 1988). Although protein purification indicated that the receptor consists of two major proteins, it now appears that at least three different subunits designated as $\alpha$, $\beta$ and $\gamma$ are necessary to build a functional $GABA_A$/BZ receptor (Pritchett et al., 1989). These three classes of subunits appeared to be composed of several variants suggesting the possibility of a very high receptor diversity (Schofield, 1989). The heterogeneity of $GABA_A$/BZ receptors was suggested several years ago on the basis of results using selective ligands or photolabeling experiments, however the high diversity of $GABA_A$ receptor subunits revealed by molecular cloning tends to indicate that this heterogeneity is far greater than previously thought (Sieghart, 1989). In order to better clarify the mechanisms of action of benzodiazepines it was important to identify brain regions and cells that contain $GABA_A$/BZ receptors and which are therefore not only potential targets for tranquillising drugs but also likely to have a reduced firing rate after GABA release.
In this chapter we will review the localisation and the sites of gene expression of $GABA_A$ receptors in the rat brain. Results obtained with three different histochemical techniques will be presented.

## QUANTITATIVE RECEPTOR AUTORADIOGRAPHY

Compared to biochemical assays sensitivity and anatomical resolution are the two main advantages of autoradiographic methods. These methods consisted originally of in vivo experiments where radioligand was injected into the animal. An in vitro approach for the microscopic localisation of receptors was developed in 1979 and enabled the mapping of binding sites for several neurotransmitters, hormones and drugs, including benzodiazepines (Young and Kuhar,1979; Kuhar et al., 1986). The subsequent use of tritium-sensitive film made the quantification of autoradiographic results much easier (Palacios et al., 1981).

We have investigated the distribution of benzodiazepine receptors in rat brain after in vitro and in vivo binding using, respectively, $^3$H-methylclonazepam (agonist) and $^3$H-flumazenil (antagonist). When benzodiazepine receptor ligands e.g. methylclonazepam, flumazenil or β-CCM (methyl ester of β-carboline-3-carboxylate) were used as radioligands, sections were incubated at 4°C for 40 min. in 170mM Tris/HCl pH 7.7 in the presence of 1nM of radioligand. They were then washed for 2 min. in cold buffer, dipped in cold water, dried and exposed to sheet film for 3 weeks. When muscimol was used as a radioligand the buffer consisted of 310mM Tris/citrate pH 7.1. In this case, sections were first preincubated in buffer at 4°C for 20 min. then incubated for 40 min. with 5nM of radioligand and washed for 1 min. in cold buffer. When TBPS was used as the radioligand, a 15 min preincubation at room temperature was carried out in 50 mM Tris/HCl pH 7.5, 200mM KCl, 1mM EDTA before a 90 min. incubation at room temperature with 20nM of radioligand. In this case sections were washed at 4°C for 10 min. The in vivo binding procedure consisted of an intraveinous injection in the tail of 20μCi of $^3$H-flumazenil in 100μl of 0.9% NaCl. The animal was sacrificed after 5min., the brain rapidly removed and frozen. The autoradiographic images obtained after in vitro binding revealed a very specific and heterogeneous distribution of binding sites in brain regions known to receive a GABAergic innervation e.g. the external plexiform layer of the olfactory bulb, ventral pallidum, hippocampal formation and substantia nigra (Fig.1-2). The distribution of the methylclonazepam binding sites is analogous to that obtained with other benzodiazepine receptor ligands, such as flunitrazepam, clonazepam, flumazenil and also β-CCM (Fig.3). Furthermore the analogy between the distribution of binding sites obtained after in vivo and in vitro binding suggested that the in vitro method is suitable to reveal receptors whose occupancy by benzodiazepine is responsible for their pharmacological effects. The binding sites

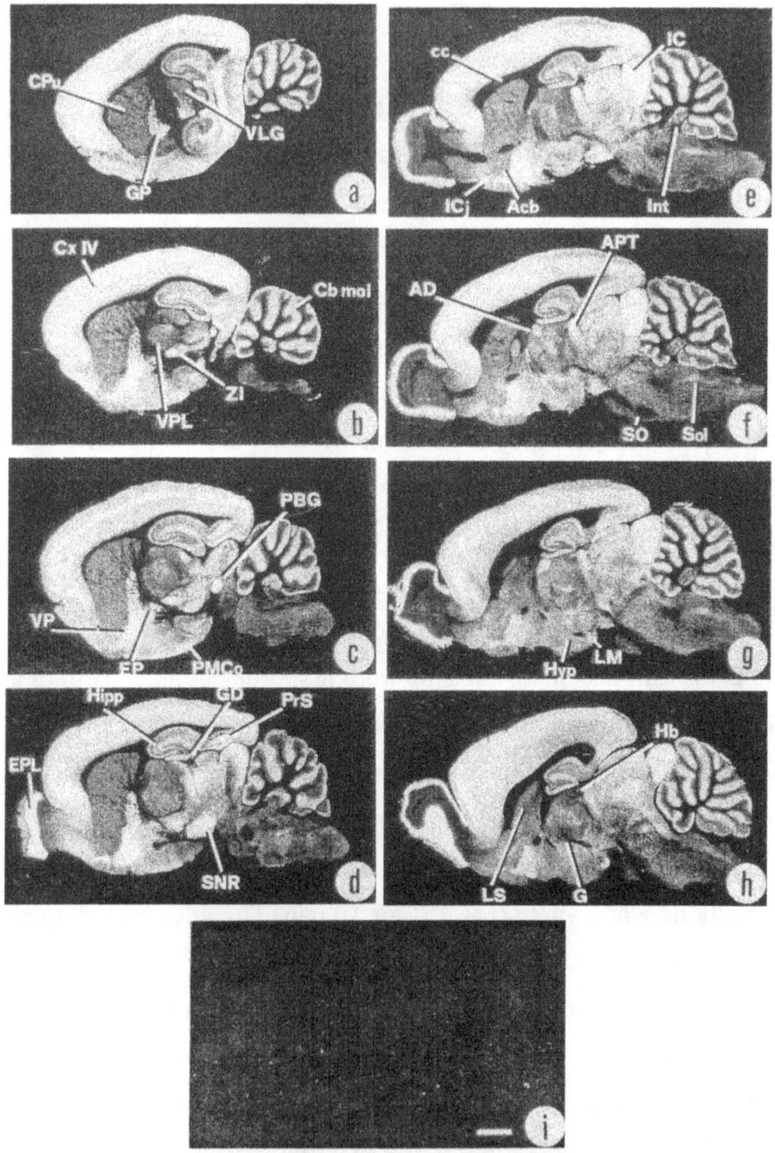


Fig.1: (a-h) Distribution of $^{3}$H-methylclonazepam binding sites on parasagittal sections of rat brain after in vitro binding. (i) Non-specific binding obtained in presence of 1$\mu$M of displacer. bar=2mm.

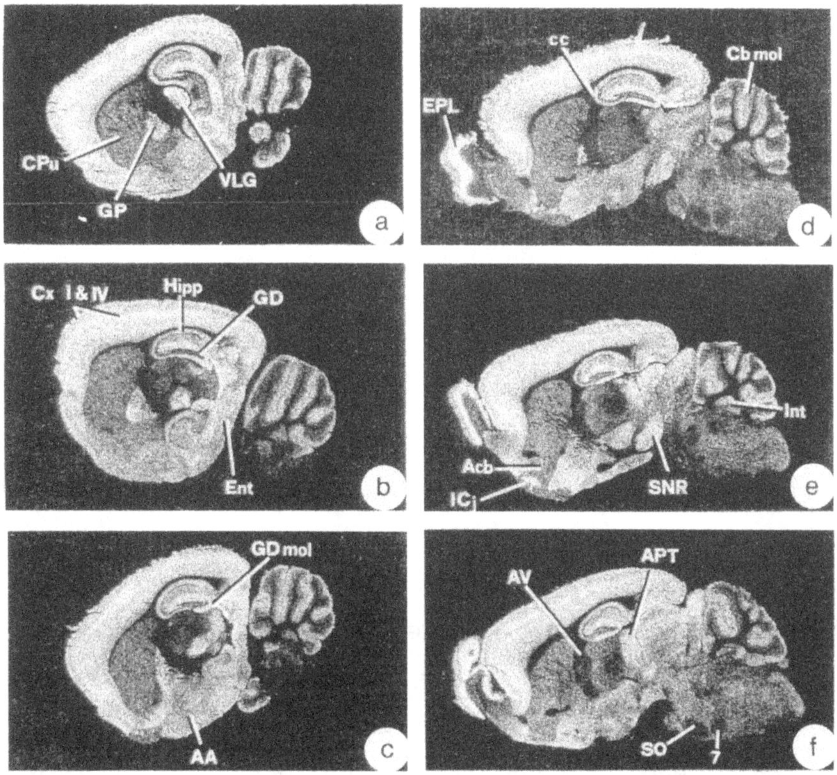


Fig.2: Distribution of $^3$H-flumazenil binding sites on parasagittal sections of mouse brain after in vivo binding.

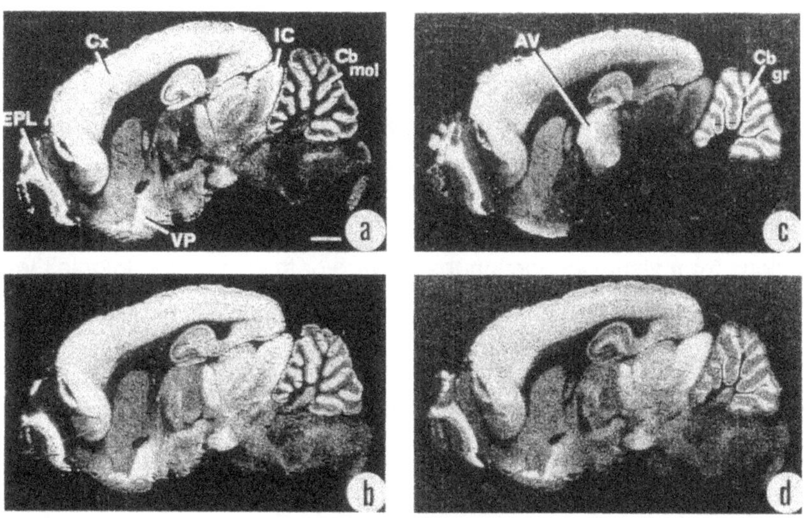


Fig.3: Distribution of binding sites for $^3$H-flumazenil (a), $^3$H-βCCM (b), $^3$H-muscimol (c) and $^{35}$S-TBPS (d) after in vitro binding. bar=2mm.

were mainly found in areas of cellular contacts supporting the idea that the radioligand is indeed located at its receptor. The distribution of benzodiazepine receptors matched almost perfectly with that of other $GABA_A$ receptor ligands, for example TBPS, with the exception of the cerebellum where the granule cell layer appears to contain more TBPS binding sites than the molecular layer (Fig.3). In the case of muscimol binding the same cerebellar pattern appeared but in general the analogy of brain distribution is less perfect. Muscimol binding sites were found in the thalamus and caudate-putamen while pallidal regions and colliculi contain much less but are rich in benzodiazepine binding sites (Fig.3). These muscimol binding sites are thought to correspond to high affinity $GABA_A$ receptors while benzodiazepine receptors are linked with low affinity $GABA_A$ receptors.The quantification of the autoradiograms was accomplished using appropriate tissue radioactivity standards and a computerised image analyser (ASBA, Wild Leitz). The radioactivity of these standards made from brain paste was determined accurately by sample combustion and were co-exposed with the tissue sections (Fig.4). Using an iodinated benzodiazepine as a radioligand we did not detect significant tissue quenching (Fig.4), a phenomenon which has been described for tritiated ligands (Kuhar and Unnerstall, 1985). The same density ratios were obtained with each radioligand. After calibration of the densitometric system, we were able to determine with a high anatomical resolution the binding characteristics of these ligands.These characteristics in brain sections were similar to those determined in synaptosomal membranes i.e. the specific binding was saturable, of high affinity and was increased by GABA. The regional affinities and binding capacities are in good agreement with those reported for membrane homogenates (Fig.6). Even though it was obvious from visual inspection that in vitro or in vivo methodologies lead to the same distribution pattern of binding sites we were able after quantification to show a good correlation for the binding site densities of a dozen brain regions after in vitro and in vivo binding (Fig.5). Autoradiographic studies have also been used to study receptor heterogeneity. While it is clear that $GABA_A$ receptors are heterogeneous, $GABA_A$/BZ receptors are divided into at least two groups, BZ1 and BZ2 receptors, since the triazolopyridazine CL 218,872 shows selectivity for the BZ1 type (Squires et al., 1979). Using in vitro competition binding experiments with this compound we

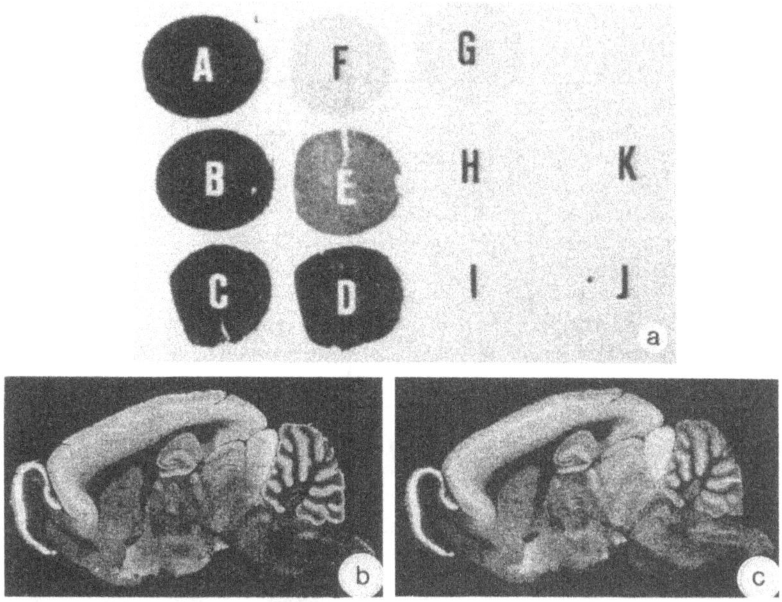

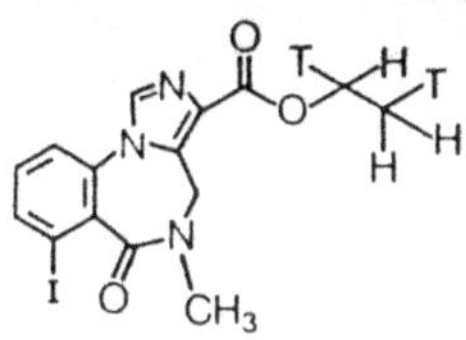

Tritium-labelled ligand

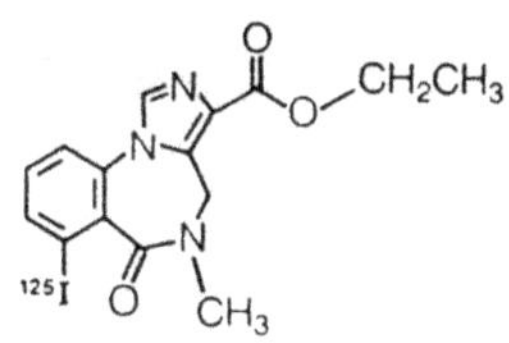

$^{125}$Iodine-labelled ligand

Fig.4: (a) Radioactive brain paste standards after 3 weeks exposure. (Activities in μCi/g tissue: A=41.5; B=21.9; C=10.2; D=5.78; E=2.88; F=1.35; G=0.59; H=0.36; I=0.16; J=0.07; K=0.04). Comparison of the distributions and densities of binding sites for a tritiated (b) or iodinated (c) benzodiazepine.

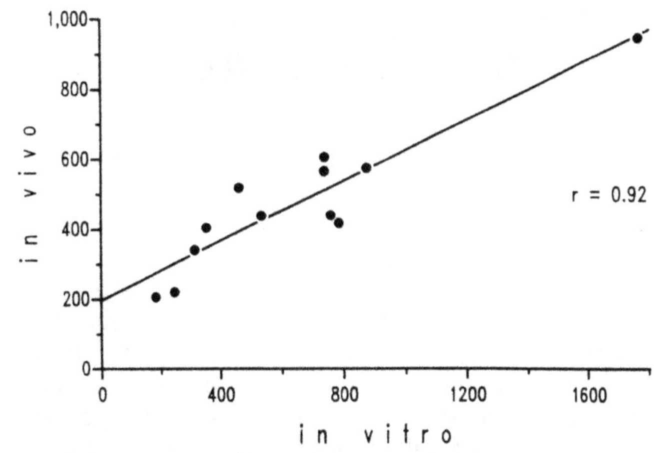


Fig.5: Comparison between regional densities obtained after in vitro binding with $^3$H-methylclonazepam and in vivo binding with $^3$H-flumazenil. These densities were obtained by densitometric measurements

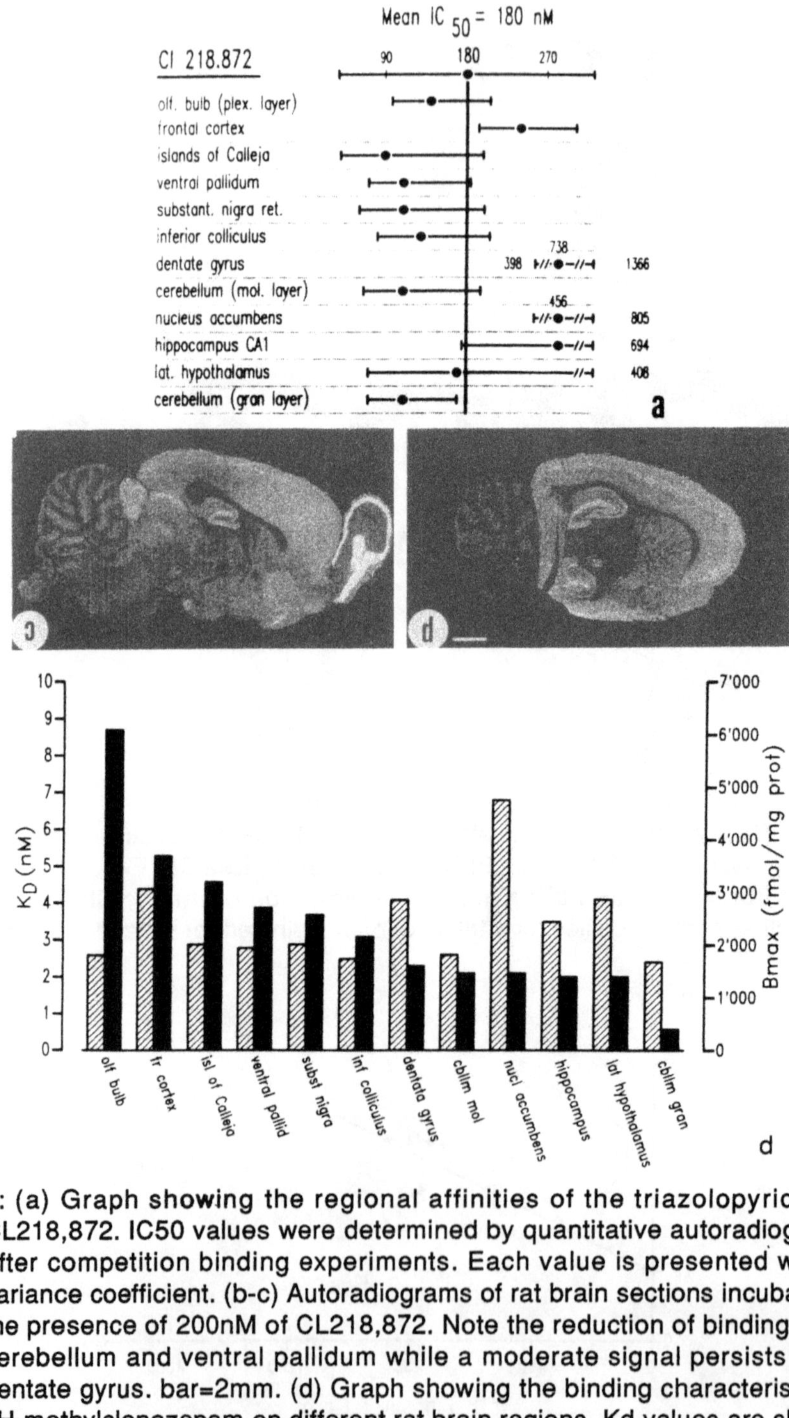


Fig.6: (a) Graph showing the regional affinities of the triazolopyridazine CL218,872. IC50 values were determined by quantitative autoradiography after competition binding experiments. Each value is presented with its variance coefficient. (b-c) Autoradiograms of rat brain sections incubated in the presence of 200nM of CL218,872. Note the reduction of binding in the cerebellum and ventral pallidum while a moderate signal persists in the dentate gyrus. bar=2mm. (d) Graph showing the binding characteristics of $^{3}$H-methylclonazepam on different rat brain regions. Kd values are showed by hatched bars, Bmax by shaded bars. These values have been obtained after quantification of the autoradiograms.

were able to detect inhibition of binding in cerebellum, substantia nigra and pallidal regions thought to be enriched in the BZ1 type of receptors (Fig.6). We were also able to determine by quantification the regional IC50 values of CL 218,872. A maximum ratio of 8 has been found between IC50 in BZ2 enriched regions like dentate gyrus and BZ1 regions (Fig.6). This compound was initially claimed to be only anxiolytic, but appears to be as sedative as the usual benzodiazepine agonists. Some hypnotics like quazepam and zolpidem have the same selectivity as CL 218,872 suggesting that occupancy of BZ1 receptors might be responsible for the sedative effects while BZ2 receptors might trigger the anxiolytic and anticonvulsant effects. So far it has not been possible so far to correlate the regional binding characteristics of a ligand with its pharmacological profile.

## IMMUNOHISTOCHEMISTRY

Despite the good anatomical resolution of radiohistochemical techniques they still have some limitations, especially when one is interested in the subcellular localisation of receptors. Benzodiazepine photoaffinity ligands and electron microscopy have been used to identify the subcellular targets of the drug but the problem of radiation spread and non-specific binding clearly limits the resolution of this radiohistochemical approach (Moehler et al., 1980;1981). To overcome these problems we used an immunohistochemical approach with antibodies directed against the $GABA_A$/BZ receptor combined with the peroxidase-anti peroxydase method (Sternberger, 1979).

Benzodiazepine receptors have been purified from rat brain by benzodiazepine affinity chromatography as previously described (Schoch et al., 1984; Richards et al., 1988). The eluted material was analysed on denaturing polyacrylamide gel electrophoresis and consisted of a mixture of proteins having apparent molecular weight from 47 to 60kDa and even larger (Fig.7). The two major bands recognised as $\alpha$ and $\beta$ by the monoclonal antibodies bd28 and bd17, respectively, were found at 49 and 57 kDa (Haering et al., 1985). This purified preparation was used for raising polyclonal antibodies. Two antisera were tested on ELISA assays and on western blots (Fig.7). One of them recognised strongly a low molecular weight band

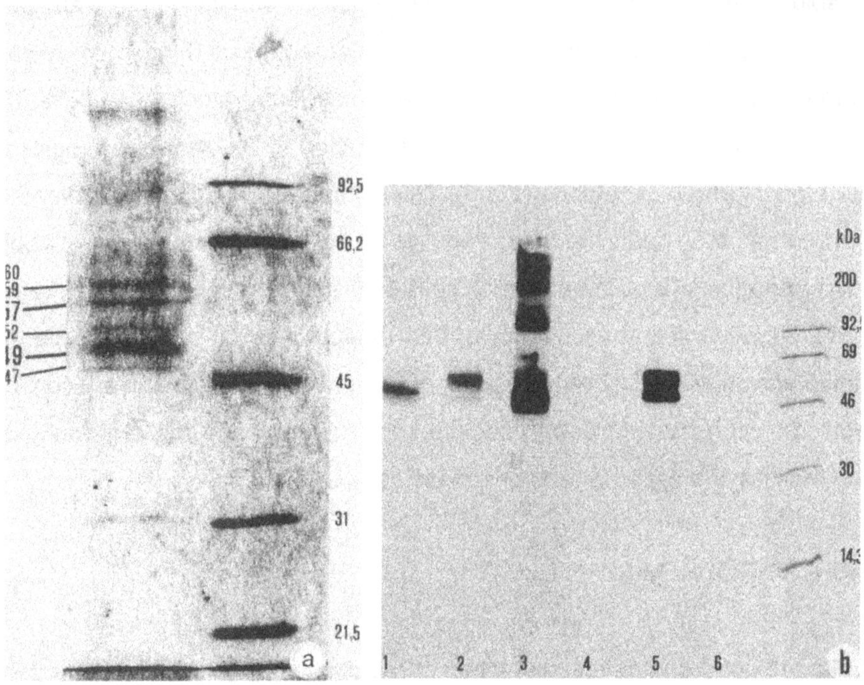


Fig.7: (a) SDS-Polyacrilamide gel of purified GABA/BZ receptors after electrophoresis and Coomassie blue staining. Apparent molecular weight of the receptor proteins are indicated on the left. (b) Immunoblot of purified GABA/BZ receptor using 1:bd28; 2:bd17; 3:Antiserum 09; 4:Preimmune serum 09; 5:Antiserum 08; 6:Preimmune serum 08.
Standard molecular weights are indicated on the right side of the figures.

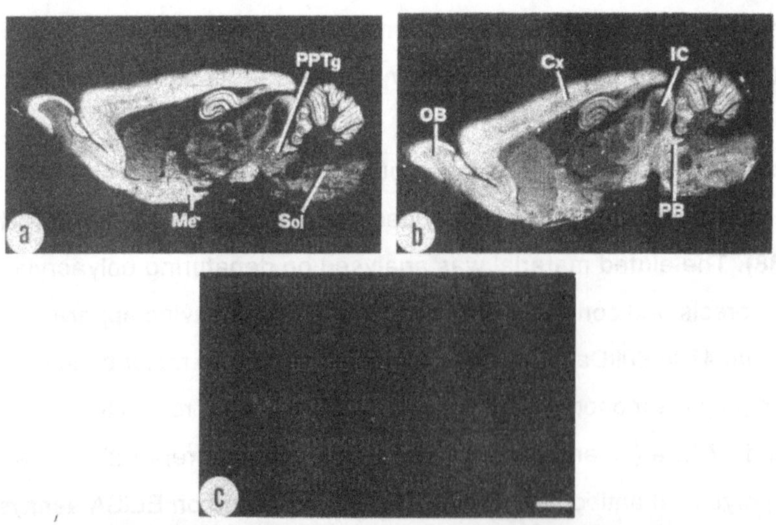


Fig.8: Distribution of the antigenic sites on parasagittal sections of rat brain revealed by the antiserum 09 (a) and 08 (b). (c) Signal obtained using preimmune serum 09 as primary antibody. bar=2mm.

which was recognised also by bd28, in addition to higher molecular weight entities that were thought to be heteropolymers containing the α subunit. The second antiserum recognised two proteins revealed by the two monoclonal antibodies and therefore thought to be α and β. Using the indirect peroxidase-antiperoxidase method, we tested these two antisera immunocytochemically. They both gave a similar distribution of immunoreactivity on rat brain section (Fig. 8). Both preimmune sera did not recognise any protein on western blots nor reveal antigenicity on brain sections (Fig.7-8). We investigated further the reaction product obtained with the α-specific antiserum (Antiserum 09 diluted at 1/2500). A strong immunoreactivity was found in the olfactory bulb, cerebral cortex, ventral pallidum, hippocampus, substantia nigra and cerebellum while a weaker reaction occured in the superior colliculus, thalamus and caudate-putamen (Fig.8). In fact most of the immunoreactivity was found in regions of cell contacts such as the external plexiform layer of the olfactory bulb or the dendritic fields of the pyramidal cells of the hippocampus (Fig.9). Even if most of the immunoreactivity has present on cellular processes such as in the substantia nigra or ventral pallidum, it was also possible to detect some reaction product around cell bodies e.g. in the Islands of Calleja or in the anterior olfactory bulb (Fig.9). In the cerebellum, the immunoreactivity was found on fibres in the molecular layer while a very dense reaction product, which probably corresponds to glomeruli, occured in the granule cell layer. At the subcellular level, receptor antigenic sites were found mainly at axo-dendritic synapses (Fig.10). The reaction product was always seen at pre- and post-synaptic membranes. We did not detect any immunoreactivity in the intracellular compartment. In the case of the granule cells of the cerebellum, a strong immunoreactivity was found at soma-somatic contacts where no GABAergic terminals occur.

The macroscopic distribution of the immunoreactivity detected with our antiserum matched the distribution of benzodiazepine binding sites with some exceptions, such as: i)the colliculi, which appear poorly antigenic compared to the amount of binding sites they contain; ii)the ventral portion of the molecular layer of the dentate gyrus which is not labelled immunocytochemically; iii)the granule cell layer of the cerebellum which seems to be more antigenic than expected from the autoradiographic study. It is likely that our antiserum detects a high proportion of the

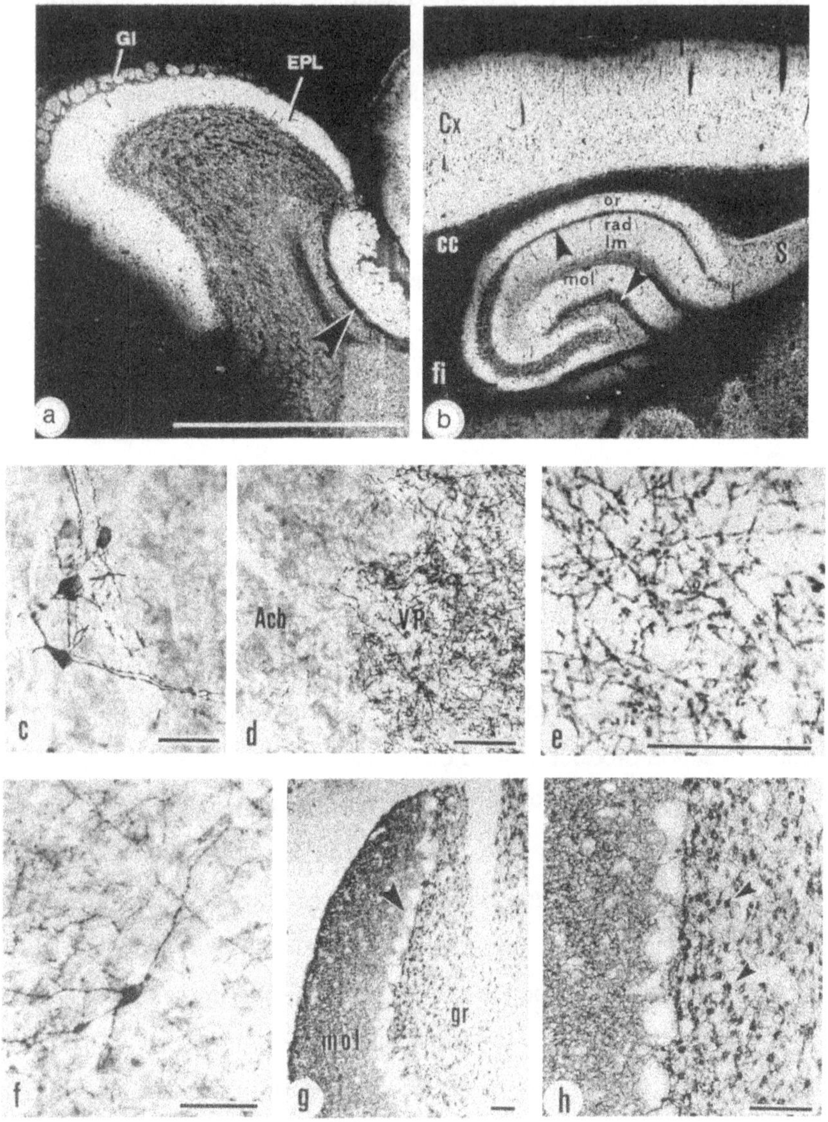


Fig.9: (a) Immunoreactivity in the olfactory bulb. Note the intense staining in the external plexiform layer as well as in the glomerular layer. (b) Immunoreactivity in the cerebral cortex and hippocampus. Note the absence of reaction product over the pyramidal and granule cell layers (arrows). bar=2mm.

Microscopic images of the distribution of the immunoreactivity in the Islands of Calleja (c), nucleus accumbens and ventral pallidum (d), substantia nigra (e), anterior olfactory bulb (f) and cerebellum (g-h). Note the absence of immunoreactivity on the Purkinje cells (g) and the intense reaction product in the granule cell layer (h). bar=50µm.

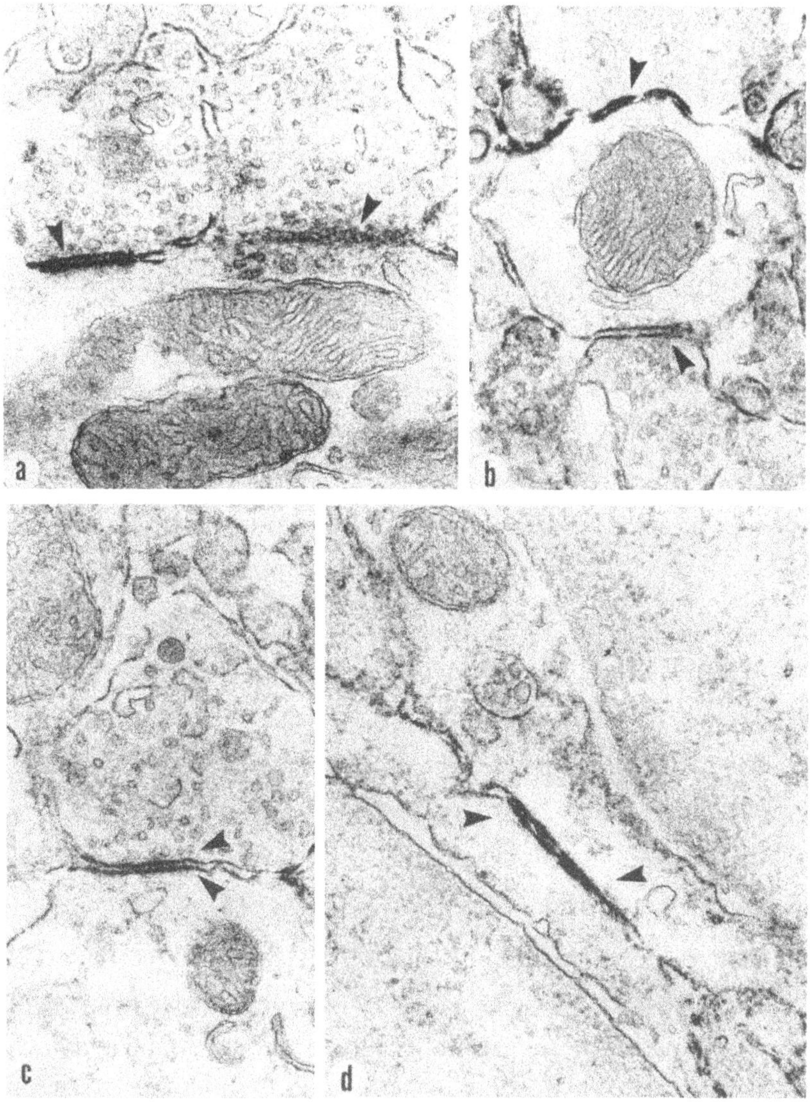


Fig.10:Electron microscopic distribution of the immunoreactivity in the cerebellum (a & d), cerebral cortex (b) and ventral pallidum (c). Immunoreactivity was found at axo-dendritic synapses both at pre and post synaptic membranes (a-c) but also at the level of soma-somatic contacts between granule cells (d). magnification # X75 000.

$GABA_A$/BZ receptors, but not all of them and we can not exclude the possibility that recognition of some $GABA_A$ receptors not linked with benzodiazepine receptors occurs, such as those in the cerebellum. These immunocytochemical results were comparable to the findings obtained using the β-specific monoclonal antibody bd17 at both the light and electron microscopic levels (Richards et al., 1987; Somogyi et al.,1989).

IN SITU HYBRIDISATION HISTOCHEMISTRY

Because of the availability of cDNA sequences for the rat $GABA_A$ receptor subunits (Malherbe et al., 1990a), it is now possible to use this final approach to study the cellular expression of these receptors. For this purpose we established a very stringent protocol using $^{35}S$-labeled cRNA probes. This protocol has been reported elsewhere and turned out to be reliable for other markers (Séquier et al., 1988;1990).

A very specific hybridisation signal was obtained whichever probe was used. The non-specific signal obtained with a sense probe was always very low (Fig.11). When tested on kidney sections, the α1 and β1 cRNA probes did not reveal any signal confirming the absence of $GABA_A$ receptor expression in this tissue (data not shown). When one of the radioactive probes was hybridised in the presence of a large excess of the other cold probe, no reduction of hybridisation signal was observed suggesting that no cross-hybridisation occured during the procedure despite the high sequence homology of the two probes (data not shown).

Surprisingly, the expression patterns of the α1 and β1 subunits visualised by autoradiography were not completely identical (Fig.11). α1 mRNA was found to be more abundant than β1 mRNA. For both subunits, a hybridisation signal was found in olfactory bulb (mitral cells), cerebral cortex, hippocampus and cerebellum. At higher magnification it was clear that despite this apparent similarity some differences existed. For example in the hippocampus, apart from the pyramidal and granule cells which express both subunits, several cells were found in stratum oriens and radiatum to be only α1 positive, while β1 positive cells were limited to the hilus. In the cerebellum, Purkinje cells as well as cells in the molecular layer clearly

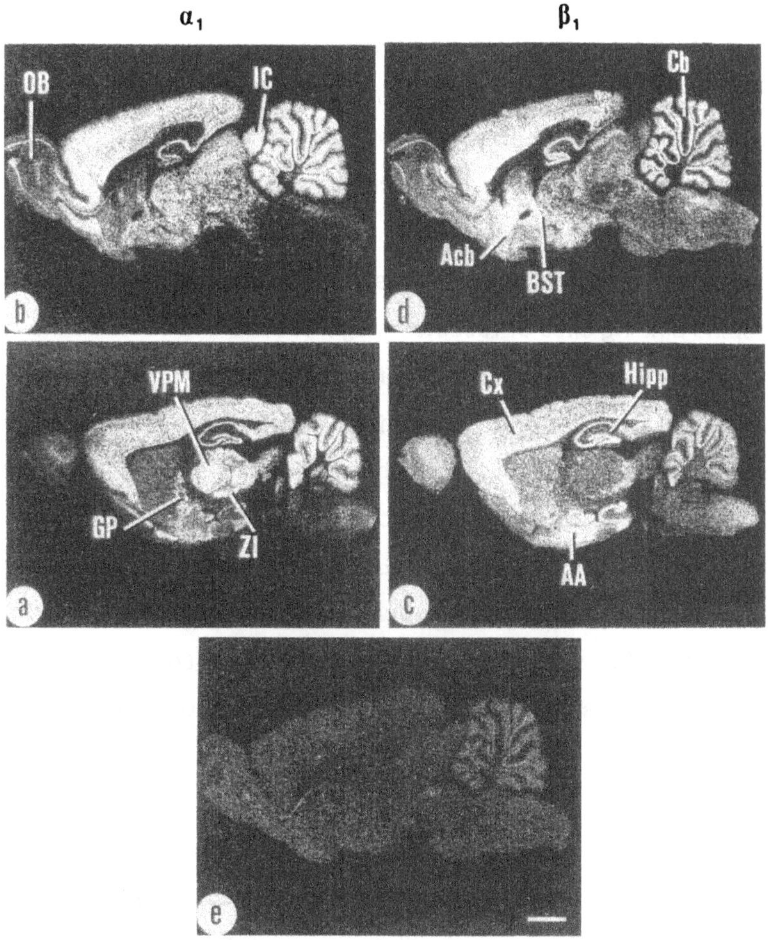


Fig.11:In situ hybridisation of the α1 subunit (a-b) and β1 subunit (c-d) on rat brain sections using cRNA probes. (e) Autoradiographic signal obtained with a sense probe as control. bar=2mm.

expressed α1 but not β1 (Fig.13). The same difference was noticed in the nucleus interpositus. More obvious differences were detected in the thalamus, inferior colliculus (Fig.11) and substantia nigra (Fig.12) and in the caudate-putamen, where a higher expression of β1 over α1 was found.

According to previous results obtained with the above mentioned approaches it is obvious that the α1 subunit is one of the main components of the $GABA_A$/BZ receptor because its mRNA is found in cells or brain regions known to contain either benzodiazepine binding sites or receptor antigenic sites and because the α subunit is considered to be the subunit to which benzodiazepines bind (Moehler et al., 1980; Sigel et al., 1983). Concerning β1, it is difficult to correlate its expression pattern with radiohistochemical and immunocytochemical results especially those obtained with the β-specific monoclonal antibody bd17 (Richards et al., 1987). These in situ hybridisation results suggested that $GABA_A$/BZ receptors are probably not composed exclusively of α1 and β1 subunits. In some cells the two subunits are clearly co-expressed and might be co-assembled, but in some others the lack of expression of one of the subunits suggests the existence of other sub-types.

## DISCUSSION

As illustrated in this chapter, histochemistry is a useful technique for determining the localisation and gene expression of a given protein in tissue, such as a neurotransmitter receptor in the brain.

According to molecular biological studies, four classes of subunit are thought to be part of the $GABA_A$ receptor (α, β, γ and δ)(Schofield et al., 1987; Pritchett et al., 1989; Shivers et al., 1989). Variants have been identified within certain classes (6α, 4β and 2γ)(Levitan et al., 1988; Ymer et al., 1989a; Khrestchatisky et al., 1989; Lueddens et al., 1990; Ymer et al., 1989b; Ymer et al., 1990) and this diversity is thought to be responsible for the pharmacological heterogeneity of the GABA/BZ receptor. Although various combinations of these subtypes have been expressed and characterised in transiently transfected cells, fully functional $GABA_A$ receptors have not yet been obtained (Verdoorn et al., 1990). Even if the co-expression of the γ2 subunit with α1 and β1 has been shown to be necessary to confer

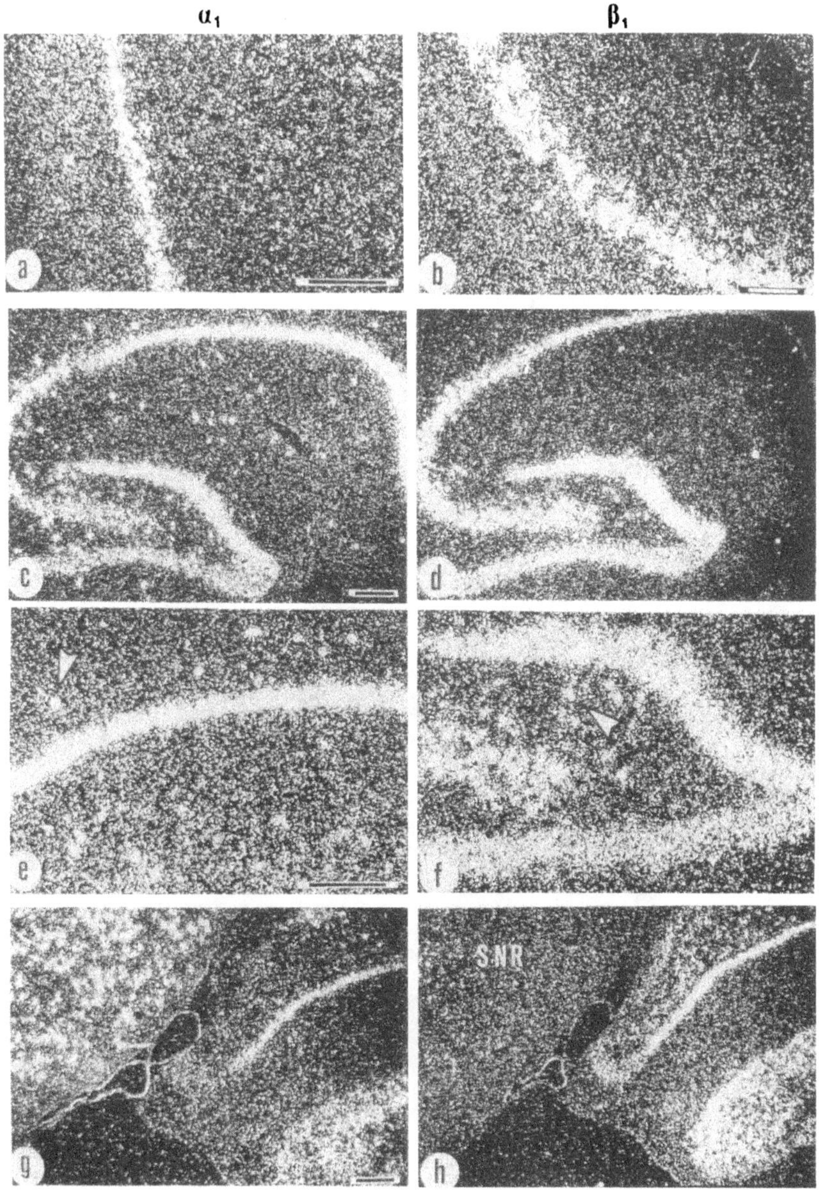


Fig.12:Cellular distribution of mRNAs coding for α1 (a,c,e,g) and β1 (b,d,f,h) subunits. Mitral cells of the main (a) and accessory (b) olfactory bulb. Pyramidal cells of the hippocampus and granule cells of the hippocampal formation (c-f). Substantia nigra (g-h). Note in the case of α1 the signal over cells in the stratum oriens and radiatum. Note also the lack of signal in the substantia nigra for β1. bar=200μm.

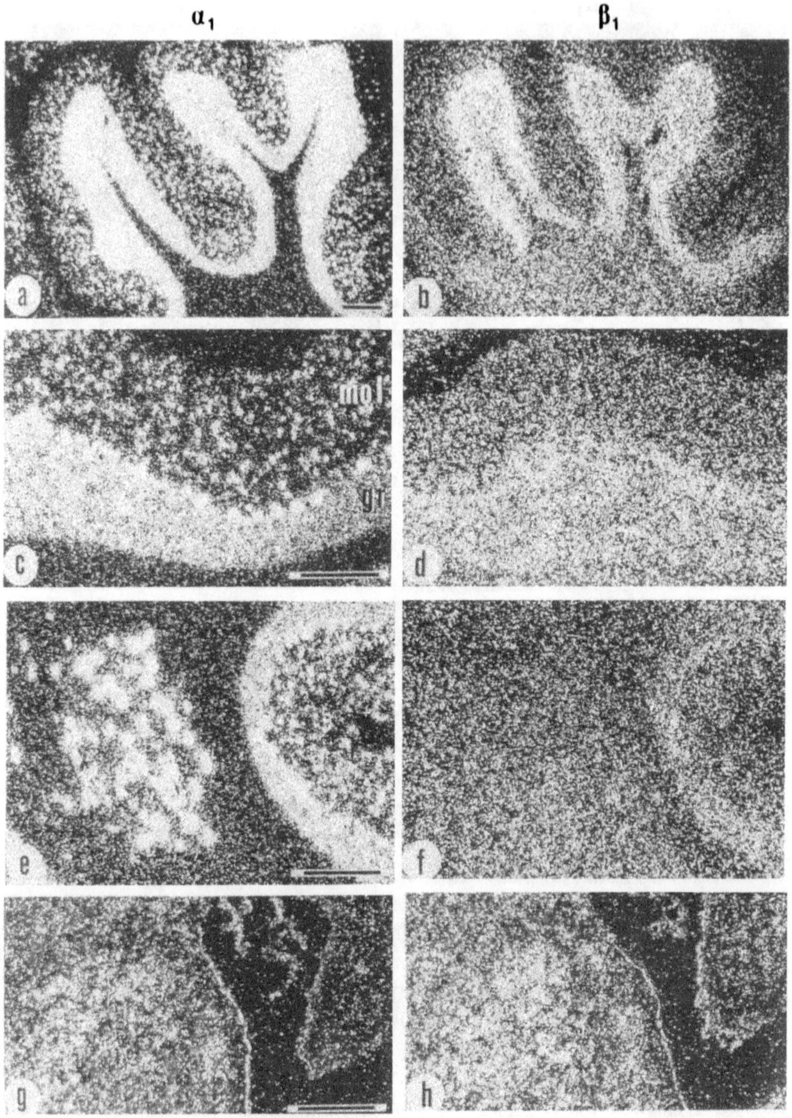


Fig.13:Cellular distribution of mRNAs coding for α1 (a,c,e,g) and β1 (b,d,f,h). Cerebellum (a-d). Nucleus interpositus (e-f). Bed nucleus of the stria terminalis (g-h). bar=200μm.

benzodiazepine potentiation, recent in situ hybridisation results suggest that not all brain regions express the three subunits (Malherbe et al.,1990b). Therefore the significance of this high molecular diversity is not clear, neither is the exact subunit composition and stoichiometry of the native GABA/BZ receptor (Schofield et al., 1990). Further investigations with the previously described histochemical methods using subunit-specific antibodies and subunit-specific oligonucleotide sequences will help to clarify the real subunit composition of the receptors in vivo as well as the possible regulation of expression of the receptors under given physiological and pathological conditions.

## REFERENCES


Costa E, Guidotti A, Toffano G (1978) Molecular mechanisms mediating the action of diazepam on GABA receptors. Brit J Psychiat 133:239-248

Haefely W, Pieri L, Polc P, Schaffner R (1981) General pharmacology and neuropharmacology of benzodiazepine derivatives. In: Hoffmeister F, Stille G (eds) Handbook of Experimental Pharmacology. Springer Verlag, Berlin Heidelberg, Vol.55/II

Haefely WE, Kyburz E, Gerecke M, Moehler H (1985) Recent advances in the molecular pharmacology of benzodiazepine receptors and in the structure activity relationships of their agonists and antagonists. Adv Drug Res 14:166-322

Haering P, Staehli C, Schoch P, Takacs B, Staehelin T, Moehler H (1985) Monoclonal antibodies reveal structural homogeneity of γ-aminobutyric acid/benzodiazepine receptors in different brain areas. Proc Natl Acad Sci USA 82:4837-4841

Khrestchatiski M, MacLennan AJ, Chiang MY, XU W, Jackson MB, Brecha N, Sternisi C, Olsen RW, Tobin AJ (1989) A novel α subunit in rat brain $GABA_A$ receptors. Neuron 3:745-753

Kuhar MJ, Unnerstall JR (1985) Quantitative receptor mapping by autoradiography:some current technical problems. Trends Neurosci 8:49-53

Kuhar MJ, De Souza EB, Unnerstall JR (1986) Neurotransmitter receptor mapping by autoradiography and other methods. Ann Rev Neurosci 9:27-59

Levitan ES, Schofield PR, Burt DR, Rhee LM, Wisden W, Koehler M, Fujita N, Rodrigues HF, Stephenson A, Darlison MG, Barnard EA, Seeburg PH (1988) Structural and functional basis for $GABA_A$ receptor heterogeneity. Nature 335:76-79

Lueddens H, Pritchett DB, Koehler M, Killisch I, Keinanen K, Monyer H, Sprengel R, Seeburg PH (1990) Cerebellar $GABA_A$ receptor selective for a behavioural alcohol antagonist. Nature 346:648-651

Malherbe P, Draguhn A, Multhaup G, Beyreuther K, Moehler H (1990a) $GABA_A$-receptor expressed from rat brain α- and β-subunit cDNAs displays potentiation by benzodiazepine receptor ligands. Mol Brain Res 8:199-208

Malherbe P,Sigel E, Baur R, Persohn E, Richards JG, Moehler H (1990b) Functional characteristics and sites of gene expression of the α1, β1, γ2-isoform of the rat $GABA_A$ receptor. J Neurosci 10:2330-2337

Moehler H, Okada T (1977) Benzodiazepine receptor:demonstration in the central nervous system. Science 198:849-851

Moehler H, Battersby MK, Richards JG (1980) Benzodiazepine receptor protein identified and visualized in brain tissue by photoaffinity label. Proc Natl Acad Sci USA 77:1666-1670

Moehler H, Richards JG, Wu JY (1981) Autoradiographic localisation of benzodiazepine receptors in immunocytochemically identified γ-aminobutyrergic synapses. Proc Natl Acad Sci USA 78:1935-1938

Palacios JM, Niehoff DL, Kuhar MJ (1981) Receptor autoradiography with tritium-sensitive film:potential for computerized densitometry. Neurosci Lett 25:101-105

Polc P, Moehler H, Haefely W (1974) The effect of diazepam on spinal cord activities: possible sites and mechanisms of action. Naunyn-Schmiedebergs Arch Exp Path Pharmak 284:319-337

Pritchett DB, Sontheimer H, Shivers BD, Ymer S, Kettenmann H, Schofield PR, Seeburg PH (1989) Importance of a novel $GABA_A$ receptor subunit for benzodiazepine pharmacology. Nature 338:582-585

Richards JG, Schoch P, Haering P, Takacs B, Moehler H (1987) Resolving $GABA_A$/benzodiazepine receptors: cellular and subcellular localisation in the CNS with monoclonal antibodies. J Neurosci 7:1866-1886

Richards JG, Schoch P, Haering P, Takacs B, Moehler H (1988) Strategies for raising antibodies against benzodiazepine receptors. In: Van Leeuven, Buijs, Pool, Pach (eds) Molecular Neuroanatomy. Elsevier Science Publishers, pp 159-190

Roberts E (1986) GABA: the road to neurotransmitter status. In: Olsen RW, Venter JC (eds) Benzodiazepine/GABA Receptors and Chloride Channels: Structural and Functional Properties. Alan R. Liss Inc., New-York, pp 1-39

Schmidt RF, Vogel ME, Zimmermann M (1967) Die Wirkung von Diazepam auf die praesynaptische Hemmung und andere Rueckenmarks-reflexe. Naunyn-Schmiedebergs Arch Exp Path Pharmak 258:69-82

Schoch P, Haering P, Takacs B, Staehli C, Moehler H (1984) A GABA/Benzodiazepine receptor complex from bovine brain: purification, reconstitution and immunological characterisation. J Receptor Res 4:189-200

Schoch P, Richards JG, Haering P, Takacs B, Staehli C, Staehelin T, Haefely W, Moehler H (1985) Co-localisation of $GABA_A$ receptors and benzodiazepine receptors in the brain shown by monoclonal antibodies. Nature 314:168-171

Schofield PR, Darlison MG, Fujita N, Burt DR, Stephenson FA, Rodrigues H, Rhee LM, Ramachandran J, Reale V, Glencorse TA, Seeburg PH, Barnard EA (1987) Sequence and functional expression of the $GABA_A$ receptor shows a ligand-gated receptor super-family. Nature 328:221-227

Schofield PR (1989) The $GABA_A$ receptor: molecular biology reveals a complex picture. Trends Pharmacol 10:476-478

Schofield PR, Shivers BD, Seeburg PH (1990) The role of receptor subtype diversity in the CNS. Trends Neurosci 13:8-11

Séquier JM, Richards JG, Malherbe P, Price GW, Mathews S, Moehler H (1988) Mapping of brain areas containing RNA homologous to cDNAs encoding $\alpha$ and $\beta$ subunits of the rat $GABA_A$ $\gamma$-aminobutyrate receptor. Proc Natl Acad Sci USA 85:7815-7819

Séquier JM, Brennand J, Barhanin J, Lazdunski M (1990) Regional expression of a MCD-peptide and dendrotoxin I-sensitive voltage-dependent potassium channel in rat brain. FEBS Lett 263:163-165

Shivers BD, Killisch I, Sprengel R, Sontheimer H, Koehler M, Schofield PR, Seeburg PH (1989) Two novel $GABA_A$ receptor subunits exist in distinct neuronal subpopulations. Neuron 3:327-337

Sieghart W (1989) Multiplicity of $GABA_A$-benzodiazepine receptors. Trends Pharmacol 10:407-411

Sigel E, Stephenson FA, Mamalaki C, Barnard EA (1983) A gamma-aminobutyric acid/benzodiazepine complex of bovine cerebral cortex J Biol Chem 258:6965-6971

Somogyi P, Takagi H, Richards JG, Moehler H (1989) Subcellular localisation of benzodiazepine/$GABA_A$ receptors using monoclonal antibodies in the cerebellum of rat, cat and monkey. J Neurosci 9:2197-2209

Squires RF, Braestrup C (1977) Benzodiazepine receptors in rat brain. Nature 266:732-734

Squires RF, Benson DI, Braestrup C, Coupet J, Klepner CA, Myers V, Beer B (1979) Some properties of brain specific benzodiazepine receptors:new evidence for multiple receptors. Pharm Biochem Behav 10:825-830

Sternberger LA (1979) Immunocytochemistry, 2nd edn. Wiley and Sons, New York

Verdoorn TA, Draguhn A, Ymer S, Seeburg PH, Sackmann B (1990) Functional properties of recombinant rat $GABA_A$ receptors depend upon subunit composition. Neuron 4:919-928

Ymer S, Schofield PR, Draguhn A, Werner P, Koehler M, Seeburg PH (1989a) $GABA_A$ receptor β subunit heterogeneity: functional expression of cloned cDNAs. EMBO J 8:1665-1670

Ymer S, Draguhn A, Koehler M, Schofield PR, Seeburg PH (1989b) Sequence and expression of a novel $GABA_A$ receptor α subunit. FEBS Lett 258:119-122

Ymer S, Draguhn A, Wisden W, Werner P, Keinanen K, Schofield PR, Sprengel R, Pritchett DB, Seeburg PH (1990) Structural and functional characterization of the γ1 subunit of the $GABA_A$/benzodiazepine receptors. EMBO J 9:3261-3267

Young WS, Kuhar MJ (1979) A new method for receptor autoradiography: [$^3$H] opioid receptors in rat brain. Brain Res 179:255-270

# REPORT ON THE DISCUSSION OF THE FOURTH SESSION : RECEPTORS

R. Couteaux
Dépt. de Cytologie, Institut des Neurosciences,
C.N.R.S., Université Pierre et Marie Curie,
7 quai Saint-Bernard, 75252 Paris-Cedex 05

The three lectures delivered during this session concerned the results of experiments carried out using different techniques, mainly immunocytochemical, on three types of receptors : cholinergic, neuropeptide and GABA/BZ. The session, which centered on these receptors, was completed in the poster session by communications (this volume) dealing with opiate receptors (Conrath et al.), substance P receptors (Zouaoui et al.) and serotonin receptors (El Mestikawy et al.)

The first lecture, given by Dr. U.J. McMahan, was a study of the formation of the postsynaptic apparatus, in which the nicotinic cholinergic receptors of the neuromuscular junction are located. In his study, Dr. McMahan analyzed the mechanisms which control the synthesis and distribution of these receptors in the skeletal muscle fibers during normal development and muscle regeneration.

The regulation of the synthesis and distribution of the nicotinic cholinergic receptors is closely linked to that of the metabolism of other proteins of the postsynaptic apparatus which also intervene in neuromuscular transmission, such as acetylcholinesterase. The various, cytochemical and structural specializations which characterize the postsynaptic apparatus form a coherent whole, in which the location and number of cholinergic receptors, although functionally of primordial importance, are subject, like the other specializations, to extremely complex neuromuscular interactions.

One of the numerous factors involved in the formation of the postsynaptic apparatus has now been made clear, as Dr. McMahan and his team were the first to show the crucial role of molecules located in the synaptic basal lamina in the forming, maintenance and regenerating of postsynaptic specializations, and also presynaptic specializations such as active zones.

In this symposium, chiefly devoted to methods, the remarkable efficiency of the strategy brought into play in this work deserves special mention. At each stage of the study, many techniques were combined, ranging from the most conventional morphological, biochemical and radioautographic procedures to the latest methods used in immunology and molecular biology. I would like here to summarize the main stages of this strategy. The work which led an essential function in neuromuscular junction organization to be assigned to

NATO ASI Series, Vol. H 58
Neurocytochemical Methods
Edited by A. Calas and D. Eugène

molecules of the synaptic basal lamina began with the experiments in muscle and nerve regeneration carried out in 1977 and 1978 by Dr. McMahan together with his colleagues L.M. Marshall and J.R. Sanes. A second series of experiments was based on the postulate that a deep homology exists between the muscle and the electric organ of Torpedo which is considerably richer in synapses than the muscle. In electric organ fractions enriched in basal lamina components the factor agrin was found, which on cultured chick myotubes induced the formation of patches that contained high concentrations of acetylcholine receptors, acetylcholinesterase and butyrylcholinesterase, three components of the postsynaptic apparatus at the neuromuscular junction. Monoclonal antibodies directed against agrin intensely stained molecules in the synaptic cleft of Torpedo, frog and chicken neuromuscular junctions, showing that the AChR-AChE-BUChE aggregating factor located in the basal lamina at the neuromuscular junction is immunologically similar to agrin. The antibodies against agrin also stained the cytoplasm of motoneuron cell bodies. These observations suggested that agrin-like molecules synthetized by motoneurons might be released by axon terminals and incorporated by the synaptic basal lamina.

The results of these experiments, which constitute a major advance in the study of the formation of the postsynaptic apparatus, raise new problems which were discussed during this session. Two of them, in particular, are not easy to solve completely : how far does the synthesis of the agrin-like molecules concentrated in the synaptic basal lamina depend on the motoneurons, and by which mechanism does agrin work in the formation of the postsynaptic apparatus ?

Dr. J.Y. Couraud undertook, in his lecture, to evaluate, mainly from the cytochemical standpoint, two immunological methods used to produce antibodies against receptors : the anti-idiotypic approach developed during the last decade, and the "anti-complementary peptide" approach, based on the molecular recognition theory recently proposed by J.E. Blalock and his collaborators.

To appreciate the results of both approaches, Dr. Couraud took as his chief example the identification of the receptors of a neuropeptide, substance P, in the central nervous system (rat spinal cord) and the peripheral nervous system (rat parotid gland cells and guinea-pig and human trachea).

The anti-idiotypic method has the advantage over the conventional procedures for anti-receptor antibody production of not requiring purified or semi-purified receptors for immunization. Anti-idiotypic antibodies are much more stable than peptide ligands like most hormone or neurotransmitter derivatives, which are generally small molecules subject to rapid hydrolysis by endogenous proteases. This stability allows protease inhibitors to be omitted from the incubation media and is also an important condition for the success of the immunocytochemical localization of the receptors. The anti-complementary peptide approach is more restricted than the more general anti-idiotypic approach, since it can only be applied to receptors for peptide ligands whose genes have already been cloned.

The micrographs obtained by light microscopy when both methods were applied to localizing substance P receptors or receptor-like material displayed better resolution than the micrographs obtained by using iodinated substance P in radioautographic experiments.

In electron microscopy, preliminary images were obtained which are encouraging but it seems premature to formulate an opinion on the value of substance P receptor localization at the ultrastructural level, specially as regards its relationships with cytoplasmic organelles.

The lecture given by Dr. J.M. Séquier concerned the localization and sites of gene expression of $GABA_A$/BZ receptors in the rat brain.

From the methodological point of view, the choice of these receptors was particularly interesting, because their identification raises very great difficulties and allowed Dr. Séquier to review all the histo- and cytochemical resources now available to overcome them.

The $GABA_A$/BZ receptor is a ligand-gated chloride channel that mediates synaptic inhibition in the mammalian brain ; however, the exact subunit composition and stoichiometry of native GABA/BZ receptors are not yet clear. The great diversity of $GABA_A$ receptor subunits revealed by molecular cloning tends to indicate that this diversity is even greater than was previously supposed. The existence of several subtypes of $GABA_A$ might explain the pharmacological heterogeneity of these receptors.

Under these conditions, the identification of cells expressing $GABA_A$/BZ receptors required several approaches : they comprised quantitative receptor radioautography, immunohisto- and cytochemistry using polyclonal antibodies against different subunits at different structural levels including the ultrastructure, and *in situ* hybridization histochemistry which enables the cells expressing the subunits of the receptor to be identified by means of cRNA probes or oligonucleotide probes.

The results obtained with these methods clearly demonstrated that they constitute useful tools for determining the location and sites of gene expression of proteins such as neurotransmitter receptors in the brain.

# EXTRACELLULAR DYES INDUCE PHOTODEGENERATION AND PHOTOPERMEABILIZATION OF VERTEBRATE AND INVERTEBRATE NEURONES

S. Picaud[1,2], H. Wunderer[1,3], L. Peichl[2] and N. Franceschini[1]
1) CNRS, Laboratoire de Neurobiologie,
Equipe de Neurocybernétique,
31 chemin J. Aiguier,
13009 Marseille,
FRANCE

Dyes can sensitize cells to light, resulting in lethal photodynamic damage (review: Spikes, 1989). Although dyes are widely used in neurobiology, their photosensitizing ability has not been investigated thoroughly on the nervous system but was only attested occasionally with electrophysiological techniques (review: Pooler, 1987). In some instances, however, the photodynamic action of dyes has been turned to advantage to produce photoinactivation of single neurones (Miller and Selverston, 1979) or of their neurites (Cohan et al, 1983; Jacob et Miller, 1985). To produce these effects, the dye was injected *intracellularly* into the neurone with a micropipette, the photoinactivation itself being subsequently achieved by a short pulse of laser light.

Recently, we observed structural changes in neurones selectively irradiated in the presence of a dye which was simply added to the *extracellular* space of the nervous tissue. First, the irradiation was found to induce classical features of neuronal degeneration (Picaud et al., 1988). Second, by using a dye exclusion test to assess cell viability, we showed that irradiated neurones can become permeable to extracellular substances (Picaud et al., 1990a,b). Thirdly, lethal damage of the irradiated neurones was confirmed by their

2) Max-Planck Institut für Hirnforschung, Deutschordenstrasse 46,
6000 Frankfurt/M71, Germany
3) Institut für Zoologie der Universität, Biologie 1, Universitätstrasse 31,
D 8400 Regensburg, Germany

NATO ASI Series, Vol. H 58
Neurocytochemical Methods
Edited by A. Calas and D. Eugène

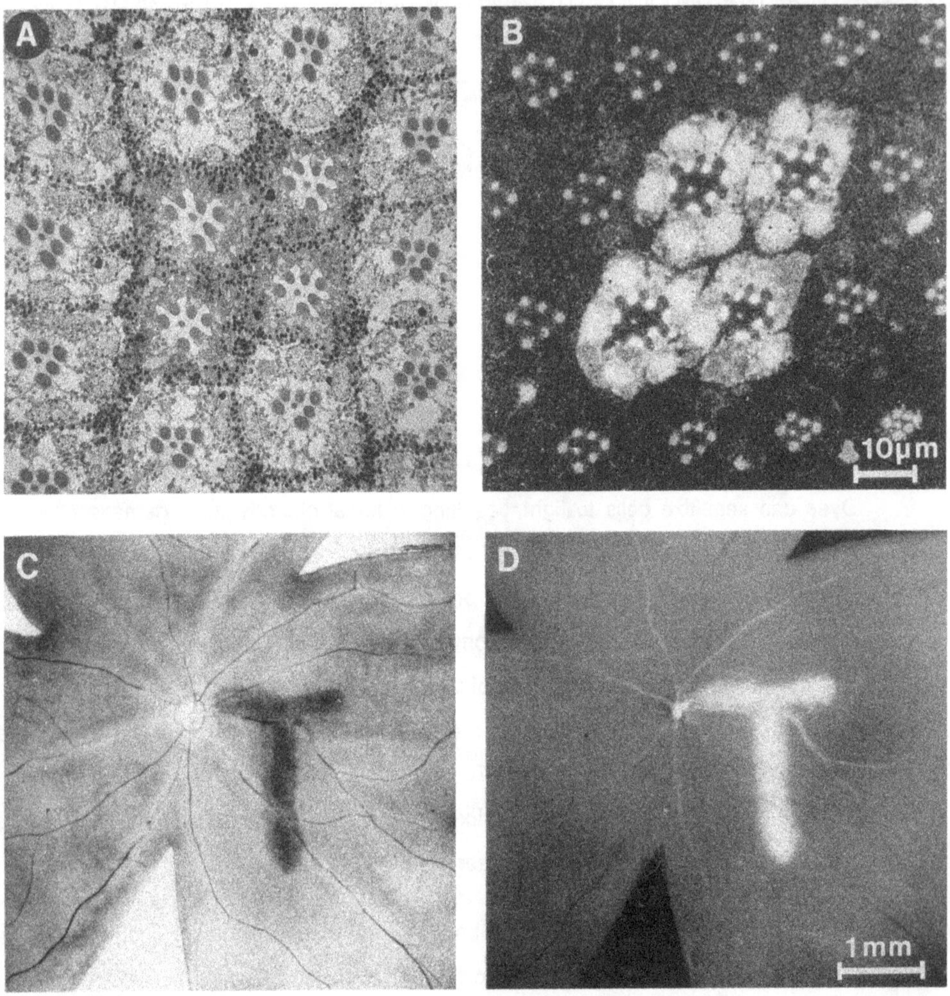


Figure 1: Dye-induced photodegeneration and photopermeabilization of retinal neurones in the housefly *Musca domestica* (a,b) and in the albino rat (c,d). (a) Low power electron micrograph of a cross-section through a treated fly retina (from Picaud et al., 1988). Receptor neurones irradiated in the presence of a dye (central patch: four ommatidia) show an increase in electron-density of their cytoplasm, whereas the receptors in surrounding ommatidia maintain their normal electron-lucent cytoplasm. (b) Photomicrograph of the fluorescence under blue excitation from a cross-section through a treated fly retina. Note that the dye Lucifer Yellow, used to assess cell permeability, stains selectively the cytoplasm of the irradiated neurones (central patch: four ommatidia) (a,b same magnification). (c,d) Photomicrographs of a flatmounted whole rat retina showing the effects of irradiation in the presence of a dye (from Picaud et al., 1990a). The T-shaped irradiated area displays a darkening of the retinal tissue under transmitted light (c) and a bright fluorescent staining with Lucifer Yellow (d). This latter staining results from the cytoplasmic labelling of the irradiated retinal neurones which became permeable to the dye (c,d same magnification).

subsequent eradication from the neural network (Picaud et al., 1988; Picaud et al, in prep.). All these observations were first made on the fly retina (Picaud et al., 1988; Picaud et al., 1990b), in which individual neurones can be observed in vivo (Franceschini et al., 1981). The same phenomena, however, were subsequently induced in non photoreceptive neurones of the rat retina (horizontal cells, bipolar cells, amacrine cells, ganglion cells) (Picaud et al., 1990a). The observed phenoma, which we termed "photodegeneration", "photopermeabilization" and "photoablation", respectively, offer interesting applications to the anatomical and physiological studies of neural networks.

A dye (Sulforhodamine 101, Rose Bengal) is applied to the extracellular space of the retina. A selected patch of neurones is irradiated *in vivo* for 30 minutes within the absorption band of the dye with the light of a conventional epi-fluorescence microscope. The following structural changes are observed depending on the survival time after the irradiation:

1) When the retina is fixed immediately after the irradiation, neurones and fibers display classical features of neuronal degeneration such as an electron-dense cytoplasm (Fig. 1a). No specific histological procedure is required to produce this cytoplasmic labelling which is also detectable with transmitted light microscopy. At the macroscopical level, the irradiated area becomes darker than the surrounding unirradiated tissue in transmitted light, the irradiated pattern appearing as "imprinted" in the nervous tissue (Fig. 1c).
2) When cell permeability is assessed with a test dye (Lucifer Yellow) applied to the extracellular space immediately after the irradiation, this latter dye is observed to permeate the irradiated neurones and label their cytoplasm (Fig.1b). At the macroscopic level, this process results in the selective staining of the irradiated area (Fig. 1d).
3) When the animal is sacrificed several days after the irradiation, glial cells are observed to phagocytose the irradiated neurones (Picaud et al., 1988; Wunderer et al., in prep; Picaud et al., in prep.). After 2 months survival time, irradiated en passant fibers are usually found to have retrogradely degenerated in the rat retina (Picaud et al., in prep.).

Various dyes (Rose Bengal, Merocyanine 540, Sulforhodamine 101, Hematoporphyrin, Procion Yellow, Lucifer Yellow) were found to produce these effects. For light to be effective, however, the irradiation wavelength had to be tuned to the specific absorption band of the dye. This requirement and the experimental parameters (low light intensity, long period of irradiation) indicated that induction of these phenomena relied on cell photosensitization by the dye, the structural changes observed indicating photodynamic damage (review: Spikes, 1989).

Photopermeabilization provides an adequate procedure for selectively labelling preselected neurones with a fluorescent dye (e.g. Lucifer Yellow cf. Fig.1b). Using the well known projection pattern of fly photoreceptor neurones, we showed that this labelling is adequate and reliable for tracing axonal paths. The fine-grained electron-dense staining resulting from neuronal degeneration is more suitable for ultrastructural studies. Since this staining does not obscure presynaptic specializations, it can be used to identify the terminals of a given, preselected neurone in order to examine its synaptic connectivity. Therefore both photopermeabilisation and photodegeneration are interesting tools for anatomical studies of the nervous system (Picaud et al., 1990b). Furthermore, since irradiated neurones happen to be removed from the neural network by glial cells in a selective and non invasive way, this kind of lesion becomes particularly interesting for physiological and developmental studies on various nervous systems and for the investigation of the neuronal bases of behaviour.

While the selectivity of photoinactivation (Miller and Selverston, 1979) relies upon a preliminary *intracellular* dye injection, the selectivity of the phenomena described here is based on optical imaging, with the interesting consequence that they are applicable from the level of single cells to that of neuronal ensembles. In the near future, coupling the dye to a vector such as an antibody homing in on a given cell population may even add a *cellular* selectivity to the *spectral* and *spatial* selectivity of irradiation. The described phenomena readily offer interesting perspectives for the functional analysis of those neural systems which can be made accessible to a light beam. These include in particular ganglions, retinae, cortices, brain slices and cell cultures....

Acknowledgments. This work was supported by the CNRS and the Fondation pour la Recherche Médicale (France), by the MPG and DFG (SFB4,ES) (Germany) and by the CODEST (EEC). S. Picaud received fellowships from the University of Marseilles, from the Ministère des Affaires Etrangères, from the MPG and from the Fondation de la Vocation. H.J. Wunderer received a post-doctoral fellowship from CNRS/MPG.

References

Cohan CS, Hadley RD, Kater SB (1983) "Zap axotomy": localized fluorescent excitation of single dye-filled neurons induces growth by selective axotomy. Brain Res., 270: 93-101.

Franceschini N, Kirschfeld K, Minke B (1981) Fluorescence of photoreceptor cells observed in vivo. Science, 213: 1264-1267.

Jacob GA, Miller JP (1985) Functional properties of individual neuronal branches isolated in situ by laser photoinactivation. Science, 228: 344-346.

Miller JP, Selverston AI (1979) Rapid killing of single neurons by irradiation of intracellularly injected dye. Science, 206: 702-704.

Picaud S, Peichl L, Franceschini N (1990a) Dye-induced "photodegeneration" and "photopermeabilization" of mammalian neurons in vivo. Brain Res., 531: 117-126.

Picaud S, Wunderer H, Franceschini N (1988) "Photo-degeneration" of neurones after extracellular dye application. Neurosci. Lett. 95: 24-30.

Picaud S, Wunderer H, Franceschini N (1990b) Dye-induced photopermeabilization and photodegeneration: a lesion technique useful for neuronal tracing. J. Neurosci. Methods 33, 101-112.

Pooler JP (1987) Photodynamic modification of excitable cell function. In: Heitz JR and Downum KR (Eds) Light-activated pesticides. Amer. Chem. Soc., Washington pp 109-121.

Spikes JD (1989) Photosensitization. In: Smith KC (Ed) The science of photobiology, 2nd edn, Plenum,pp. 79-110.

# A METHOD TO STUDY PROJECTIONS OF GABAERGIC NEURONS USING DELETION OF THE GABA IMMUNOSTAINING BY METALLIC IONS

Isabelle COURNIL, Pierre MEYRAND and Maurice MOULINS
Laboratoire de Neurobiologie et Physiologie Comparées
Université de Bordeaux I/ CNRS
place Peyneau
33120 Arcachon
France

GABAergic transmission plays an important role in the function of both the vertebrate and invertebrate nervous system. However, the physiological effects of discrete GABAergic neurons on neural circuits are not so well defined. Immunocytochemistry has been used to localize GABA in neural somata, fibers and terminals, and this maps its distribution throughout whole nervous systems. However, to understand how a particular GABAergic neuron or group of neurons influences a target neuron or neuronal network, it is necessary to devise anatomical methods that allow the identification of the appropriate GABA containing neurons and enable their characterization in electrophysiological studies. It is difficult to use the simple observation of immunostained preparations to resolve the projections of individual axons when they pass through regions of ganglionic neuropil, where labelled axon terminals interfere with the staining of fibers, or through the junction between nerves where axons may ramify into several branches. To overcome these limitations, we present here an original method that is specific to GABAergic neuronal systems.

## MATERIALS AND METHODS

The method is illustrated by using the well-characterized Crustacean stomatogastric nervous system (Selverston and Moulins, 1987). Experiments were performed using the lobster *Homarus gammarus*. The stomatogastric nervous

NATO ASI Series, Vol. H 58
Neurocytochemical Methods
Edited by A. Calas and D. Eugène

systems were dissected free from the stomach of the animal, and placed *in vitro* in physiological saline.

Axonal backfilling of nickel was performed using 1 M nickel chloride solution in distilled water. This solution was applied to the cut end of a nerve, in a small vaseline pool and left for 48 hours at 4°C in a humid chamber. All these nickel chloride labelled preparations were then rinsed in saline and processed for GABA immunocytochemistry.

GABA immunolabelling was performed as in Cournil et al (1990). Briefly, nervous tissue was fixed three hours in 2% glutaraldehyde with 0.1 M sodium cacodylate buffer containing 2% sodium chloride at pH 7.4. This was followed by Bouin acetic fixation for 6 hours, after which the preparation was rinsed and embedded in paraffin wax. Ten micron serial sections were cut, immunostained and visualized with the peroxidase-antiperoxidase method. The GABA antiserum supplied by M. Geffard (Bordeaux, France) is diluted 1:1000. Its specificity has been tested previously (Cournil et al, 1990). GABA immunostaining was also performed on control preparations without nickel filling.

## THE METHOD AND ITS APPLICATION

Metals such as cobalt and nickel are routinely used as axonally transported neuronal tracers, in order to map neuronal pathways. We have recently demonstrated that filling an identified GABA neuron cell body with nickel (using iontophoretic injection) completely deletes GABA immunostaining in the filled neuron (Cournil et al, 1991). We have used this effect of nickel on GABA immunostaining as a new tool to analyze GABAergic projections within a nervous system. First, a precise control mapping of the GABA immunoreactive neurons must be done. Second the metallic salt is introduced into some neurons. The subsequent GABA immunoreaction will only label those neurons which are not filled with metal, and the morphology of the metal filled GABA neurons can be deduced from the comparison with the control mapping.

This method is presented using an invertebrate preparation. In the stomatogastric nervous system from Crustacean it has been shown that GABA is involved in the control of activity of the neuronal networks found in the stomatogastric ganglion. This control can be performed by ten GABAergic neurons projecting via the stomatogastric nerve (stn) to the networks, whose cell bodies of origin were suspected to be located in more anterior ganglia. The oesophageal ganglion (OG), which is one of these anterior ganglia, contains two GABA immunoreactive somata and their axons in the medial part of the ganglion, six GABA positive fibers (Fig. 2B) and a neuropile area with GABAergic terminals labelled in the posterior part of the ganglion (Fig 1A).

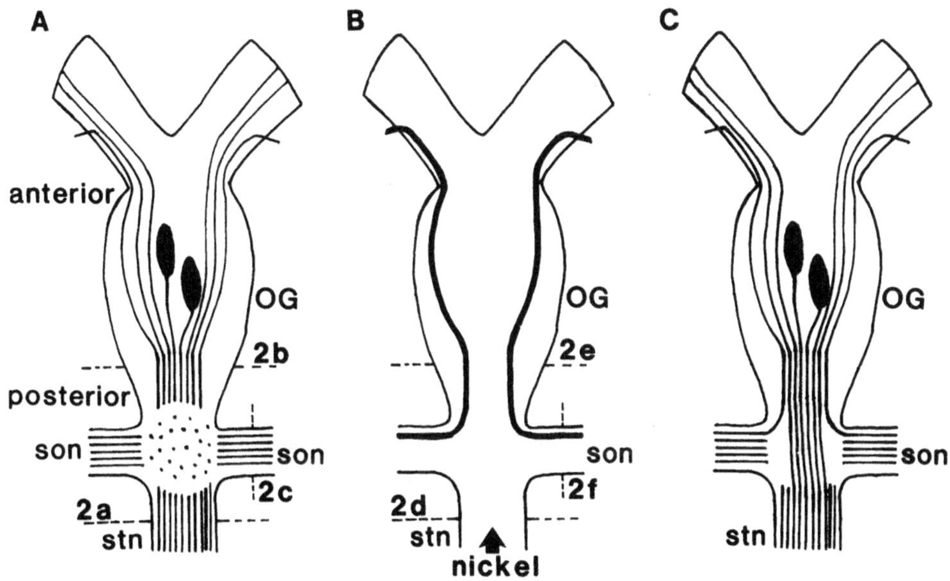


Figure 1. Schematic drawings of the GABA staining within the oesophageal ganglion and posterior nerves before and after nickel axonal transport. A: OG which contains two GABA immunoreactive neuron somata located in the middle part of the ganglion, presents in its posterior part, a neuropile (stippled area) with GABA immunoreactive terminals and three efferent nerves: the stn with ten GABA immunoreactive axonal fibers and the two bilateral son's with five labelled fibers. 2a, 2b and 2c are the locations of sections shown in figure 2 A, B and C. B: The GABA immunostaining is represented in the OG and the three nerves described in A, after a nickel axonal transport from the cut stn (arrow). Only two fibers are labelled in OG and project respectively into each son. 2d, 2e and 2f are the locations of sections shown in figure 2 D, E and F. C: From the comparison of GABA mapping as presented in A (staining of all the GABAergic neurons found in the preparation) with the staining of the neurons which are not filled with nickel, the possible organization for the projections of the fibers found in OG is the following : six fibers are going into the stn and two fibers are projecting respectively into each son.

The ganglion is connected at its posterior end to three nerves (Fig. 1A). These include two superior oesophageal nerves (sons), each of which contains five GABA immunoreactive fibers (Fig. 2C), and the stomatogastric nerve (stn), attached to the stomatogastric ganglion, which has ten GABA immunoreactive fibers (Fig. 2A). The purpose of this study was to identify which GABAergic neurons of the oesophageal ganglion are projecting to the stomatogastric ganglion via the stomatogastric nerve. The exact axonal pathways of all the GABA-immunopositive oesophageal neuronal fibers is impossible to follow as they cross the oesophageal ganglion neuropile area. The "classical" double labelling method of Lucifer Yellow backfilling combined to the GABA immunodetection applied to the preparation permitted to identify only two neurones, whose somata sit in the oesophageal ganglion which project into the stomatogastric nerve. It is shown here, how the method of deletion of the GABA immunostaining by nickel has been conclusive to identify in the oesophageal ganglion the GABAergic fibers projecting to the stomatogastric ganglion. Nickel was backfilled from the cut end of the stomatogastric nerve and the preparation was stained for GABA (Fig. 1B). There was no more staining on the sectionned nerve because all the axons, included those from GABAergic neurones have been filled with nickel (Fig. 2D), however two labelled axons were found in the middle part of the oesophageal ganglion (Fig. 2E). They are the axons of neurons which are not filled with nickel, these are traced into each superior oesophageal nerve (Fig. 2F) and consequently they are not projecting into the stomatogastric nerve. This experiment means that six GABAergic neurons within the oesophageal ganglion project into the stomatogastric nerve. Two of them have their soma in the oesophageal ganglion and it has been shown (Cournil et al, 1990) that the four others cell bodies are located in more anterior ganglia.

In conclusion this metal deletion technique is perfectly adapted for neuroanatomical tract tracing when combined with GABA immunocytochemistry. The resulting "tracing by substraction" enables the precise determination of GABAergic projections in invertebrate nervous systems and will be similarly useful for understanding GABAergic neuronal organization in vertebrate brains.

## REFERENCES

Cournil I, Meyrand P, Moulins M (1990) Identification of all GABA immunoreactive neurons projecting to the lobster stomatogastric ganglion. J Neurocytol 19:478-493.

Cournil I, Meyrand P, Moulins M (submitted) A method to determine projections of GABA immunoreactive neurons. J Neurosci Methods

Selverston AI, Moulins M (eds) (1987) The Crustacean stomatogastric system. A model for the study of central nervous system. Springer Verlag, Berlin Heidelberg

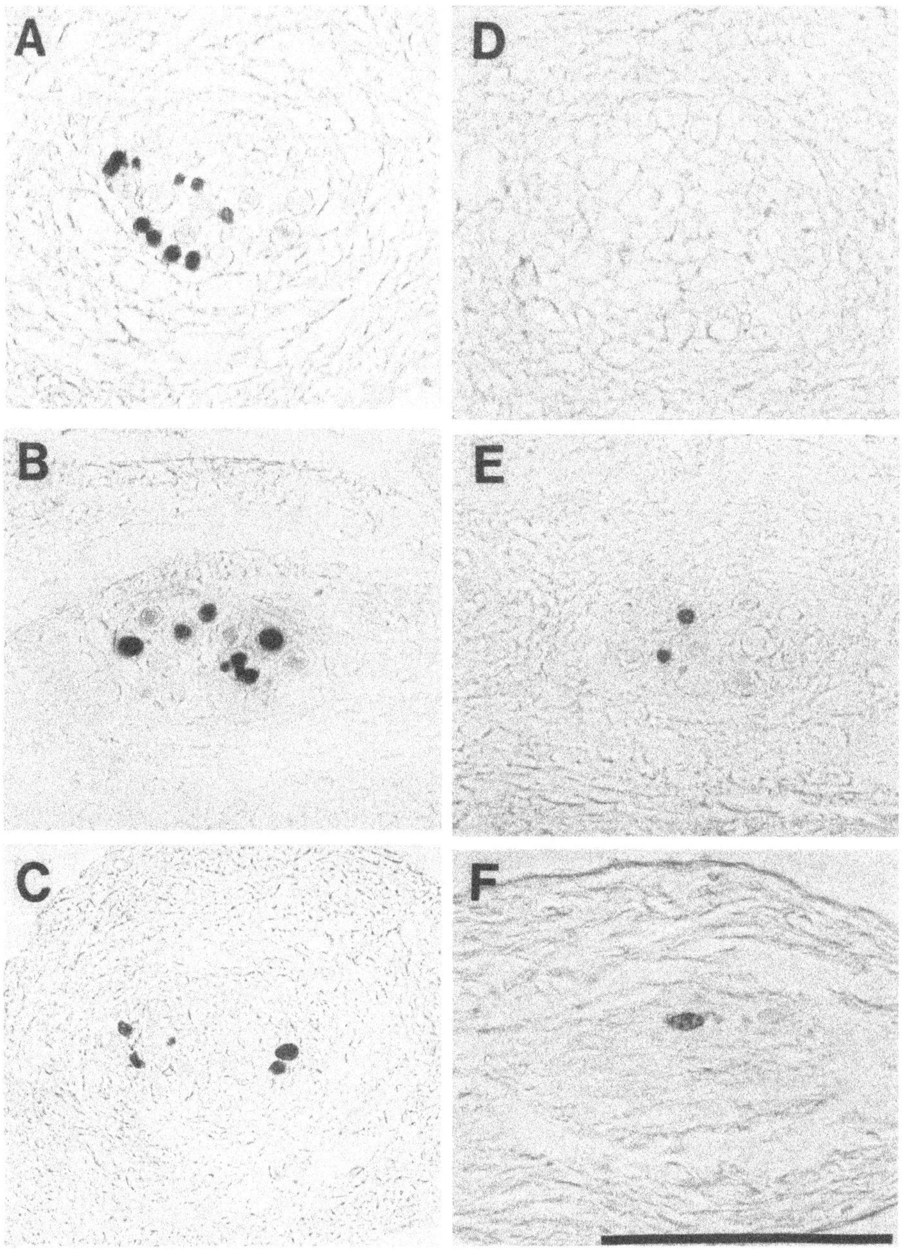


Figure 2. GABA immunostaining in oesophageal ganglion and posterior nerves before (A,B and C) and after (D, E and F) nickel axonal transport from the stomatogastric nerve.
Into the OG eight stained axonal fibers are found in the middle part of the ganglion (B). In the three nerves efferent from the posterior part of the ganglion, five labelled axonal fibers are found in each son (C) and ten fibres in the stn (A). After the nickel axonal transport from the cut stn there is no more staining in this nerve (D) but two fibers are still immunoreactive within the OG (E) and one fiber is labelled in each son (F). Calibration bar: 100 µm

# ULTRASTRUCTURAL IMMUNOCYTOCHEMICAL LOCALIZATION OF CHOLINE ACETYLTRANSFERASE IN SYMPATHETIC GANGLIA

Margaret R. Matthews, A. Ribeiro-da-Silva* & A. Claudio Cuello*
Department of Human Anatomy
University of Oxford
South Parks Road
Oxford OX1 3QX
United Kingdom

In the central nervous system the demonstration of neurones and nerve terminals immunoreactive for choline acetyltransferase (ChAT) has been successfully achieved by a number of groups, using several different primary antibodies, and has already almost the status of a routine procedure. In the peripheral nervous system the immunolocalization of this transmitter-synthesizing enzyme has not been entirely straightforward, as the antibodies and procedures which are effective in central nervous tissue do not always prove so efficacious in the periphery. Lindh et al. (1986) and Landis et al. (1987) have recently demonstrated immunolocalization of the enzyme in nerve terminal networks in peripheral autonomic ganglia at the light microscope level. We here report its immunocytochemical, ultrastructural localization, in nerve fibres and nerve endings of sympathetic ganglia of the rat.

Young adult male Wistar rats were perfusion-fixed under intraperitoneal Equithesin anaesthesia with a mixture of 4% paraformaldehyde, 0.1% glutaraldehyde and 15% saturated picric acid in 0.1M phosphate buffer, pH 7.4; next, with fixative without glutaraldehyde; and finally with 10% sucrose, w/v, in the same phosphate buffer, each for 30 min. Coeliac-superior

*Department of Pharmacology and Therapeutics, McGill University, 3655 Drummond Street, Montreal, PQ, Canada H3G 1Y6

NATO ASI Series, Vol. H 58
Neurocytochemical Methods
Edited by A. Calas and D. Eugène

mesenteric and superior cervical ganglia were dissected out, equilibrated in 30% sucrose-phosphate buffer, frozen in liquid nitrogen and cut at 50 µm on a cryotome. Slices were treated for 15-30 min with 1% sodium borohydride, rinsed, then immunostained free-floating in the following sequence of antibodies, with intervening rinses:

1. Rat monoclonal antibody to porcine choline acetyltransferase, ChAT 11, 72h at 4°C
2. Mouse anti-rat IgG, 2h at room temperature (ca. 22°C)
3. Rat monoclonal bi-specific antibody to mouse IgG and horseradish peroxidase (HRP), McC10, overnight at 4°C

The last two antibody treatments were repeated once more, then dilute horseradish peroxidase (HRP) solution, 5 µg/ml, was applied and allowed to bind to the bispecific antibody for 2h at 22°C. Bound HRP was detected by a diaminobenzidine reaction, intensified by nickel and cobalt. After osmication and flat-embedding in Epon, slices were thin-sectioned for electron microscopy and examined with or without further light contrasting.

Ultrathin sections showed labelling of myelinated and unmyelinated nerve fibres and nerve endings in the ganglion slices (Fig. 1a). In the coeliac-superior mesenteric ganglion there was clear evidence of heterogeneity of nerve terminals. Some nerve endings remained unlabelled in the immediate neighbourhood of well-labelled endings (Fig. 1b). Labelled synaptic terminals in this ganglion contained either regular, spherical synaptic vesicles or irregular, pleomorphic synaptic vesicles. Both these types of labelled nerve endings could contain large dense-cored vesicles, some of which showed evidence of immunolabelling over their cores (Fig. 1b). In the superior cervical ganglion virtually all the labelled nerve terminals contained regular, spherical synaptic vesicles.

It is not clear why the transmitter-synthesizing enzyme choline acetyltransferase should be more difficult to demonstrate immunocytochemically in the peripheral than in the central nervous system. One may speculate that the enzyme might become

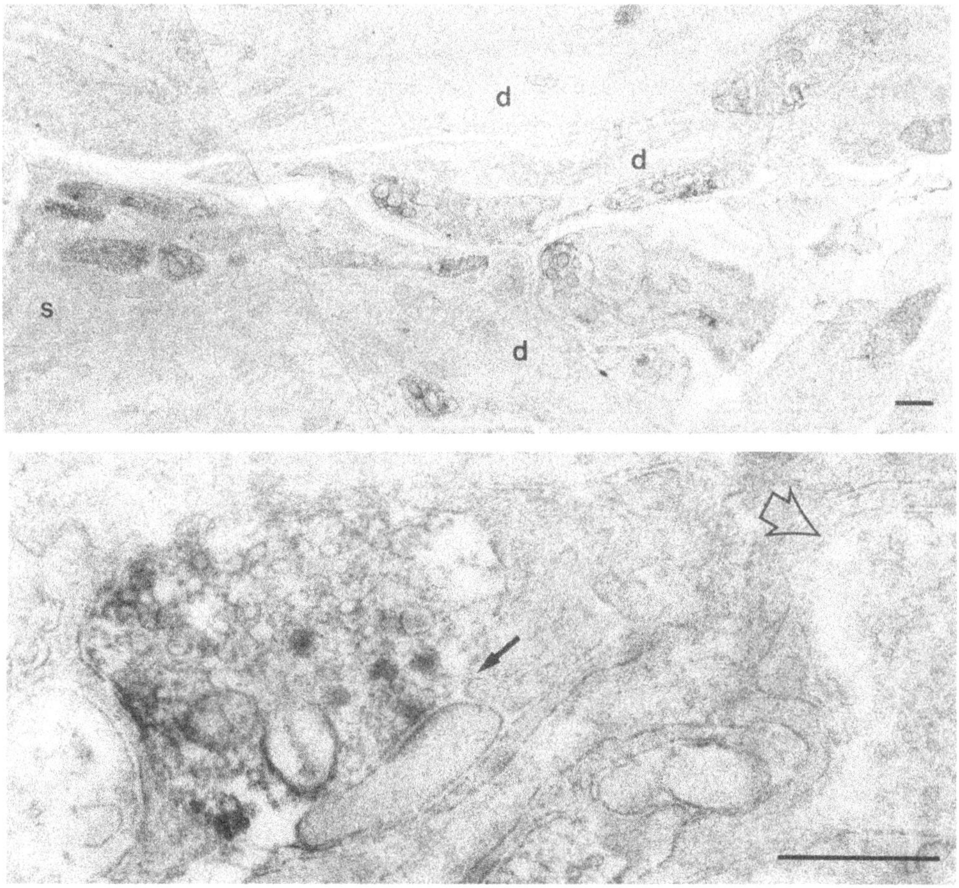


Fig. 1. ChAT-immunoreactive nerve terminals in the coeliac-superior mesenteric ganglion. Specimens osmicated but otherwise uncontrasted, apart from the end-product of the immunoperoxidase reaction. Scale bars indicate Ø.5 µm. a: An area at low magnification showing numerous immunolabelled nerve terminals in a region containing several dendrites (e.g. d) and part of a neurone soma (s). b: View at high magnification showing a ChAT-immunoreactive nerve terminal in synaptic contact (arrow) with a small spine-like profile. End-product of the immunoreaction coats the external surfaces of small synaptic vesicles, large dense-cored vesicles and mitochondria in the labelled terminal, and is also seen over the cores of some, but not all, of the large dense-cored vesicles. On the right (open arrow) is seen part of an unlabelled nerve terminal, containing strongly pleomorphic synaptic vesicles.

modified by some form of post-translational processing during its transport to the periphery. The presence of reticular collagen in peripheral nerves and ganglia may also be a complicating factor. A particular value of this immunocytochemical localization of ChAT in peripheral ganglia is in the opportunity which it offers for double-labelling studies, in order to investigate the coexistence of transmitters and the connectivity of distinctive pathways. Such further studies are in progress.

## References

Leblanc GG, Trimmer BA, Landis SC (1987) Neuropeptide Y-like immunoreactivity in rat cranial parasympathetic neurons: coexistence with vasoactive intestinal peptide and choline acetyltransferase. Proc natl Acad Sci USA 84:3511-3515

Lindh B, Staines W, Hökfelt T, Terenius L, Salvaterra PM (1986) Immunohistochemical demonstration of choline acetyltransferase-immunoreactive preganglionic nerve fibres in guinea pig autonomic ganglia. Proc natl Acad Sci USA 83:5316-5320

# A DUAL IMMUNOPEROXIDASE LABELLING METHOD TO STUDY THE ULTRASTRUCTURAL RELATIONSHIPS BETWEEN CHOLINE ACETYLTRANSFERASE- AND NEUROPEPTIDE Y-CONTAINING NEURONS IN THE RAT STRIATUM

**J. Vuillet[1,2], R. Dimova[3], A. Nieoullon[2] and L. Kerkerian-Le Goff[2]**

1. Centre de Microscopie Electronique, Faculté de Médecine, Bd Jean-Moulin, 13385 Marseille cedex 5 ; 2. Unité de Neurochimie, Laboratoire de Neurosciences Fonctionnelles, CNRS, 31 Chemin J.-Aiguier, 13402 Marseille cedex 9, France

In the striatum, neuropeptide Y (NPY) and choline acetyltransferase (ChAT) are to be found in populations of aspiny neurons with distinct morphological features. Indeed, immunocytochemical studies using selective antibodies have indicated that NPY-labelled neurons are medium-sized type IV aspiny neurons (Vuillet et al., 1989b), whereas ChAT-labelled neurons belong mainly to the category of large neurons in the striatum (Bolam et al., 1984). Neurochemical studies in combination with lesion experiments and morphological studies have provided evidence that both NPY- and ChAT-containing neuronal populations represent interneurons, receive afferents from extrinsic origin (Chang, 1988; Vuillet et al., 1989a; 1989b), and provide synaptic input to medium-sized spiny neurons that are predominantly striatal output neurons (Izzo and Bolam., 1988; Vuillet et al, 1989b). These data suggest that both cholinergic and NPY interneurons in the striatum may play an integrative role which may be of particular relevance in the control of sensorimotor function. The question of whether or not synaptic interactions exist between those NPY and cholinergic interneurons then appears of great interest for further functional resolution of the striatal network. Two major difficulties emerge to answer this question : both types of neurons represent less than one percent of the total striatal neuron population; in addition, besides the paucity of NPY terminals, both ChAT and NPY axonal boutons form predominantly symmetrical type synapses that are difficult to characterize. Then study of the ultrastructural relationships between NPY and ChAT neurons requires the use of a dual immunolabelling approach allowing both selection of labelled neurons at light microscopic level and clearcut visualization of membrane specializations. Combination of immunoperoxidase with radioimmunocytochemical or post-embedding immuno-gold methods does not seem to respond to those criteria.

3. Regeneration Research Laboratory, Bulgarian Academy of Sciences, 1431 Sofia, Bulgaria

NATO ASI Series, Vol. H 58
Neurocytochemical Methods
Edited by A. Calas and D. Eugène

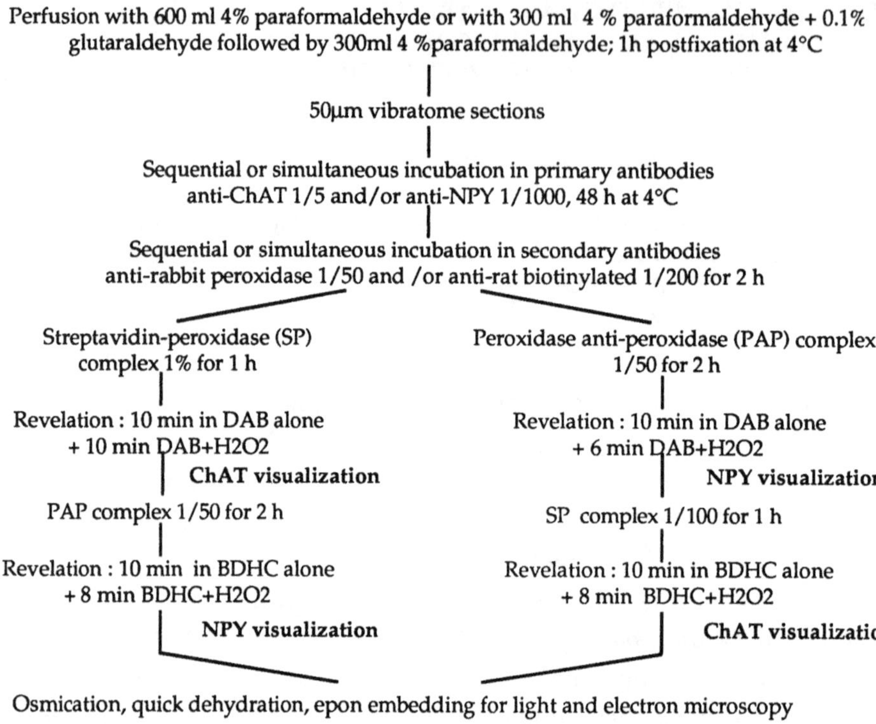


**Figure 1 : Dual immunoperoxidase labelling procedure for simultaneous detection of ChAT and NPY immunoreactivities.** All incubations were carried out in either Tris saline or phosphate buffer 0.1 M at pH 7.4, except for BDHC solutions which were performed in 0.01M phosphate buffer at pH 6.8.

This fact prompted us to develop a dual immunoperoxidase labelling method to simultaneously localize at the light and electron microscopic level NPY and ChAT immunoreactivities on the same striatal sections. The approach was to combine peroxidase-anti-peroxidase and biotin-streptavidin labelling methods using 3-3'-diaminobenzidine (DAB) and benzidine dihydrochloride (BDHC) as distinct chromogens. Preliminar data showed that the diffuse DAB end-product is easily distinguishable from BDHC crystalline aggregates but also that the use of BDHC is, in our hands, to be limited to the visualization of cell bodies and dendrites. Then, to study the reciprocal ultrastructural relationships between NPY- and ChAT-containing neurons, the two following reverse methods described in detail in fig. 1 were used in a complementary manner : 1) combination of NPY/BDHC with ChAT/DAB labellings to examine whether NPY neurons constitute targets for cholinergic nerve terminals; 2) combination of ChAT/BDHC with NPY/DAB labellings to determine whether, reciprocally, cholinergic neurons are contacted by NPY nerve terminals.

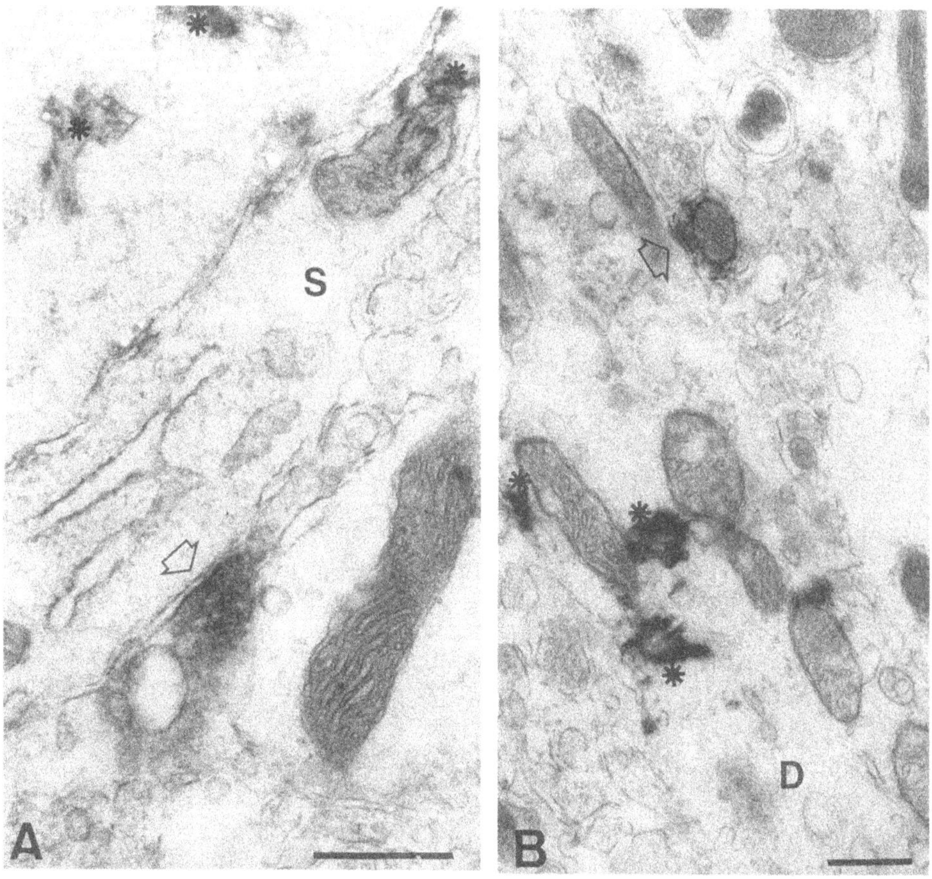


**Figure 2 : Axosomatic and axodendritic associations of ChAT immunoreactive terminals (DAB-labelled) with NPY neurons (BDHC-labelled).** A. A ChAT-immunoreactive terminal forms a synaptic contact of the symmetrical type (open arrow) on the soma (S) of a NPY neuron (asterisks: BDHC crystals). B. A ChAT-immunoreactive terminal forms a symmetrical synaptic contact (open arrow) on the shaft of a BDHC-labelled (asterisks) NPY-immunoreactive dendrite (D). Bars: 500 nm.

Results show DAB-labelled ChAT nerve terminals in contact with BDHC-labelled NPY cell bodies and dendrites (Fig. 2). Conversely, DAB-labelled NPY terminals were found to contact BDHC-labelled ChAT perikarya (Fig. 3).

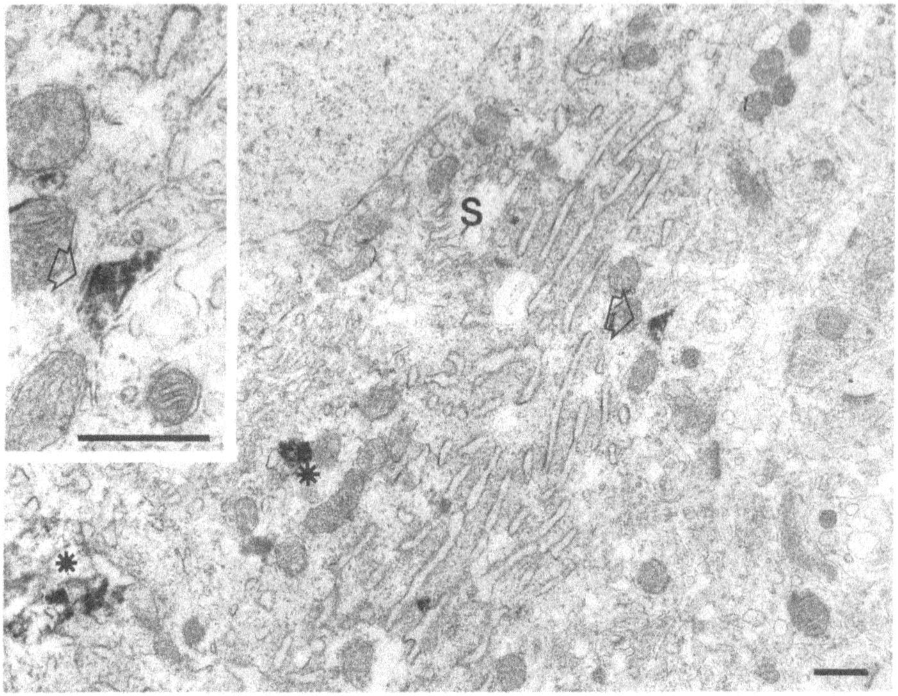


**Figure 3 : Example of axosomatic association of a NPY-immunoreactive terminal with a ChAT-positive neuron in the rat dorsal striatum.** A ChAT-immunoreactive soma (S) labelled with BDHC crystals (asterisks) is contacted by a DAB-labelled NPY-immunoreactive terminal (open arrow). Inset : High magnification of the axosomatic association visualized above, which forms a synaptic contact of the symmetrical type (open arrow). Bars: 500 nm.

In both cases, the synaptic specializations formed by the intensely labelled boutons were characterized by accumulation of vesicles surrounded by DAB end-product at presynaptic level, by thickening of pre- and postsynaptic membranes and sometimes by dense reactive material in the synaptic cleft. Classification of these contacts as symmetrical type synapses was possible due to the fact that BDHC labelling of perikarya and dendrites did not present any background and led us examine postsynaptic membrane densities on serial sections. These observations emphasize the usefulness of dual immunoperoxidase labelling method using DAB and BDHC as chromogens to study synaptic interactions in particular when the neuronal populations examined are poorly represented. In addition, data reveal reciprocal synaptic interactions between NPY and cholinergic neurons in the striatum, providing morphological support for the existence of a functional link between these two striatal interneuron populations. This interaction is to consider with particular interest in the balance between the activity of cholinergic neurons and the activity of nigrostriatal dopaminergic afferent fibers (see Izzo and Bolam, 1988). Previous studies on striatal circuitry have led to the conclusion that medium-sized spiny neurons provide the basic network of the striatum since they receive information of extrinsic origin (from the cortex, the substantia nigra at least) and from other striatal intrinsic neurons, and since they make contact within and outside the striatum (Bolam, 1984). However, it must be kept in mind that those neurons represent quantitatively the major striatal neuron population, and their inputs and outputs are quantitatively easy to identify due to their extensive arborizations, dendritic spines and axon collaterals. These characteristics have probably led to an underestimation of the striatal synaptic interactions involving the striatal aspiny interneurons that represent minority populations with long, but predominantly few, dendritic arborization. Recent advances in immunocytochemical investigations allow to show that aspiny neurons keep up multiple relationships with the other striatal systems. In particular, we have contributed to show that NPY aspiny interneurons also receive afferents from extrinsic origin (from the cortex and the substantia nigra), from intrinsic neuronal populations (such as GABA neurons, and, as presently reported, cholinergic neurons), and contact intrinsic neuronal populations representing striatal output neurons and other interneurons such as the cholinergic ones (Aoki and Pickel, 1989; Kubota et al., 1988; Vuillet et al, 1989a; 1989b; 1990). Then, accurate immunocytochemical investigations using dual labelling methods at the electron microscopic level, by allowing visualization of the sparse but diverse synaptic interactions involving striatal NPY neurons and more generally aspiny interneurons, provide morphological support for a fundamental integrative role of these neurons in the striatal network.

**Acknowledgements**

This work was supported by grants from the Faculty of Medicine in Marseille (Université Aix-Marseille II) and from CNRS.

Aoki C. and Pickel V.M. (1989) Neuropeptide Y in the cerebral cortex and the caudate-putamen nuclei : Ultrastructural basis for interactions with GABAergic and non-GABAergic neurons. J. Neurosci. 9 : 4333-4354.

Bolam J.P. (1984) Synapses of identified neurons in the neostriatum. In D. Evered and M. O'Connor (eds) Ciba Foundation Symposium 107 : Functions of the Basal Ganglia. London : Pitman, pp. 30-47.

Bolam J.P., Wainer B.H. and Smith A.D. (1984) Characterization of cholinergic neurons in the rat neostriatum. A combination of choline acetyltransferase immunocytochemistry, Golgi-impregnation and electron microscopy. Neurosci. 12 : 711-718.

Chang H.T. (1988) Dopamine-acetylcholine interaction in the rat striatum. A dual-labeling immunocytochemical study. Brain Res. Bull. 21 : 295-304.

Izzo P.N. and Bolam J.P. (1988) Cholinergic synaptic input to different parts of spiny striatonigral neurons in the rat. J. Comp. Neurol. 269 : 219-234.

Kubota Y., Inagaki S., Shimada S., Kito S., Eckenstein F. and Tohyama M. (1987) Neostriatal cholinergic neurons receive direct synaptic inputs from dopaminergic axons. Brain Res. 413 : 179-184.

Vuillet J., Kerkerian L., Kachidian P., Bosler O. and Nieoullon A. (1989a) Ultrastructural correlates of functional relationships between nigral dopaminergic or cortical afferent fibers and neuropeptide Y-containing neurons in the rat striatum. Neurosci. Lett. 100 : 99-104.

Vuillet J., Kerkerian L., Salin P. and Nieoullon A. (1989b) Ultrastructural features of NPY-containing neurons in the rat striatum. Brain Res. 477 : 241-251.

Vuillet J., Kerkerian-Le Goff L., Kachidian P., Dusticier G., Bosler O. and Nieoullon A. (1990) Striatal NPY-containing neurons receive GABAergic afferents and may also contain GABA : An electron microscopic study in the rat. Europ. J. Neurosci. 2 : 672-681.

# CHARACTERIZATION OF MICROGLIA AND BRAIN MACROPHAGES

N. DELHAYE-BOUCHAUD, C. FARDEAU, I. RIVA, S. CARON and C. BOUCHAUD
Institute of Neurosciences CNRS UA 1199 Department of Cytology
University P. & M. Curie
75252, Paris Cedex O5 France

The central nervous system contains two populations of phagocytic cells: the non-resident phagocytes derived from circulating precursor cells and known as brain macrophages, and the ramified microglia, the origin of which remains controversial, that are considered as the resident macrophage population. These two classes of phagocytes are recruited in a different manner in response to local injury or infection of the brain, depending on the severity of brain lesion or of blood vessels damage (see reviews in Perry and Gordon 1988 and in Streit et al. 1988).

After viral infections or traumatic lesions such as stab wounds affecting the vascular supply, the non resident macrophages are mobilized from blood and play a major part in the removal of degenerating neural tissue. After non traumatic degenerative processes, the ramified or «resting» microglia that lack phagocytic properties undergo a series of morphological changes and transform into proliferating ameboid microglia acting as scavenger cells. The number of reactive microglia also increases markedly in the normal developing brain, in areas where cell death or axon loss naturally occur; it has been suggested that developing reactive microglia subsequently differentiate into adult resting microglia.

Like other macrophages, brain macrophages can be recognized by taking advantage of their phagocytic properties (Bocchini et al. 1988). However after peripheral administration of various markers such as carbon particles, trypan blue, silver protein complex (figs. 1, 2), or fluorescent latex beads, the only regions where these prelabeled cells can be detected in the adult brain are the circumventricular organs (for instance the subfornical organ, the choroid plexus, the median eminence, the area postrema) deprived of a blood-brain barrier (Wislocki and Leduc, 1952)

The cerebral ventricles contain polymorphic motile cells (Kolmer cells) resembling free macrophages, mainly located on the surface of the choroid plexuses (Chamberlain, 1974). This localisation is favourable for transmission and scanning electron microscopic studies (figs. 3 , 4).

Features of brain macrophages and microglia have been studied morphologically

NATO ASI Series, Vol. H 58
Neurocytochemical Methods
Edited by A. Calas and D. Eugène

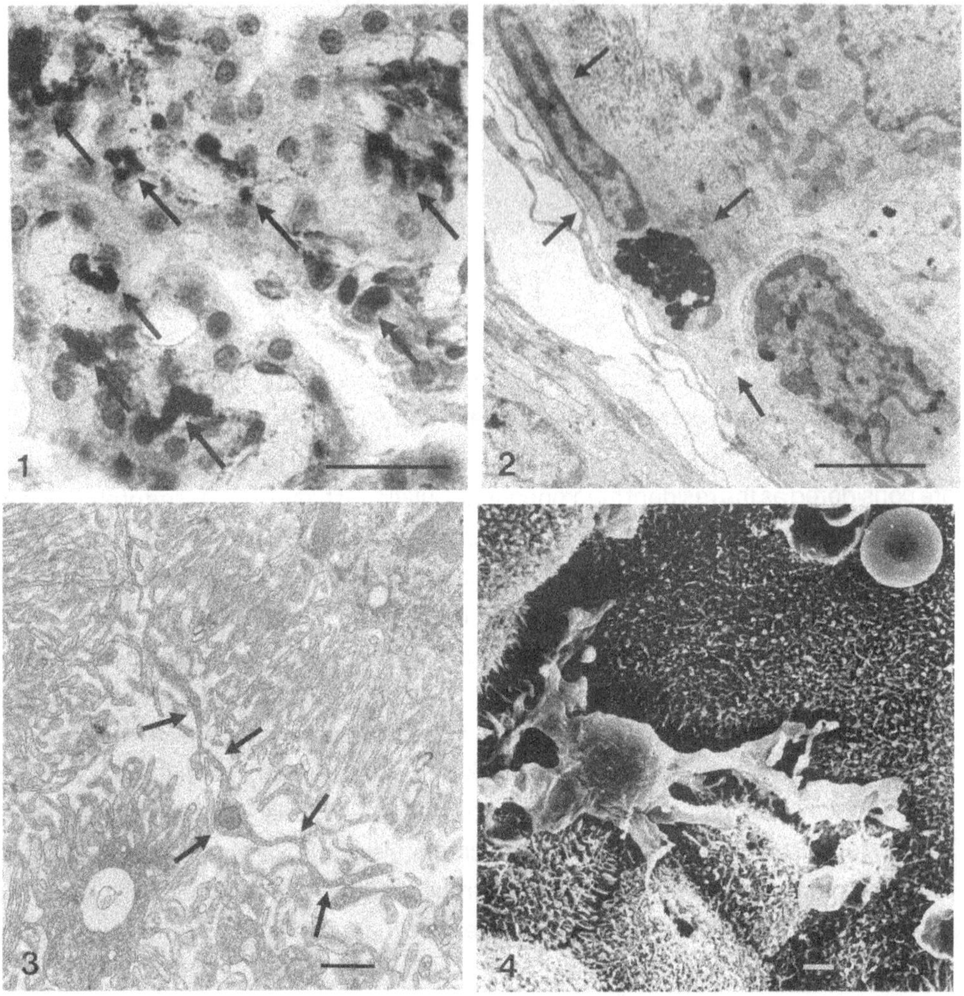

Figs. 1-4 .- Identification of rat brain macrophages (arrows);

*phagocytosis* of the silver-protein complex in two circumventricular organs: the choroid plexus (fig. 1) and the subfornical organ (fig. 2, in a vacuole of sequestration); *electron microscopy* (fig.3, T.E.M., fig. 4, S.E.M.): intraventricular Kolmer cells on the choroid plexus. Note in fig. 4 the macrophage-erythrocyte proximity.

Scales : 1: 30 µm; 2-4: 1 µm

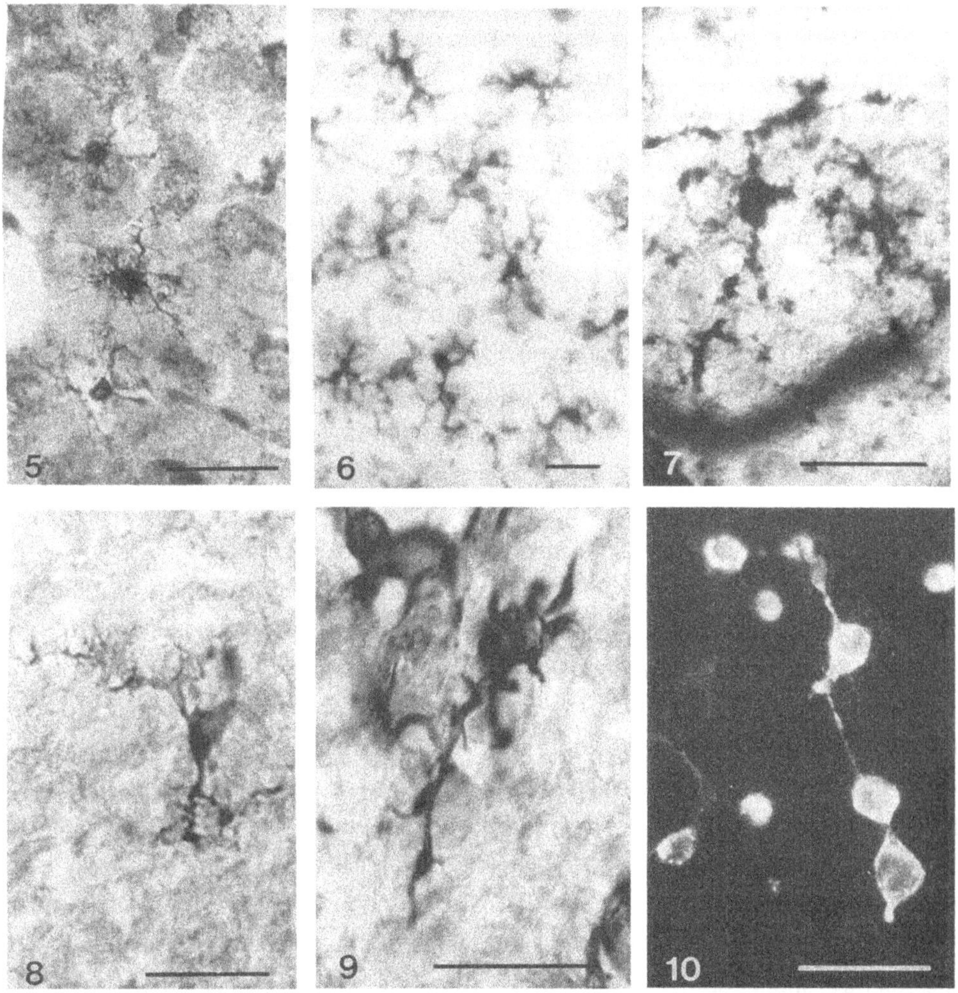


Figs. 5-10.-Identification of microglia by histological and cytochemical techniques.

Fig. 5: Rio Hortega's method. Rat cortex;

Figs. 6-7: microglia revealed by its nucleoside diphosphatase activity. Rabbit (fig. 6) and rat (fig. 7) retina;

Figs. 8-9: B4 isolectin binding of microglia in the rat cerebellum (fig. 8) and in the staggerer mouse cerebellum (fig. 9);

Fig. 10: immunostaining of microglia by anti Mac 1 monoclonal antibody in a culture of rat cerebellar cortex.

Scale: 20 μm

both at the light and electron microscopic level. Microglial cells were first described by del Rio Hortega (1932), as a distinct third neuroglial element, using the classical technique of silver carbonate impregnation (fig. 5). The resting microglia have a small nucleus, a scanty cytoplasm and few slender and extensively branched processes. These processes are covered with tiny protrusions, giving the cell a very typical aspect. At the ultrastructural level, the intraparenchymatous microglia are characterized by a dark protoplasm containing numerous lysosomes and lipid bodies.

In recent years more sophisticated procedures have been developed for recognition of microglia, both in the adult and developing brain.

In addition to the classical enzyme histochemical methods for lysosomal enzymes such as acid phosphatases (Imamoto et al. 1982), or for non-specific esterases (Koski et al. 1976), methods for the detection of thiamine pyrophosphatase or nucleoside diphosphatase (Murabe and Sano, 1982) along the outer surface of the plasma membrane have been used (figs. 6, 7).

Another method for the visualization of microglial cells uses conjugates of two types of plant lectins: the B4 isolectin derived from *Griffonia simplicifolia* (Streit et al, 1988) and *Ricinus communis* agglutinin-1 ( Mannoji et al, 1986 ) which bind respectively $\alpha$-D and $\beta$-D galactose containing glycoconjugates in the membranes (figs. 7, 8).

Finally, antibodies have been raised that specifically label microglia and macrophages (fig. 10). Resting microglia in the adult, and all stages of microglia in the developing brain express the Fc receptor as well as the type 3 complement receptor (Mac 1) and high levels of a macrophage-specific membrane glycoprotein of unknown function (Springer et al, 1979, Austyn and Gordon, 1981). Antisera to ferritin also positively stain resting and activated microglia (Kaneko et al, 1989). In the normal brain, expression of major histocompatibility complex (MHC) antigens is absent or very low in resting microglia; on the contrary, under pathological conditions, there is a marked upregulation of both class I and class II antigens. This is also true for the CD4 antigen and for the human leucocyte antigen, the expression of which is increased on reactive microglia.

In conclusion, until recently, there has been some difficulty visualizing brain macrophages and microglial cells, especially because of the functional plasticity of the latter. The classical del Rio Hortega's method suffers from the fact that silver impregnation remains hazardous and specific only for resting microglia.

Conjugated lectins and enzyme histochemical methods reliably label monocytes, brain macrophages and resting, reactive or phagocytic microglia, but they do not stain other glial or neuronal elements. In addition they also label cerebral and retinal micro-vessels, confirming a common embryological origin for microglia and endothelial cells.

The main advantage of the antibodies is a selective labeling of macrophages and all physiological stages of microglia .

Austyn J M, Gordon S (1981) F 4/80, a monoclonal antibody directed specifically against the mouse macrophage. Eur J Immunol : 805-815

Bocchini V, Rebel G, Masarelli R, Schuber F, Muller C D (1988) Latex beads phagocytosis capacity and ecto-NAD$^+$ glycohydrolase activity of rat brain microglial cells in vitro. Int J Devel Neurosc 6: 525-534

Chamberlain J G (1974) Scanning electron microscopy of epiplexus cells (macrophages) in the fetal rat brain. Am J Anat 139: 443-447

Imamoto K, Fujiwara R, Nagai T, Maeda T (1982) Distribution and fate of macrophagic ameboid cells in the rat brain. Arch Histol Jap 45: 505-518

Kaneko Y, Kitamoto T, Tateishi J, Yamaguchi K (1989) Ferritin immunohistochemistry as a marker for microglia. Acta Neuropathol 79: 129-136

Koski J R, Poplack DG, Blaese RM, (1976) In methods in cell mediated and tumors immunity. 359-362 Acad Press New-York

Mannoji H, Yeger H, Becker L E (1986) A specific histochemical marker (lectin Ricinus communis agglutinin-1) for normal human microglia, and application to routine histopathology. Acta Neuropathol 71: 341-343

Murabe Y, Sano Y (1982) Morphological studies on neuroglia: VI Postnatal development of microglial cells. Cell Tissue Res 225: 469-485

Perry V H, Gordon S (1988) Macrophages and microglia in the nervous system. TINS 11: 273-277

Rio Hortega P del, (1932) In Cytology and cellular pathology of the nervous system. 2: 482-534 PB Hoeber Inc New-York

Springer T, Galfré G, Secher DS, Milstein C (1979) Mac 1: a macrophage differentiation antigen identified by monoclonal antibody. Eur J Immunol 9:301-306

Streit W J, Graeber M B, Kreutzberg GW (1988) Functional plasticity of microglia: a review. Glia 1: 301-307

Streit W J, Kreutzberg GW (1987) Lectin binding by resting and reactive microglia. J Neurocytol 16: 249-260

Wislocki G B, Leduc E H (1952) Vital staining of the hematoencephalic barrier by silver nitrate and trypan blue, and cytological comparisons of neurohypophysis, pineal body, area postrema, intercolumnar tubercle and supraoptic crest. J Comp Neurol 96: 371-413

Acknowledgments: This work was supported by the Association Française contre les Myopathies and the Fondation de France.

# VISUALIZATION AND QUANTIFICATION OF 5-$HT_{1A}$ RECEPTORS IN THE RAT BRAIN BY RADIOIMMUNOHISTOCHEMISTRY

S. El Mestikawy, M. Riad, L. Matthiessen[1], H. Gozlan, M. Hamon, G. Daval[1] and D. Vergé[1].
Neurobiologie Cellulaire et Fonctionnelle
INSERM U288, CHU Pitié-Salpêtrière
91 bd de l'Hôpital
75013 Paris

The localization of serotoninergic 5-$HT_{1A}$ receptor binding sites has been extensively studied in the rat central nervous system (CNS) by autoradiography using radioactive ligands such as $[^3H]$8-OH-DPAT (Pazos and Palacios, 1985; Vergé et al., 1986). 5-$HT_{1A}$ binding sites are mainly found in limbic areas (hippocampus, septum, amygdala) which are concerned with the control of mood. Interestingly, some selective agonists (ipsapirone, buspirone) exhibit marked antidepressant and anxiolytic properties (Traber et Glaser, 1987). The raphe nuclei, where are concentrated the somas of 5-HT neurons, are also rich in 5-$HT_{1A}$ binding sites.

However, the visualization of the receptor protein at a better level of resolution could afford further information about its localization on the target cells of serotoninergic terminals as well as on the raphe cell bodies. Unfortunately, autoradiography using reversible ligands is not a convenient method for this purpose. Furthermore, for 5-$HT_{1A}$ sites as for other receptors (R) functioning with G proteins, only the complex R-G exhibits a high affinity for agonists, and consequently their radioactive derivatives (notably $[^3H]$8-OH-DPAT) can label the receptor coupled with G, but not those which are uncoupled. Conversely, all 5-$HT_{1A}$ receptors are expected to be recognized by specific antibodies and visualized at a cellular level.

We therefore used an approach based upon anti-receptor antibodies raised against a synthetic polypeptide (El Mestikawy et al., 1990). A radioimmunohistochemical technique has been developed for the visualization and the quantification of G-coupled plus uncoupled 5-$HT_{1A}$ receptors in the rat brain.

[1]Département de Cytologie, Institut des Neurosciences, CNRS UA 1199, Université Pierre et Marie Curie, 7 Quai Saint-Bernard, 75005 Paris

NATO ASI Series, Vol. H 58
Neurocytochemical Methods
Edited by A. Calas and D. Eugène

## Experimental procedures

A 26 amino-acids peptide corresponding to a specific portion (residues 243-268) of the third cytoplasmic loop of the rat 5-$HT_{1A}$ receptor (Albert et al., 1990) was coupled to bovine serum albumin and injected every 4 weeks to male New-Zealand rabbits. The serum used in the present study was collected 4 weeks after the 7th booster injection.

The immunoautoradiographic experiments were performed as previously described (El Mestikawy et al., 1990). Briefly, adult male Sprague-Dawley rats were anaesthetized with chloral hydrate and sacrificed by an intracardiac perfusion of 200 ml of 0.9% NaCl containing sodium nitrite (1 g/l). The brain was quickly removed and frozen at -30°C. Coronal or horizontal sections (20 µm thick) were cut at -15°C in a cryomicrotome and thaw-mounted onto gelatin-coated slides. Sections were then preincubated for 30 min in PBS (50 mM $Na_2HPO_4/NaH_2PO_4$, 0.9% NaCl, pH 7.4) containing 3% (w/v) bovine serum albumin (BSA), and washed with PBS alone for 5 min. After a 30 min incubation in PBS containing 1% BSA and the antiserum (final dilution 1/1000), sections were washed in PBS (3 x 5 min) and dipped in the same buffer supplemented with [$^{35}$S]IgG-anti-rabbit IgG (570 Ci/mmol, Amersham) for 2 h. They were then washed 3 times in PBS (10 min each) and once in distilled water (15 s), dried with cold air and exposed to β-max films (Amersham) at -80°C for 48 h.

The autoradiographic labelling of 5-$HT_{1A}$ sites with 2 nM [$^{3}$H]8-OH-DPAT or 50 pM [$^{125}$I]BH-8-MeO-N-PAT was performed as previously described by Vergé et al. (1986) and Gozlan et al. (1988), respectively. Autoradiograms were generated on $^{3}$H-hyperfilms (Amersham) with an exposure of 2 months for [$^{3}$H]8-OH-DPAT and 3 days for [$^{125}$I]BH-8-MeO-N-PAT.

For the labelling of 5-$HT_{1A}$ receptors in the rat immature cerebellum, similar experiments were performed with sagittal sections of the cerebellar vermis of 8 day-old rats.

Densitometric measurements were carried out with a home-made computer-assisted image analyser.

## Results and discussion

The autoradiograms obtained from sections exposed to the antiserum and then to [$^{35}$S] IgG-anti-rabbit IgG were remarkably similar to those from sections labelled by the radioactive agonists [$^{3}$H]8-OH-DPAT and [$^{125}$I]BH-8-MeO-N-PAT (Fig. 1). In

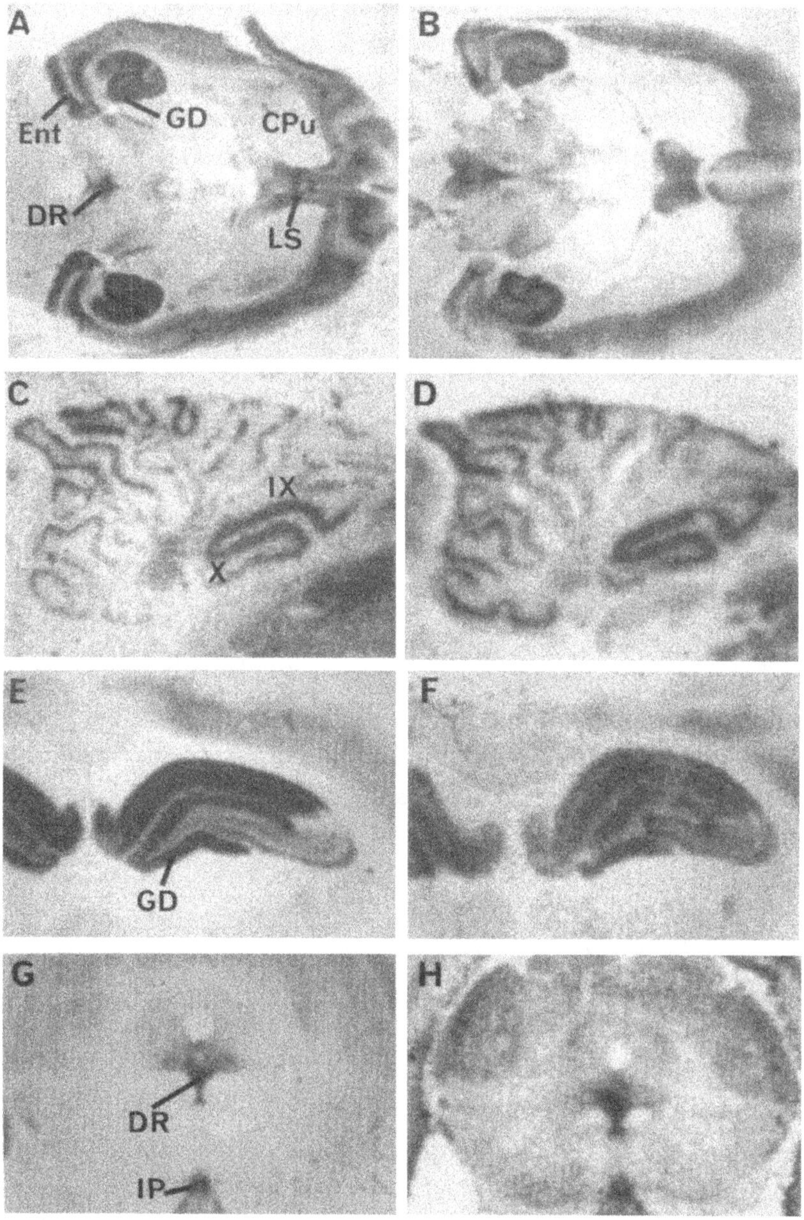


Figure 1: Autoradiograms from sections exposed to the radioligands [$^{125}$I]BH-8-MeO-N-PAT (A,C) or [$^{3}$H]8-OH-DPAT (E,G) or to the antiserum (B,D,F,H). A,B: horizontal sections from adult rat brains; C,D: sagittal sections of cerebellar vermis from a 8-day old rat; E-H: coronal sections from adult rat brains at the level of the hippocampus (E,F) and the dorsal raphe nucleus (G,H). CPu, caudate putamen; DR, dorsal raphe nucleus; Ent, entorhinal cortex; GD: gyrus dentatus; IP, interpeduncular nucleus; LS: lateral septum; IX, X: cerebellum, lobules IX and X.

the adult rat, the lateral septum, the entorhinal cortex (Fig. 1A,B), the hippocampus (Fig. 1A,B,E,F), the dorsal raphe nucleus (Fig. 1A,B,G,H) and the interpeduncular nucleus (Fig. 1G,H) were highly labelled with the two techniques. In contrast, the striatum (Fig. 1A,B) and the substantia nigra (not shown) were devoid of 5-$HT_{1A}$ receptors. Interestingly, the 5-$HT_{1A}$ binding sites revealed by radioactive agonists in the cerebellum of young rats (Daval et al., 1987) but not in the adult animals could also be visualized by radioimmunohistochemistry (Fig. 1C,D). In sections of cerebellar vermis from a 8-day-old rat, a strong labelling was observed with the two methods in all the lobules, probably in the molecular layer (Vergé et al., 1990). In this latter example, it was obvious that the distribution of 5-$HT_{1A}$ receptors was similar not only qualitatively, but also quantitatively, the posterior lobules IX and X being more intensely labelled than the anterior lobules.

By contrast, the labelling by [$^{35}$S]IgG-anti-rabbit IgG in the hippocampus appeared less intense than in the dorsal raphe nucleus, whereas [$^{3}$H]8-OH-DPAT binding was similar in these two areas (Vergé et al., 1986) (Fig 1E-H).

Densitometric measurements on autoradiograms confirmed that the ratio of labelling in hippocampus versus dorsal raphe nucleus was close to 1 with the radioactive agonist, and markedly lower with the antiserum (Table 1). It can therefore be proposed that, in the hippocampus, nearly all 5-$HT_{1A}$ receptors are present in the high affinity state recognized by radiolabelled agonists as well as by the antibodies. In contrast, in the dorsal raphe nucleus, a significant amount of receptors could be labelled only by the anti-receptor antibodies. Consequently, a population of "spare" receptors could exist in the raphe but not in the hippocampus. This hypothesis is supported by the observation that receptor alkylation by N-ethoxy-2-ethoxy-1,2-dihydroxyquinoline (EEDQ) completely suppresses the functional response to 5-$HT_{1A}$ agonists in the hippocampus but not in the dorsal raphe nucleus (Yocca et al., 1990). Such a functional difference between 5-$HT_{1A}$ receptors in the raphe nuclei and in the hippocampus is not unexpected, since they exert different actions upon the serotoninergic system. In the raphe nuclei, 5-$HT_{1A}$ sites are located on somas and/or dendrites of 5-HT neurons (Vergé et al., 1986). They control the firing of 5-HT neurons (Sprouse and Aghajanian, 1987) and their stimulation decreases the 5-HT release in terminal fields. In contrast, agonists binding at postsynaptic sites on the target cells in the hippocampus mimic the action of 5-HT.

These results emphasize the potential use of radioimmunohistochemistry for the quantitative evaluation of 5-$HT_{1A}$ receptors whether or not they are coupled to G-protein. Of course, further investigations are needed to firmly establish the validity of measurements of optical densities on autoradiograms as a method of quantifying densities of 5-$HT_{1A}$ receptor proteins on brain sections (see Correa et al., 1988).

Table 1: Quantification of autoradiograms from sections exposed to [$^3$H]8-OH-DPAT or to the antiserum.

| structure | [$^3$H]8-OH-DPAT | [$^{35}$S]IgG-anti-rabbit IgG |
|---|---|---|
| gyrus dentatus (GD) | 0.293 ± 0.030 | 0.253 ± 0.011 |
| dorsal raphe n.(DR) | 0.261 ± 0.024 | 0.444 ± 0.020 |
| GD/DR | 1.12 | 0.57 |

Optical densities (± S.E.M.) were measured on 4 to 6 sections from 4 rats.

References


Albert PR, Zhou QY, Van Tol HHM, Bunzow JR, Civelli O (1990) Cloning, mRNA tissue distribution, and functional expression of the rat 5-$HT_{1A}$ receptor gene. J Biol Chem 265: 5825-5832

Correa FMA, Guilhaume SG, Saavedra JM (1988) Quantitative radioimmunohistochemical method using [$^{125}$I]-protein A to measure the content of methionine enkephalin in discrete rat brain areas. J Histochem Cytochem 36:1379-1386

Daval G, Vergé D, Becerril A, Gozlan H, Spampinato U, Hamon M. (1987) Transient expression of 5-$HT_{1A}$ receptor binding sites in some areas of the rat CNS during postnatal development. Int J Devl Neurosci 5:171-180

El Mestikawy S, Riad M, Laporte AM, Vergé D, Daval G, Gozlan H, Hamon M (1990) Production of specific anti-rat 5-$HT_{1A}$ receptor antibodies in rabbits injected with a synthetic peptide. Neurosci Lett 118:189-192

Gozlan H, Ponchant M, Daval G, Vergé D, Ménard F, Vanhove A, Beaucourt JP, Hamon, M. (1988) [$^{125}$I]Bolton-Hunter-8-methoxy-2(N-propyl-N-propylamino)-tetralin as a new selective radioligand of 5-$HT_{1A}$ sites in the rat brain. In vitro binding and autoradiographic studies. J Pharmacol Exp Ther 244:751-759

Pazos A, Palacios JM (1985) Quantitative autoradiographic mapping of serotonin receptors in the rat brain. I. Serotonin-1 receptors. Brain Res 346:205-230

Sprouse JS, Aghajanian GK (1987) Electrophysiological responses of serotonergic dorsal raphé neurons to 5-$HT_{1A}$ and 5-$HT_{1B}$ agonists. Synapse 1:3-9

Traber J, Glaser T (1987) 5-$HT_{1A}$ receptor-related anxiolytics. Trends Pharmacol Sci 8:432-437

Vergé D, Daval G, Marcinkiewicz M, El Mestikawy S, Gozlan H, Hamon M (1986) Quantitative autoradiography of multiple 5-$HT_1$ receptor subtypes in the brain of control or 5,7- dihydroxytryptamine-treated rats. J Neurosci 6:3474-3482

Vergé D, Matthiessen L, Bailly Y, Daval G, Kia K, Gozlan H., Hamon M (1990)Localization of 5-$HT_{1A}$ serotonin binding sites in the cerebellum of young rats. Neurochem. Int. in press

Yocca, FD, Iben, I, Meller, E (1990) Lack of receptor reserve (RR) at postsynaptic 5-$HT_{1A}$ receptors linked to adenylate cyclase (AC) in rat hippocampus (HIP). Soc Neurosci 16:1034 (Abst. 427.4)

# ANTI-IDIOTYPIC ANTIBODIES AGAINST OPIOID RECEPTORS: FIRST ULTRASTRUCTURAL EVIDENCE FOR A PRESYNAPTIC INHIBITION OF OPIOIDS IN THE DORSAL HORN OF THE RAT SPINAL CORD

M. Conrath[1], D. Zouaoui[1], E. Apartis[1], F. Herbrecht[2], A. Cupo[2]
1-Institut des Neurosciences
CNRS URA 1199, Université Paris 6
7 Quai St Bernard, 75005 Paris
France

## Introduction

Enkephalins (ENK), belong to the opioid peptides and are thus potential endogenous ligands for opiate receptors. They are present in numerous cells (perikarya and terminals) of the dorsal horn of the spinal cord (Hökfelt et al., 1977; Glazer and Basbaum, 1983; Aronin et al., 1981) and have been shown to inhibit the response of dorsal horn neurons to noxious stimulation (Duggan et al., 1977; Zieglgansberger and Tulloch, 1979). Both a postsynaptic and a presynaptic mechanism of action have been proposed. The postsynaptic hypothesis is now strongly supported by electrophysiological data (references above) and morphological results showing that ENK-immunoreactive varicosities are found presynaptic to dendrites and cell bodies of the dorsal horn . Furthermore, Ruda et al. (1984) have demonstrated that about half of the spino-thalamic projection in laminae I and V of the dorsal horn are contacted by ENK-immunoreactive varicosities.

As to the presynaptic hypothesis, it derives from electrophysiological and opiate receptor binding studies. In particular, dorsal rhizotomy leads to a 30-60% decrease in mu and delta opiate receptors in the dorsal horn (Hiller et al., 1978; Zajac et al., 1989). This result suggests that part of the opiate receptors are localized on the primary afferent terminals. However, anatomical support for such a presynaptic mechanism (axo-axonic contacts between primary afferent terminals and ENK-containing varicosities) was so far virtually non existent.

An ultrastructural study of the localization of the opiate receptors, which would allow for the identification of the **target cells**, would bring about arguments in favor -or in disfavor- of the pre and/or the postsynaptic mechanism of action of the ENK.

The anti-idiotypic approach (see Couraud et al., this issue), which permits to obtain antibodies recognizing the receptor without its purification, allows a study under the electron microscope (EM).

---

2- Institut de Pharmacologie Moléculaire et Cellulaire, Sophia Antipolis, 660 Route des Lucioles, 06560 Valbonne, France

NATO ASI Series, Vol. H 58
Neurocytochemical Methods
Edited by A. Calas and D. Eugène

## Methods

Two monoclonal anti-idiotypic antibodies (named 5 and 51) were used in this study. They were obtained by immunizing rats with an anti-DADLE (the DADLE being a potent opiate agonist, delta and, to a lower extend, mu specific). The antibodies were respectively tested for their anti-receptor and anti-idiotypic activities. They are capable to bind to membrane preparations from the NG 108/15 cell line or from rat striatum and recognize, with immunocytochemical methods, surface antigen of the NG 108/15 cell line (Cupo and Kaldy., 1986). Their anti-idiotypic activity, measured by their ability to prevent the ligand/anti-DADLE interaction, reaches 100% and 60% for antibodies 5 and 51 respectively.

An indirect immunocytochemical technique was used in this study, with horseradish peroxidase (HRP) as a marker. Experiments are performed on 50 m vibratome sections from Wistar rats previously perfused with a 4% paraformaldehyde solution. The sections were preincubated for 30 minutes in 1% normal goat serum before the incubation in the solution of antibodies (1:1000) is achieved overnight at 4°C. Ascitic fluid precipitate by 40% ammonium sulfate and purified by DEAE-A52 chromatography were used. Following an extensive washing, the HRP was revealed by diaminobenzidin and hydrogen peroxide.

Blocks of the dorsal horn were dissected out, postfixed in 2% $OsO_4$ and embedded in araldite. Ultrathin sections were then examined with a Philips EM 300 microscope and we chose to focus our attention on lamina II, where numerous afferent fibres are known to terminate.

## Results - Discussion

Under the light microscope, the labelling is very dense in the dorsal horn (laminae I and II), around the central canal (lamina X) and in discrete regions of the ventral horn, probably corresponding to the region where there are groups of motoneurons (lamina IX). The results obtained are found to be similar with the two antibodies. It has been previously shown in the rat, that these antibodies do bind in other regions of the brain known to be rich in opiate receptors, eg in the striatum, the hippocampus and the subfornical organ (Conrath and Cupo, 1989). A non-immune serum does not label tissue sections. Preincubation of the sections with DADLE or synthetic ENK, before the addition of the antibodies does not prevent the labelling, even if high concentrations of peptidase inhibitors are added to the incubation media. This finding is in accordance with previous results obtained with an anti-substance P anti-idiotypic antibody (Conrath et al., 1988) and may most probably -at least partly- be accounted

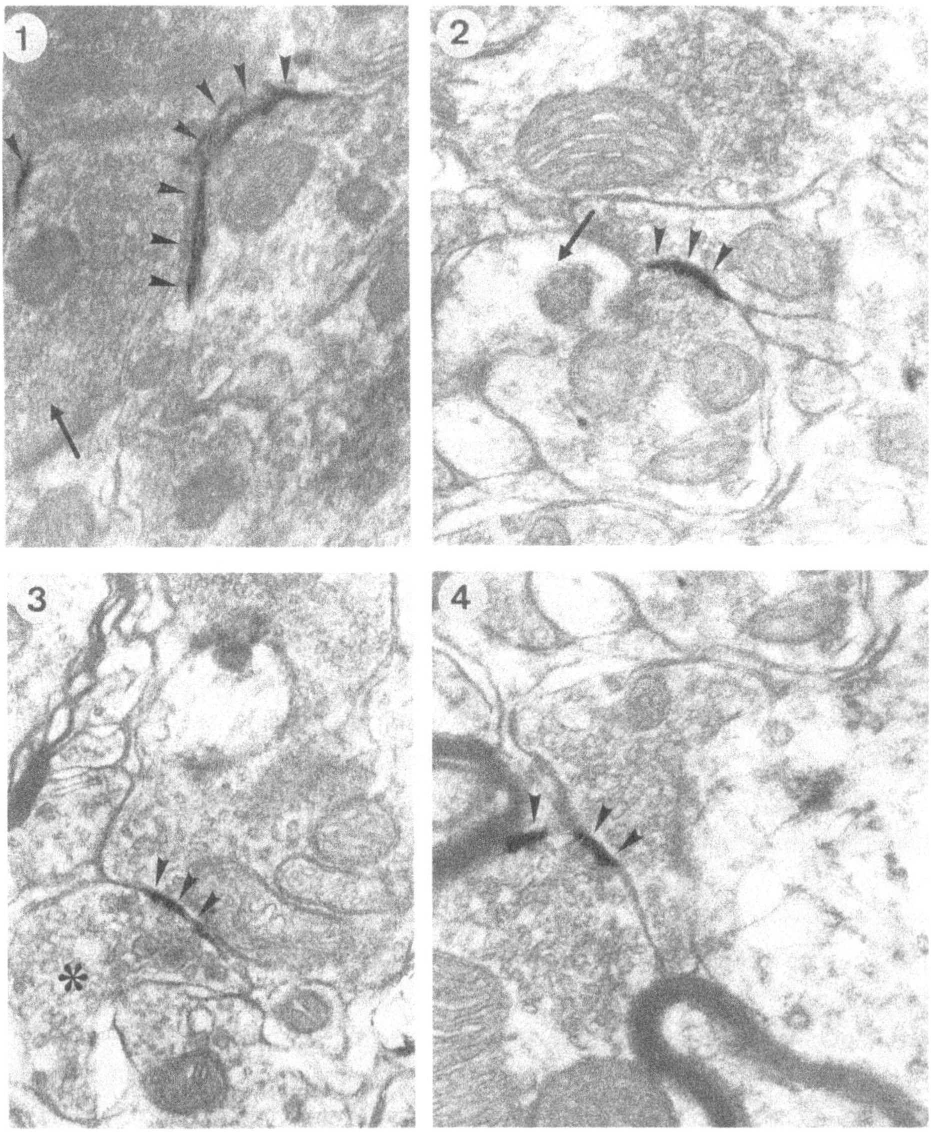


Figures: Immunocytochemical labelling (arrow heads) obtained with two anti-idiotypic antibodies (5: 1 and 2; 51: 3 and 4) in lamina II of the dorsal horn of the rat spinal cord. Some unlabelled synapses are pointed by arrows.
1: the labelling is localized between an axon terminal and a dendrite.
2 to 4: the precipitate takes place at appositions between two axon terminals containing synaptic vesicles.
3: one of the axon terminal (asterix) contains also some several large granular vesicles.

for, first by the different sizes of the molecules in competition, and secondly by the fact that they each may bind to a slightly different region (epitope) of the receptor. By contrast, the preincubation of sections (from fresh tissue) with the anti-idiotypic antibodies (for 2 hours at room temperature) leads to a 20% decrease in the binding of $^{125}$I DADLE. This was measured by means of a quantitative autoradiography.

Under the EM, part of the immunoprecipitate is localized inside the cell bodies or processes. The other part of the reaction takes place on the plasma membranes as well as inside the intermembranous space. Half of the reaction occurs at the site of appositions, either between a varicosity and a dendrite (Fig. 1), or between a dendrite and an unidentified process. Most striking is the observation that about half of the membranous labelling is localized between 2 axon terminals (Figs. 2-4), which are clearly identified by the presence of numerous synaptic vesicles. On the micrograph 3, in addition to small synaptic vesicles, one terminal shows numerous large granular vesicles, known to contain neuropeptides. In this case it may be the presynaptic opiate containing terminal or the postsynaptic afferent terminal; both could indeed contain a variety of peptides. The reaction extends to about 200-250 nm, a length close to that of synapses in the neighbouring. This can be checked by comparing the labelled with the non labelled synaptic differentiations seen in the same micrographs (Figs. 2-4). Because of the high density of the immunoprecipitate, synaptic differentiations, if present, cannot be seen. Preliminary results with 10 nm colloidal gold, have shown that the reaction can be localized on the postsynaptic differentiations. It has to be noticed that in the case where the labelling is localized at the site of axo-dendritic appositions (Fig. 1), the mean length of the reaction is about 1 m, i.e. which is longer than that of a synaptic differentiation.

These results provide the first direct anatomical argument in favor of a presynaptic mechanism for the action of endogenous opiates in the dorsal horn of the rat spinal cord. Moreover, they confirm the postsynaptic action of opiates. The mechanism of release (i.e. synaptic or not) remains however obscure. From our results, synapses seem more conceivable at the site of axo-axonic interactions rather than at the site of axo-dendritic interactions. Whether the presynaptic axons contain ENK will have to be verified. While it could well be the case for the postsynaptic mechanism of action, it is highly unprobable for the presynaptic mechanism, since ENK-varicosities do not contact axon terminals (see references in the Introduction). Two hypotheses can be proposed: (i) ENK could be released at distance from its target cells; (ii) another endogenous opiate may exert this presynaptic inhibition.

**References**

Aronin N, Difiglia M, Liotta AS, Martin JB (1981) Ultrastructural localization and biochemical features of immunoreactive Leu-enkephalin in monkey dorsal horn. J Neurosci 1: 141-145.

Conrath M, Cupo A (1989) Immunocytochemical demonstration of brain delta opiate receptors using anti-idiotypic antibodies. Advances in the Biosciences, vol 75, Pergamon Press, London, 273-275

Conrath M, Couraud JY, Pradelles P (1988) Anti-idiotypic antibodies as a tool for cytochemical identification of substance P receptors in the central nervous system. J Histochem Cytochem 36: 1397-1401

Cupo A, Kaldy P (1986) Monoclonal anti-idiotypic antibodies which recognized the binding site of delta receptor: fine specificity of the anti-idiotypic antibodies. Prog in Opioid Res, Holaday JW, Law PY and Herz A, eds, 25-28

Duggan AW, Hall JG, Headley PM (1977) Enkephalins and dorsal horn neurons of the cat: effects an responses to noxious and inoxious skin stimuli. Br J Pharmacol 61: 399-408

Glazer E, Basbaum AI (1983) Opioid neurons and pain modulation: an ultrastructural analysis of enkephalin in cat superficial dorsal horn. Neurosci 10: 357-376

Hiller JM, Simon EJ, Crain SM, Peterson ER (1978) Opiate receptors in culture of fetal mouse dorsal root ganglia (DRG) and spinal cord: predominance in DRG neurites. Brain Res 145: 396-400

Hökfelt TA, Ljundahl A, Terenius L, Elde R, Nilsson G (1977) Immunocytochemical analysis of peptides possibly related to pain and analgesia: enkephalin and substance P. Proc Natl Acad Sci USA 74: 3081-3085

Ruda MA, Coffield J, Dubner R (1984) Demonstration of postsynaptic opioid modulation of thalamic projection neurons by the combined techniques of retrograde horseradish peroxidase and enkephalin immunocytochemistry. J Neurosci 4: 2117-2132

Zajac JM, Lombard MC, Peschanski M, Besson JM, Roques BP (1989) Autoradiographic study of mu and delta opioid binding sites and neutral endopeptidase-24.11 in rat after dorsal root rhizotomy. Brain Res 477: 400-403

Zieglgansberger W, Tulloch IF (1979) The effects of methionine and leucine enkephalin on spinal neurons of the cat. Brain Res. 167: 53-64

# ANTIBODIES TO COMPLEMENTARY PEPTIDE FOR LABELLING SUBSTANCE P RECEPTORS IN THE DORSAL HORN OF THE RAT SPINAL CORD FOR BOTH LIGHT AND ELECTRON MICROSCOPY

D. Zouaoui[1], S. Maillet[2], J.Y.Couraud[2], M. Conrath[1]
1- Institut des Neurosciences
CNRS UA 1199, Université Paris 6
7 Quai St Bernard, 75005 Paris
France

**Introduction** Several questions remain unanswered concerning the role of substance P (SP) in the central nervous system, particularly in the dorsal horn of the spinal cord, where it was long considered as a transmitter or co-transmitter of pain information. It seems now that SP is not present in all nociceptive-responding cells of the dorsal ganglia (Leah et al., 1985), but is present in numerous other cells including descending tract and local circuit neurons (Barber et al., 1979; Hökfelt et al., 1977). Without an adequate method of tagging SP receptors with labels visible in the electron microscope (EM), their ultrastructural disposition remains unknown, in particular, their pre- or postsynaptic localization. Antibodies to complementary peptides (CP) may be promising labels of SP receptors visible in both light and electron microscopes.

**Methods** A complementary peptide was prepared as described by Couraud et al. (this issue) with a 3' to 5' reading of the RNA complementary to that encoding for SP. Experiments were performed on 4% paraformaldehyde perfused Wistar rats. 50 m vibratome sections of the cervical spinal cord were processed for immunocytochemistry. Sections were preincubated in normal goat serum before incubation in the anti-CP antibodies (1:2000). Anti-rabbit IgG coupled to either horseradish peroxidase (HRP) or 1nm colloidal gold were used as second antibody. The HRP was revealed by the diaminobenzidine method; colloidal gold was intensified by silver (Chan Palay, 1987). Blocks of the dorsal horn were embedded in araldite for examination in the EM.

2- Service de Pharmacologie et d'Immunologie, DSV/DRIPP, CEN·Saclay, 91190, Gif sur Yvette, France

NATO ASI Series, Vol. H 58
Neurocytochemical Methods
Edited by A. Calas and D. Eugène

**Results - Discussion** The general distribution of the immunoreaction in the cervical spinal cord fits well with that observed by conventional autoradiography using $^{125}$I-SP (see Couraud et al., this issue), and resolution is further improved since a labelling could be observed in numerous cell bodies of the dorsal horn (figure 1).

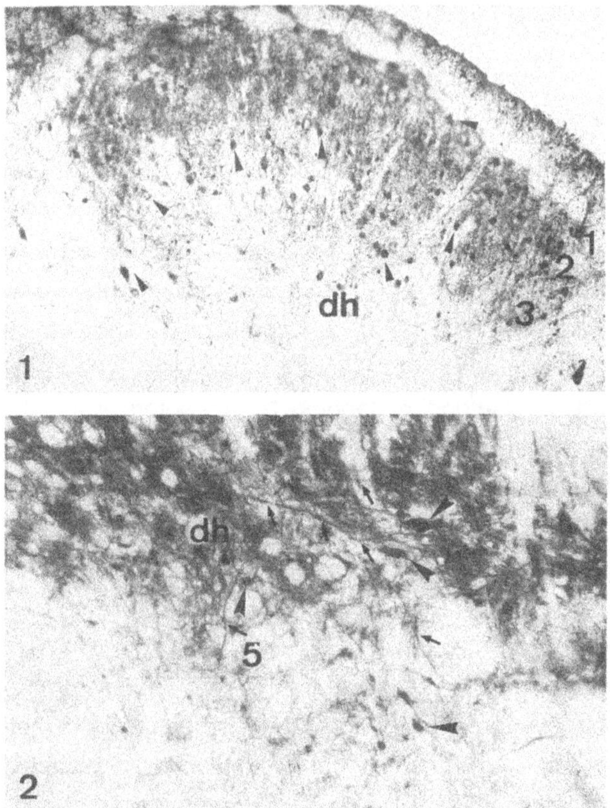


Figure 1: Immunoreactivity in the dorsal horn of the cervical spinal cord obtained with the anti-CP antibodies (1 nm colloidal gold intensified with silver).
1: General view of dorsal horn laminae 1 to 4. Many immunoreactive fibres and cell bodies (arrow heads) are observed in laminae 1 and 2. The density of IR elements decreases in laminae 3 and 4.
2: Immunoreactivity in lamina 5 of the dorsal horn. Numerous immunoreactive cell bodies (arrow heads) are observed. Long immunoreactive processes can be followed (small arrows).

Controls showed that no labeling was obtained with a non-immune serum (not shown). A two-hour preincubation of the sections with $10^{-3}$ M SP (figure 2: 2) or neurokinin (NK) A (in the presence of $10^{-3}$ M peptidase inhibitors: benzamidin, bacitracin, leupeptin, chymostatin) largely prevented the immunostaining Figure 2:1). By contrast, preincubation with NKB under the same experimental conditions did not modify the staining. These results suggested that the anti-CP recognized SP and NKA binding sites.

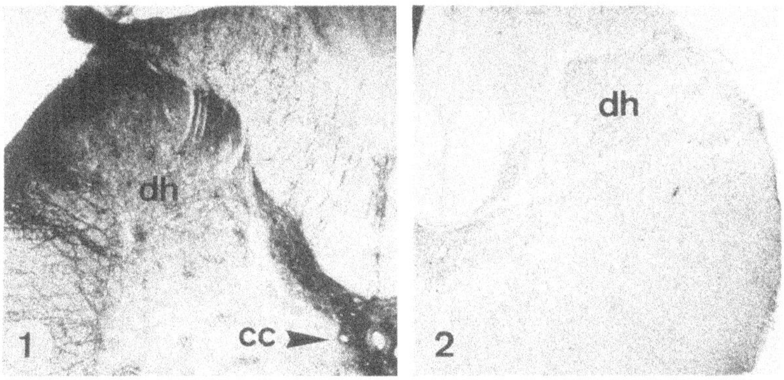


Figure 2: 1: Immunoreactivity with anti-CP in the dorsal horn (dh) and around the central canal (cc).
2: Preincubation of the section in $10^{-3}$ M SP, in the presence of peptidase inhibitors, before addition of the anti-CP, almost totally prevented the immunoreaction.

It is worth further investigating whether these binding sites represent NK1 and NK2 receptors. It has been shown that NK1 and NK2 receptors, which present some similarities (see Quirion and Dam, 1988), are both present in the spinal cord (Ninkovic et al., 1985). In the dorsal horn, many immunoreactive (IR) cell bodies are observed in laminae 1 to 3 (figure 1: 1). The density of IR cell bodies in the neck of the dorsal horn is relatively low. Numerous IR dendrites are observed (figure 1: 2). These IR cell bodies are most likely contacted by SP-containing terminals, since they probably synthesize SP-like receptors (or at least SP binding sites). The wide distribution of immunoreactive cells in the dorsal horn is in argument with data showing that SP is not systematically associated with pain transmission and is probably not related to all modalities of pain stimuli.

In the EM, we have examined laminae 1 and 2. Half of the immunoprecipitate was localized in cell bodies and dendrites (figure 3: 1), on free ribosomes, rough endoplasmic reticulum and neurotubules. The other half of the immunoreaction occurred on plasma membranes, mostly at appositions between an axon terminal

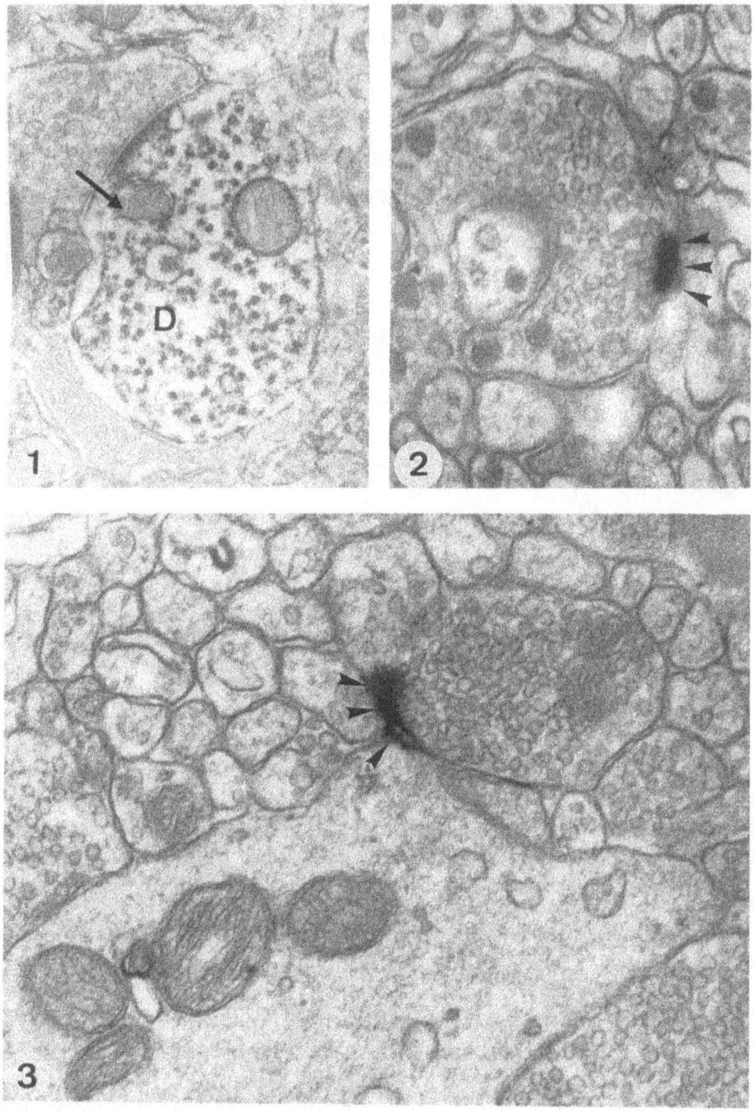


Figure 3: Immunoreactivity obtained with the anti-CP antibodies in the EM in the two superficial layers of the dorsal horn.
1: The reaction (HRP labelling) is localized in a dendrite (D) postsynaptic (arrow indicates the polarity of the synapse) to an unlabelled axon terminal.
2 and 3: Immunoreactivity (silver-intensified colloidal gold) is localized on plasma membranes (arrow heads), at appositions between axon terminals and dendrites (3) or unidentified processes (2). The reactive length (about 300 nm) is congruent with the size of a synapse. Synaptic differentiation cannot however be visualized (see text).

and a dendrite (figure 3: 2 and 3), thus suggesting that SP mainly acts postsynaptically. Synapses might however be hidden by the immunoprecipitate which extends along 300 nm (mean length of a synaptic differentiation), and fills out the intermembranous space.

These preliminary results strongly suggest that anti-CP antibodies could be a useful label of SP receptors, or at least SP binding sites, visible in both light and electron microscopes.

## References

Barber RP, Vaughn JE, Slemmon JR, Salvaterra PM, Roberts E, Leeman SE (1979) The origin, distribution and synaptic relationships of substance P axons in rat spinal cord. J Comp Neurol 184:331-352.

Chan-Palay V (1987) Somatostatin immunoreactive neurons in the human hippocampus and cortex shown by immunogold/silver intensification on vibratome sections: Coexistence with neuropeptide Y neurons, and effects in Alzheimer type dementia. J Comp Neurol 260:201-223.

Hökfelt T, Ljungdahl A, Terenius L, Elde R, Nilsson G (1977) Immunohistochemical analysis of peptide pathways possibly related to pain and analgesia: enkephalin and substance P. Proc Natl Acad Sci USA 74:3081-3085.

Leah JD, Cameron A, Snow PJ (1985) Neuropeptides in physiologically identified mammalian sensory neurons. Neurosciences Lett 56:257-263.

Ninkovic M, Beaujouan JC, Torrens Y, Saffroy M, Hall MD, Glowinski J (1985) Differential localization of tachykinin receptors in rat spinal cord. Eur J Pharmacol 106:463.

Quirion R, Dam TV (1988) Multiple neurokinin receptors: recent developments. Regul Peptides 22:18-25.

# COMPARATIVE ULTRASTRUCTURAL LOCALIZATION OF NEUROPEPTIDE mRNAs USING TRITIATED AND BIOTINYLATED OLIGODEOXYNUCLEOTIDE PROBES

**Alain Trembleau, Marc Landry, Michelle Fèvre-Montange* and André Calas**
Institut des Neurosciences, Département de Cytologie
URA CNRS 1199
Université Pierre et Marie Curie
7 quai Saint-Bernard
75252 Paris Cedex 05

Hybridization histochemical techniques have been widely developed during the last decade, in particular for two main applications : first the localization and the cartography of the cells expressing a given mRNA, and second the quantitative studies of variations of mRNA contents in different physiological or experimental conditions.

Several groups develop now a third application of *in situ* hybridization (ISH) : the comparative subcellular localization of various nucleic acid sequences including ribosomal RNA or corresponding DNA (Binder et al. 1986 ; Binder 1987 ; Escaig-Haye et al. 1989 ; Thiry and Thiry-Blaise 1989), viral nucleic acids (Puvion-Dutilleul and Puvion 1989 ; Wolber et al. 1989), and messenger RNAs (Webster et al. 1987 ; Harris and Croy 1988 ; Brangeon et al. 1989 ; Guitteny and Bloch 1989 ; Morel et al. 1989 ; Tong et al. 1989 ; Trembleau et al. 1990). These challenges require some improvements of classical ISH, more especially the development of both high resolutive markers and their detection at the electron microscopic (EM) level.

For this purpose, we attempted to develop and to compare two hybridocytochemical methods using tritiated and biotinylated oligodeoxynucleotide probes to localize at the EM level oxytocin and vasopressin mRNAs in the rat hypothalamus.

We used for this work normal male Sprague-Dawley rats weighing 200-220 g. The rats were fixed by intra-aortic perfusion of aldehyde fixatives prepared in 0.1M phosphate buffer saline (PBS). The fixative solutions contained either 4% paraformaldehyde (PFA) or 4% PFA and 0.1 to 0.5% glutaraldehyde (GA). The brains were then removed, postfixed in 4% PFA, and 50 μm vibratome sections were cut. These slices were kept in sterile PBS.

Two synthetic oligodeoxynucleotide probes were used (Trembleau et al. 1988 ; Trembleau et al. 1990) : a 25-mer oxytocin (OT) probe and a 41-mer vasopressin (VP)

---

*Laboratoire d'Anatomie Pathologique, Faculté de Médecine Alexis Carrel, rue Guillaume Paradin, 69372 Lyon cedex 8

NATO ASI Series, Vol. H 58
Neurocytochemical Methods
Edited by A. Calas and D. Eugène

probe. These probes were labeled at their 3' extremity with $^{3}$H-dCTP or Biotin-21-dUTP by using terminal transferase. Briefly, 2 pmol of OT probe were labeled with 1.2 μCi of $^{3}$H-dCTP (Amersham, 69 Ci/mmol) and 10 pmol of the VP probe were labeled with 500 pmol of Biotin-21-dUTP. The reaction was stopped after a 1-hour (radioactive labeling) or a 2-hour (non-radioactive labeling) incubation and the labeled oligodeoxynucleotides were purified by ethanol/sodium acetate precipitation (Trembleau et al 1990).

The floating vibratome sections were incubated for 2 hours at 37°C in a sterile prehybridization buffer containing 4X SSC (1X SSC = 0.15M NaCl, 0.015M trisodium citrate), 1X Denhardt solution (0.02% BSA, 0.02% ficoll, 0.02% polyvinylpyrrolidone), and 100 μg/ml salmon sperm DNA. Then the sections were incubated overnight at 37°C in a sterile hybridization buffer (50% deionised formamide, 5X SSC, 1X Denhardt, 125 μg/ml tRNA, 100 μg/ml salmon sperm DNA) containing 0.5nM of the tritiated OT probe or 7-10nM of the biotinylated VP probe. Hybridization step was followed by several washings in SSC buffer (2X SSC : 5x20 min, 1X SSC : 2x1hour, 0.5X SSC : 2x1hour). All incubations were carried out in sterile culture dishes.

Then the radioactive slices were postfixed with osmium tetroxide, embedded in Epon, cut into semithin and ultrathin sections subsequently processed for radioautography (Boyenval and Fischer 1976). Semithin and ultrathin radioautographed sections were developed after 3 to 5 months exposure.

The biotinylated VP probe was revealed by a technique (Trembleau et al. in press) modified from the method of Arai et al (1988). Briefly, the floating sections were incubated first with an avidin-biotinylated-peroxidase complex (ABC, Vector) diluted in PBS containing 0.1% triton (PBS-triton) for 1 hour at room temperature, second with a biotinylated anti-avidin antibody (Vector) (1/200 in PBS-triton) overnight at 4°C, and third with the ABC complex in PBS-triton one hour at room temperature. The slices were rinsed 3x10 min in PBS after each incubation. Peroxidase was revealed by incubating the sections with 0.025% 3,3'-diaminobenzidine (DAB) and 0.006% hydrogen peroxide in 50 mM Tris-HCl buffer (pH 7.6). Finally, slices were postfixed with osmium tetroxide, embedded in Epon, cut in semithin and ultrathin sections. Ultrathin sections were contrasted with uranyl acetate and observed with an electron microscope.

*Results obtained with the tritiated OT probe* : A cellular labeling, clearly distinguishable from the very low background, was observed in some magnocellular neurons of the hypothalamus both on semithin (fig. A) and ultrathin sections (fig. B). Silver grains were mainly distributed over the cytoplasm of the positive neurons. Their

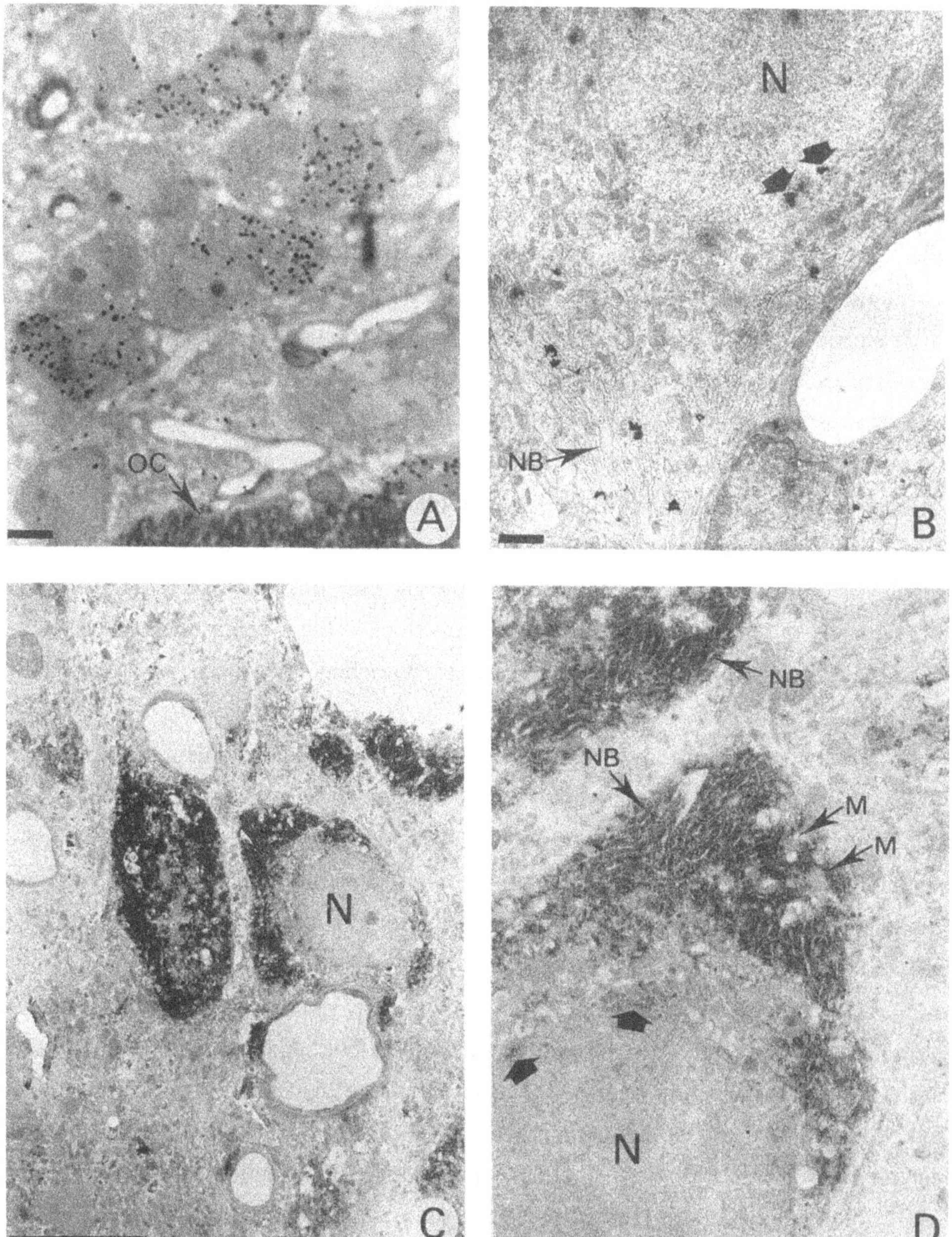


**Figure legends** : **A - B** : Localization of OT mRNAs, detected with a tritiated oligodeoxynucleotide, on semithin (A) and ultrathin (B) sections. **C - D** : Ultrastructural localization of VP mRNAs detected with a biotinylated oligodeoxynucleotide. At a high magnification (D), the detail of the cytoplasmic labeling is visualized. OC : optic chiasma ; N: nucleus ; M : mitochondria ; NB : Nissl Bodies. Large arrows show the nuclear envelop associated labeling. Bars = 10 μm (A,C) and 1 μm (B,D).

nucleus was generally not labeled, but a few silver grains were sometimes localized near the nuclear envelop (fig. B). In the cytoplasm, numerous silver grains were localized close to the endoplasmic reticulum (ER) of the Nissl bodies (fig. B). Finally, perinuclear areas containing a lot of mitochondria and no Nissl bodies seemed to be less labeled (fig. B).

*Results obtained with the biotinylated VP probe* : Before their inclusion in Epon, the labeled vibratome slices displayed positive neurons in the hypothalamus, particularly in supraoptic and paraventricular nuclei (data not shown). Moreover, the high signal to noise ratio allowed to distinguish easily the labeled neurons from the non-labeled ones in these magnocellular areas. At the ultrastructural level, positive neurons displayed an heterogeneous labeling in their cytoplasm (fig. C,D). In this compartment, a strong labeling was restricted to large peripheral areas containing a lot of endoplasmic reticulum. No labeling was observed inside the cisternae of the ER (fig. D). By contrast, entire perinuclear areas (where the ER is less abundant) displayed only a weak and scattered labeling. In addition, this type of labeling was occasionally observed close to the nuclear envelop, in the cytoplasmic compartment. No DAB precipitate was found in the nuclei ; mitochondria were never labeled (fig. C,D).

Our two pre-embedding hybridocytochemical methods, which are consistent with a good mRNA preservation and accessibility, allow the detection of hypothalamic neuropeptide mRNAs at the electron microscopic level with both a good specificity and sensitivity. Moreover, oligodeoxynucleotides which are able to diffuse easily through the whole thickness of the vibratome slices (Trembleau et al 1990), seem to be probes of choice for these approaches.

The cytological preservation is better with our radioautographic method than with the enzymatic one. Two reasons could explain that : first we added 0.1% GA to the fixative solution when we used the tritiated probe (this addition reduced the signal but permitted otherwise to maintain a good level of labeling) ; second we had to use a detergent (triton X100) to facilitate the penetration of detection systems of biotin (ABC, biotinylated-anti-avidin antibody). Fixation with GA would have decreased too much the signal with the biotinylated probe. In any case, even with the enzymatic method, the cytological preservation is sufficient to identify most of the cellular organelles.

The resolution of the tritium is not very high and does not permit to localize precisely the labeled structures. For example, we cannot say if a positive RE cisterna is labeled inside or outside its compartment. Moreover, it is impossible to know if the origin of the silver grains localized close to the nuclear envelop observed in this work and by other authors (Guitteny and Bloch 1989) is intra- or extranuclear. DAB precipitate gives a better resolution and allows a more precise localization of mRNAs : the labeling of the ER given by the biotinylated VP probe is always localized around but never inside these organelles. Mitochondria (sometimes labeled by silver grains when using a tritiated probe) are never labeled by the biotinylated probe. Finally, the labeling associated with the nuclear envelop is clearly localized with this method *outside* the nucleus. Other markers like ferritin or colloidal gold would theoretically give a better resolution. A few works describe the use of these markers by preembedding (Webster et al. 1987 ; Harris and Croy 1988 ; Brangeon et al. 1989) or postembedding (Binder 1987 ; Webster et al. 1987 ; Escaig-Haye et al. 1989 ; Thiry and Thiry-Blaise 1989 ; Puvion-Dutilleul and Puvion 1989 ; Wolber et al. 1989) methods for the detection of abundant nucleic acids (virus genomes, ribosomal RNAs, abundant mRNAs). However, the sensitivity of these techniques seems presently to be not sufficient for the detection of moderate levels of mRNAs.

In conclusion, radioautographic and enzymatic hybridocytochemistry seem to give complementary informations concerning the subcellular localization of neuropeptide mRNAs. Pre-embedding *in situ* hybridization using tritiated oligodeoxynucleotides give the best cytological preservation, but requires long time exposures. Pre-embedding ISH using biotinylated oligodeoxynucleotides permits to obtain rapidly sensitive labelings, but with a less preserved cytology.

With further improvement of their sensitivity, these techniques will probably constitute powerful tools for comparative studies of subcellular localization of exonic or intronic sequences belonging to mRNAs encoding various proteins or peptides in different physiological or experimental conditions.

**Acknowledgments** : We thank Marie-Jeanne Brisorgueil and Danièle Roche for technical assistence, and Paulette Cloup for photographic work. This work was supported by the Fondation pour la Recherche Médicale

**Litterature references**

ARAI H, EMSON PC, AGRAWAL S, CHRISTODOULOU C, GAIT MJ (1988) *In situ* hybridization histochemistry : localisation of vasopressin mRNA in rat brain using a biotinylated oligonucleotide probe. Mol Brain Res 4 : 63-69

BINDER M, TOURMENTE S, ROTH J, RENAUD M, GEHRING WJ (1986) *In situ* hybridization at the electron microscope level : localization of transcripts on ultrathin sections of lowicryl K4M-embedded tissue using biotinylated probes and protein A-gold complexes. J Cell Biol 102 : 1646-1653

BINDER M (1987) *In situ* hybridization at the electron microscope level. Scanning Microscopy 1 : 331-338

BOYENVAL J, FISCHER J (1976) Dipping technique. J Microscop Biol Cell 27 : 115-120

BRANGEON J, NATO A, FORCHIONI A (1989) Ultrastructural detection of ribulose-1,5-bisphosphate carboxylase protein and its subunit mRNAs in wild-type and holoenzyme-deficient Nicotiana using immunogold and *in situ* hybridization techniques. Planta 177 : 151-159

ESCAIG-HAYE F, GRIGORIEV V, FOURNIER JG (1989) Détection ultrastructurale d'ARN ribosomal par hybridation *in situ* à l'aide d'une sonde biotinylée sur coupes ultrafines de cellule animale en culture. C R Acad Sci, Paris 309, III : 429-434

GUITTENY AF, BLOCH B (1989) Ultrastructural detection of vasopressin messenger RNA in normal and Brattleboro rat. Histochemistry 92 :277-281

HARRIS N, CROY RRD (1988) Localization of mRNA for pea legumin : *in situ* hybridization using a biotinylated cDNA probe. Protoplasma 130 : 57-67

MOREL G, DIHL F, GOSSARD F (1989) Ultrastructural distribution of growth hormone (GH) mRNA and GH intron I sequences in rat pituitary gland : effects of GH releasing factor and somatostatin. Mol Cel Endocrinol 65 : 81-90

PUVION-DUTILLEUL F, PUVION E (1989) Ultrastructural localization of viral DNA in thin sections of Herpes simplex virus type 1 infected cells by *in situ* hybridization. Eur J Cell Biol 49 : 99-109

THIRY M, THIRY-BLAISE L (1989) *In situ* hybridization at the electron microscope level : an improved method for precise localization of ribosomal DNA and RNA. Europ J Cell Biol 50 : 235-243

TONG Y, ZHAO HF, SIMARD J, LABRIE F, PELLETIER G (1989) Electron microscopic autoradiographic localization of prolactin mRNA in rat pituitary. J Histochem Cytochem 37 : 567-571

TREMBLEAU A, CALAS A, FEVRE-MONTANGE M (1988) Combination of immunocytochemistry and *in situ* hybridization with synthetic oligonucleotide probes to localize simultaneously vasopressin, oxytocin and their mRNAs in hypothalamic magnocellular neurons. Bull Assoc Anat 72 : 101-106

TREMBLEAU A, FEVRE-MONTANGE M, CALAS A (1990) Ultrastructural localization of oxytocin mRNA in the rat hypothalamus by *in situ* hybridization using a synthetic oligonucleotide. Mol Brain Res 8 : 37-45

TREMBLEAU A, FEVRE-MONTANGE M, LANDRY M., CALAS A Techniques d'hybridation *in situ* ultrastructurale avant inclusion. Les éditions INSERM (G. Morel Ed) Techniques en Microscopie Electronique : cryo-méthodes, Immunocytochimie, Radioautographie, Hybridation *in situ.* (in press)

WEBSTER H de F, LAMPERTH L, FAVILLA JT, LEMKE G, TESSIN D, MANUELIDIS L (1987) Use of a biotinylated probe and *in situ* hybridization for light and electron microscopic localization of $P_0$ mRNA in myelin-forming Schwann cells. Histochemistry 86 : 441-444

WOLBER RA, BEALS TF, MAASSAB HF (1989) Ultrastructural localization of Herpes simplex virus RNA by *in situ* hybridization. J Histochem Cytochem 37 : 97-104

# USE OF *IN SITU* HYBRIDIZATION AND IMMUNOHISTOCHEMISTRY TO STUDY THE EFFECTS OF INTRANIGRAL INJECTIONS OF COLCHICINE ON ENKEPHALIN- AND SUBSTANCE P-CONTAINING NEURONS OF THE LIMBIC FOREBRAIN

Pierre-Alain BOYER, Alain TREMBLEAU, Philippe MAILLY and Michel ARLUISON.
Université P. & M. CURIE, Institut des Neurosciences,
Dept. cytologie, UA 1199 CNRS,
7 quai St. Bernard 75252 PARIS Cedex 05

## INTRODUCTION

The fact that the expression of genes encoding neuropeptides precursors and/or the levels of peptides in neurons may be regulated by nerve afferents has been well demonstrated in the neostriatum. In this material, it was shown that a chronic treatment with a dopaminergic antagonist resulted in a strong increase in the levels of enkephalins (ENK), correlated with an enhancement of preproenkephalin A (PPA) mRNAs (Chou et al. 1984; Tang et al. 1983; Normand et al. 1987; Romano et al. 1987; Le Moine et al. 1990). Likewise, similar changes in the levels of ENK and PPA mRNAs have been observed after unilateral dopaminergic denervation of the striatum by means of 6-hydroxydopamine (Thal et al. 1983; Young et al. 1986). In this last experimental model, by contrast, the levels of substance P (SP) and the mRNAs encoding this peptide were decreased (Voorn et al. 1987; Young et al. 1986). Furthermore, these data are substantiated by the fact that both ENK and SP containing neurons of the striatum receive direct inputs from dopaminergic nerve fibers (Kubota et al. 1986a,b).

We have been interested in the experimental work of Kamata et al. (1987) who have shown that intrastriatal injections of colchicine resulted in an ipsilateral circling behavior analogous to that induced by the degeneration of the nigrostriatal dopaminergic pathway. This similitude has led us to examine further the effects of intranigral injections of colchicine on neostriatal peptidergic neurons by studying both the levels of ENK and SP mRNAs by *in situ* hybridization (ISH), and the levels of these peptides by immunohistochemistry (IHC). In addition, these parameters were also examined in various limbic regions of the forebrain, including the ventral striatum.

NATO ASI Series, Vol. H 58
Neurocytochemical Methods
Edited by A. Calas and D. Eugène

## MATERIAL AND METHODS

**Experimental protocole**: Wistar rats anesthetized with sodium pentobarbital (60 mg/kg) received a double injection of colchicine (0,5 μl; 20 mg/ml) into the lateral and medial regions of the left substantia nigra (A: 3,7; L: 2; H:1,5 and A: 3,7; L:3; H:2 according to Paxinos and Watson, 1986) but the dopaminergic neurons of the ventral tegmental area were also destroyed by the neurotoxic.

**Tissue preparation**: after 48 hours the rats were perfused intracardially with 300 ml of 1% (for HIS) or 4% (for IHC) formaldehyde in 0,1M phosphate buffer. After 1 hour postfixation, the brains were rinsed in phosphate buffer saline (PBS). After overnight immersion in 25% saccharose, the brains were frozen at -40°C and cut at 20 μm with a cryostat. Then, the sections were sticked on gelatin coated slides to be studied by ISH.

***In situ* hybridization**: the expression of SP and PPA was examined by ISH using a 45-mer for SP (Genofit) and a 33-mer for PPA (J.M. Felix, Strasbourg) oligodeoxynucleotide probes. These probes were enzymatically labelled at the 3' end, using $^{35}$S-dATP. After 1h30 of pre-hybridization, the sections were incubated overnight at 37°C with the labelled probe diluted in the hybridization buffer (about 0,5 nM), as described by Trembleau et al. (1990). After several washes at different stringence and temperature conditions, the sections were dried and exposed 5 to 7 days against a X-ray film. The autoradiograms were analysed with a microcomputer/densitometer system converting 64 levels of grey in false colors. The sections were subsequently dipped in K5 emulsion (Ilford) and exposed one month at 4°C. After development in Kodak D19, the sections were stained with cresyl-violet and mounted in Eukitt to be observed in light microscopy.

**Immunohistochemistry**: the fixed brains were cut at 50 μm with a vibratome. Tissue sections were pretreated with 1% hydrogen peroxide for 30 min before being incubated overnight in the cold with the primary antibody to Met-enkephalin (M.Hamon) or to SP (P.Vernier) diluted to 1:1000 or 1:5000 respectively. Subsequently, the tissue sections were incubated for one hour at 20°C with biotinylated secondary antibodies to rabbit immunoglobulins (1:250) and then incubated for one hour in the avidin-biotin pre-formed complex (Vectastain kit, Vector). The peroxidase activity was revealed using diaminobenzidine as chromogen. Finally, the brain sections were mounted in buffered glycerol to be observed in light microscopy.

## RESULTS

The injection of colchicine into the substantia nigra of rats induced an ipsilateral rotation similar to that observed after intrastriatal administration; this effect was still observable 48h after the injection.

When the ISH technique was applied to the neostriatum of these rats, it was observed that the levels of both SP and PPA mRNAs were decreased as compared to the uninjected side, the levels of PPA mRNA being only slightly decreased in the dorsal region of the caudate-putamen (Figs. 1-2).

In addition, other powerful effects of intranigral injections of colchicine were noted in some limbic nuclei of the forebrain. Thus, we have observed that the levels of SP mRNAs were strongly enhanced in the medial amygdaloid nucleus (AM) and the ventromedial nucleus of the hypothalamus (HVM) ipsilateral to the injection (Figs. 3-4). In the latter nucleus, the levels of PPA mRNAs were also found increased on the injected side in several experimental animals. Furthermore, the quantity of PPA mRNA appeared generally enhanced also in the piriform cortex and/or olfactory tubercle ipsilateral to the injection of colchicine (Fig. 2).

The up- or down- regulation of the expression of genes governing the synthesis of ENK or SP reported above may tentatively be explained by the blockade of dopaminergic projections originating in the ventral mesencephalon. This possibility is supported by the fact that 1) intranigral injections of colchicine results in an ipsilateral circling behavior similar to that observed after unilateral destruction of the nigrostriatal pathway by 6-hydroxydopamine; 2) the levels of SP mRNAs in neostriatal neurons are depressed in both types of intranigral injections. However, the lowered expression of the PPA gene which is observed in the ipsilateral neostriatum is not in agreement with the similitude of both experimental models. Furthermore, it is worth noting that, if the primary olfactory areas and the amygdala are known as being richly innervated by dopaminergic nerve fibers, the HVM lacks almost complete this kind of innervation (see Beckstead et al., 1979) and indirect effects must be researched in this case.

On the other hand, intranigral injections of colchicine have remote effects on the perikaryal immunoreactivity for ENK and SP (as well as for other neuropeptides) in various limbic forebrain nuclei which does not always correspond to the changes in the levels of mRNA described above. In fact, we have observed that the ENK-like immunoreactivity was strongly enhanced on the injected side in numerous nerve cell bodies of the central nucleus of the amygdala, HVM and lateral/perifornical areas of the hypothalamus, whereas the SP-like immunoreactivity was also highly increased in numerous perikarya of the AM, HVM and dorsolateral

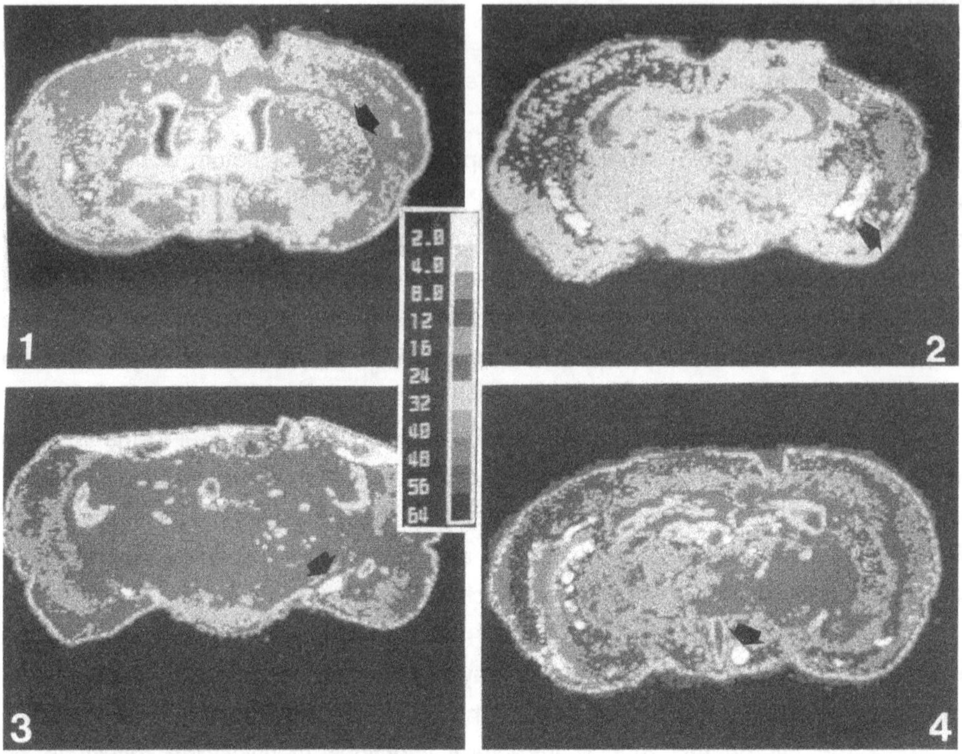


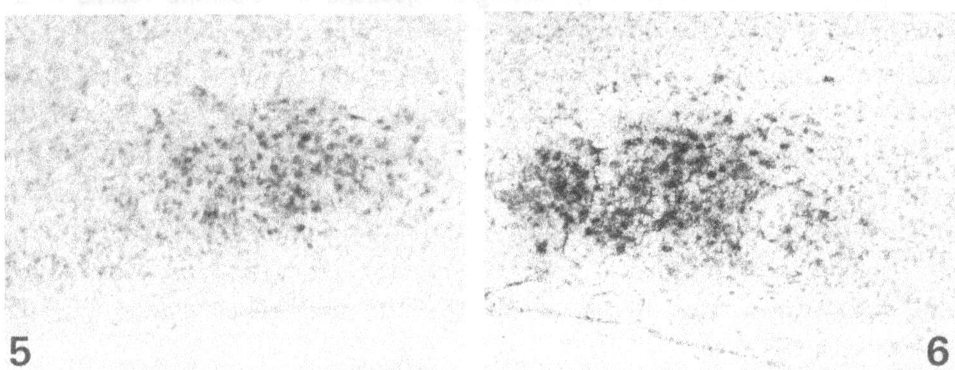


**Figs. 1-4**: Densitometric analysis of the effects of intranigral injections of colchicine (2x O.5 μg) on the expression of two genes encoding the precursors of substance P (1, 3, 4) and enkephalins (2) in the rat forebrain (injected side on the right). The levels of SP mRNAs are clearly decreased in the neostriatum ipsilateral to the injection (Fig. 1, arrow) whereas there is only a weak reduction of the pre-proenkephalin A mRNAs. The latters, however, appear increased in the piriform cortex (Fig. 2, arrow). By contrast, the levels of SP mRNAs are strongly enhanced in the medial amygdaloid nucleus (Fig. 3, arrow) and the ventromedial nucleus of the hypothalamus (Fig. 4, arrow).

**Figs. 5-6**: Histoautoradiographic detection of SP mRNAs in the neurons of the medial nucleus of the amygdala of one experimental rat. The intranigral injection of colchicine results in a strong increase of the levels of SP mRNAs (Fig. 6) as compared to the uninjected side (Fig. 5).

hypothalamus. Thus, a correlation between the enhanced peptide immunoreactivity and the stimulation of the corresponding gene occurs only in the AM and HVM in the case of SP and in the HVM in the case of ENK. In other forebrain nuclei sensitive to colchicine, the increased immunoreactivity for ENK or SP which has been observed, may be due to a blockade of the axonal transport of neuropeptides by colchicine; however, more indirect effects of colchicine are also possible.

## REFERENCES


BECKSTEAD RM, DOMESICK VB, NAUTA WJH (1979) Efferent connection of the substantia nigra and ventral tegmental area in the rat. Brain Res. 175: 191-217

CHOU J, TANG J, YANG HYT, COSTA E (1984) Increase of striatal Met-enkephaline-arg-phe (YGGFMRF) content elicited by long-term treatment with haloperidol. J.Pharmacol. Exp. Ther. 229: 171-174

KAMATA K, OKUYAMA S, HASHIMOTO S, KUMURA MA, AIHARA H, KASUYA Y (1987) Atrophy of the striatum and motor disturbance induced by colchicine. Brain Res. 421: 353-357

KUBOTA Y, INAGAKI S, KITO S (1986a) Innervation of substance P neurons by catecholaminergic terminals in the neostriatum. Brain Res. 375: 163-167

KUBOTA Y, INAGAKI S, KITO S, TAKAGI H, SMITH AD (1986b) Ultrastructural evidence of dopaminergic input to enkephalinergic neurons in rat striatum. Brain Res. 367: 374-378

LE MOINE C, NORMAND E, GUITTENY AF, FOUQUE B, TEOULE R, BLOCH B (1990) Dopamine receptor gene expression by enkephalin neurons in rat forebrain. Proc. Natl. Acad. Sci. USA. 87: 230-234

NORMAND E, POPOVICI T, FELLMANN D, BLOCH B (1987) Anatomical study of enkephalin gene expression in the rat forebrain following haloperidol treatment. Neurosci.Lett. 83: 232-236

PAXINOS G, WILSON C (1986) The rat brain in stereotaxic coordinates. Sydney: Academic Press

ROMANO GJ, SHIVERS BD, HARLAN RE, HOWELLS RD, PFAFF DW (1987) Haloperidol increases proenkephalin mRNA levels in the caudate-putamen of the rat: a quantitative study at the cellular level using in situ hybridization. Mol. Brain Res. 2: 33-41

TANG F, SCHWARTZ JP, COSTA E (1983) Increase of proenkephalin mRNA and enkephalin in content of rat striatum after daily injection of haloperidol for two to three weeks. Proc.Natl.Acad.Sci. 80: 3841-3844

THAL LJ, SHARPLESS NS, HIRSCHHORN ID, HOROWITZ SG, MAKMAN MH (1983) Striatal met-enkephalin concentration increase following nigrostriatal denervation. Biochem.Pharmacol. 32: 3297-3301

TREMBLEAU A., CALAS A., FEVRE-MONTANGE M. (1990) Ultrastructural localization of oxytocin mRNA in the rat hypothalamus by in situ hybridization using a synthetic oligonucleotide. Mol. Brain Res. 8: 37-45.

VOORN P, ROEST G, GROENEWEGEN HJ (1987) Increase of enkephalin and decrease of SP immunoreactivity in the dorsal and ventral striatum of the rat after mid-brain 6-hydroxydopamine lesions. Brain Res. 412: 391-396

YOUNG III WS, BONNER TI, BRANN MR (1986) Mesencephalic dopamine neurons regulate the expression of neuropeptide mRNAs in the rat forebrain. Proc.Natl.Acad.Sci. 83: 9827-9831

# REPORT ON THE POSTER SESSION

J. Taxi
Dépt. de Cytologie, Institut des Neurosciences,
C.N.R.S., Université Pierre et Marie Curie,
7 quai Saint-Bernard, 75252 Paris-Cedex 05

The 18 posters showed at the workshop gave a stimulating illustration of several topics treated in the lectures, and also of several others. Ten out of the 18 posters are presented as short papers in this book: they correspond to underlined authors in the following report.

Two posters concerned the use of intracerebral grafts as a tool for analyzing the phenotypic plasticity of neurons and glial cells. One of them (Alvarado-Mallart and Sotelo) showed the evolution of embryonic cerebellar tissue grafted in the cerebellum of mutant mice with Purkinje cell deficiency; the other (Jousselin-Hossaja) dealt with the fate of chromaffin tissue of young mouse adrenal medulla grafted in the hippocampus of mice of the same strain and weight. In both models the cytology of the grafted cells and of their connections were analyzed by electron microscopy, combined with cytochemistry. The prospection of different sites of graft implantation should provide a great variety of environmental conditions, especially as regards the different possible intercellular connections. Thus the intracerebral grafts appear very suitable for testing the balance between environmental and internal factors in the phenotypic expression of neurons, glia and even other cells, like chromaffin cells.

A second group of four posters illustrated several staining methods, whose the chemical basis is not completely understood. In the first poster of this group, Diaz and Pécot-Dechavassine showed the great interest of combining Koelle's histochemical method for acetylcholinesterase with a silver impregnation to study the changes occuring in the neuromuscular synapses under experimental conditions. In the second poster, presented by P.Derer, a new fluorescent dye was used to overcome certain limitations of the Golgi method. It is noteworthy that this dye works on materials already fixed by aldehydes, and is effective in the superficial zones of the tissues; however, the staining may require several weeks. In the third poster, Picaud et al. irradiated an extracellularly introduced dye to obtain a very fast degeneration of a group of neurons. This original use of dye, which has already been applied to the fly and rat retinas, opens up new possibilities for investigating nervous pathways. The last poster of this set, by Cournil et al., described another hodological method, which is restricted to the GABAergic pathways of invertebrate ganglia; in this method, the retrograde axonal transport of nickel and cobalt ions results in the selective disappearance of the anti-

NATO ASI Series, Vol. H 58
Neurocytochemical Methods
Edited by A. Calas and D. Eugène

GABA immunoreactivity. The origin of GABA-ergic pathways can therefore be determined by comparing normal and experimentally treated ganglia.

Several posters dealt with various uses of immunocytochemistry, including technical improvements. Thus Matthews et al. emphasized the efficiency of amplifying the immunological reaction when several antibodies were successively used after the first specific antibody, which in his work was an anti-choline acetyltransferase applied to sympathetic ganglia. The activity of several horseradish peroxidase (HRP) molecules, instead of just one or two, as in current procedures, increases the amount of final precipitate. This method seems applicable to all weak immunological reactions, provided that a set of bispecific antibodies to both a given animal species and HRP is available. Riche, Menini, Wiklund, Silva-Barrat and Naquet showed an example of the use of an immunohistochemical reaction, in this case for glutamate decarboxylase, to test the variations in the level of a given neurotransmitter or of a related molecule in experimental conditions.

M. Derer, Grynszpan-Winograd, Portier and Bizzini compared the plasticity of chromaffin cells in culture to their *in vivo* development by using both an anti-peripherin, an antibody against a specific molecule of the peripheral neurons, and another neuronal marker, the tetanus toxin.

The possible contribution of the fast-freeze "fixation" followed by the freeze-substitution of a chemical fixative was convincingly illustrated by Nicolas for the immunolocalization of labile tissue components. The method is of general use to retain diffusible components *in situ*, but unfortunately it only works up to a depth of 15 micrometers from the surface of the tissue block.

The simultaneous localization on the same section by immunocytochemistry of two presumed neurotransmitters or neuromodulators is an excellent tool for learning more about the regulatory connections in the nervous system. This is possible using different techniques to visualize the antibodies specific to neurotransmitters, because it enables the end products to be easily differentiated by their ultrastructural appearance, as in the study of the rat striatum presented by Vuillet et al.

Another method of double detection can be used for neurotransmitters which can be specifically taken up in the form of radiolabelled molecules and possibly submitted to the retrograde axonal transport. Radioautography of the labelled molecule (a catecholamine or serotonin) is then carried out, after the tissue sections have also been treated by immunocytochemistry for localizing another neurotransmitter or neuropeptide. By this way, both hodological and neurochemical data can be provided, as illustrated by Araneda and Calas. It is obvious that a large number of double detections will obviously be carried out in the near future for more precise analysis of nervous system connectivity.

Different histochemical reactions may also be used on the same type of cell to distinguish different functional stages, as it was done by Delhaye-Bouchaud et al. with enzyme histochemistry, lectins and immunocytochemistry of specific molecules designed to distinguish activated and resting forms of brain macrophages.

Three posters concerned the localization of receptors. In this field,localization and quantification by radiolabelled ligands are now currently used. Unfortunately, the limitations of this method render it generally unsuitable for electron microscopy. The aim of the work described in the poster presented by El Mestikawy et al. was to obtain a better estimate of the whole population of the 5-HT-1A receptors, of which only the fraction bound to G-protein reacts with the best specific ligand (8-0H-DPAT) .Accordingly, an antibody was raised against a synthetic peptide that reproduced a short sequence of the as yet unpurified receptor; thanks to this antibody, it became possible to determine the entire amount of receptor molecules present in a tissue section. This meant also that electron microscope localizations with immunoperoxidase became feasible.

The second poster concerned anti-idiotypes, selected antibodies to an anti-ligand, whose certain idiotypes have binding properties resembling those of the receptor. Thus the anti-idiotypes are thought to be able to recognize the agonist receptor .To test this possibility, anti-idiotypes to anti-enkephalins were used by Conrath et al. for localizing opiate receptors in the dorsal horn of the rat spinal cord using light and electron microscopy, including postembedding detection with colloidal gold. This is a very promising approach.

The third poster dealing with receptor histochemistry was the most revolutionary. Presented by Zouaoui et al., it describes an application of a still eagerly controversial method: the use of an antibody raised against an anti-sense peptide obtained by translating of the RNA complementary to the mRNA of the receptor (see Couraud's lecture). This poster advances interesting data concerning the value of the method, since the results obtained converge with those obtained by the radioautography of bound ligands, for instance. Moreover, controls were run by exhaustion of the antibody's reactivity with the antigen, a control which is not possible with the anti-idiotype method. The "anti-sense" method therefore seems another promising tool for receptor localization, that would include the use of electron microscopy, provided that different immunological verifications still in progress are conclusive.

Both the last posters were devoted to in situ hybridization, which marks the introduction of molecular biology into histochemistry. One poster was methodological (Trembleau et al.) and the other illustrated an application to the localization in the rat striatum of mRNAs encoding for precursors of neuroactive substances under experimental conditions (Boyer et al.). In their methodological study, Trembleau et al. compared the definition, sensitivity and other characteristics of several procedures for mRNA localization using oligonucleotide probes. The probes were labelled with tritium, digoxigenin or biotin. Although the limitations of definition in radioautography are well known, it remains of interest, because it permits quantitation. As regards digoxigenin and biotin, they were revealed by complexes containing alkaline phosphatase, which give a good definition in light microscopy. For electron microscopy, HRP was used for the final reaction. Prog ress in the definition of localization can be expected from markers of small size, like colloidal gold. Another limitation of the method may be the mobility of the oligonuclear probes. Freeze fixation and substitution (see Nicolas' poster above) might provide valuable controls of localization obtained by current fixations, despite the thinness of the samples fixed.

## AUTHOR INDEX

## SUBJECT INDEX

# NATO ASI Series H

Vol. 1: **Biology and Molecular Biology of Plant-Pathogen Interactions.** Edited by J.A. Bailey. 415 pages. 1986.

Vol. 2: **Glial-Neuronal Communication in Development and Regeneration.** Edited by H.H. Althaus and W. Seifert. 865 pages. 1987.

Vol. 3: **Nicotinic Acetylcholine Receptor: Structure and Function.** Edited by A. Maelicke. 489 pages. 1986.

Vol. 4: **Recognition in Microbe-Plant Symbiotic and Pathogenic Interactions.** Edited by B. Lugtenberg. 449 pages. 1986.

Vol. 5: **Mesenchymal-Epithelial Interactions in Neural Development.** Edited by J.R. Wolff, J. Sievers, and M. Berry. 428 pages. 1987.

Vol. 6: **Molecular Mechanisms of Desensitization to Signal Molecules.** Edited by T.M. Konijn, P.J.M. Van Haastert, H. Van der Starre, H. Van der Wel, and M.D. Houslay. 336 pages. 1987.

Vol. 7: **Gangliosides and Modulation of Neuronal Functions.** Edited by H. Rahmann. 647 pages. 1987.

Vol. 8: **Molecular and Cellular Aspects of Erythropoietin and Erythropoiesis.** Edited by I.N. Rich. 460 pages. 1987.

Vol. 9: **Modification of Cell to Cell Signals During Normal and Pathological Aging.** Edited by S. Govoni and F. Battaini. 297 pages. 1987.

Vol. 10: **Plant Hormone Receptors.** Edited by D. Klämbt. 319 pages. 1987.

Vol. 11: **Host-Parasite Cellular and Molecular Interactions in Protozoal Infections.** Edited by K.-P. Chang and D. Snary. 425 pages. 1987.

Vol. 12: **The Cell Surface in Signal Transduction.** Edited by E. Wagner, H. Greppin, and B. Millet. 243 pages. 1987.

Vol. 13: **Toxicology of Pesticides: Experimental, Clinical and Regulatory Perspectives.** Edited by L.G. Costa, C.L. Galli, and S.D. Murphy. 320 pages. 1987.

Vol. 14: **Genetics of Translation. New Approaches.** Edited by M.F. Tuite, M. Picard, and M. Bolotin-Fukuhara. 524 pages. 1988.

Vol. 15: **Photosensitisation. Molecular, Cellular and Medical Aspects.** Edited by G. Moreno, R.H. Pottier, and T.G. Truscott. 521 pages. 1988.

Vol. 16: **Membrane Biogenesis.** Edited by J.A.F. Op den Kamp. 477 pages. 1988.

Vol. 17: **Cell to Cell Signals in Plant, Animal and Microbial Symbiosis.** Edited by S. Scannerini, D. Smith, P. Bonfante-Fasolo, and V. Gianinazzi-Pearson. 414 pages. 1988.

Vol. 18: **Plant Cell Biotechnology.** Edited by M.S.S. Pais, F. Mavituna, and J.M. Novais. 500 pages. 1988.

Vol. 19: **Modulation of Synaptic Transmission and Plasticity in Nervous Systems.** Edited by G. Hertting and H.-C. Spatz. 457 pages. 1988.

Vol. 20: **Amino Acid Availability and Brain Function in Health and Disease.** Edited by G. Huether. 487 pages. 1988.

# NATO ASI Series H

Vol. 21: **Cellular and Molecular Basis of Synaptic Transmission.**
Edited by H. Zimmermann. 547 pages. 1988.

Vol. 22: **Neural Development and Regeneration. Cellular and Molecular Aspects.**
Edited by A. Gorio, J. R. Perez-Polo, J. de Vellis, and B. Haber. 711 pages. 1988.

Vol. 23: **The Semiotics of Cellular Communication in the Immune System.**
Edited by E. E. Sercarz, F. Celada, N. A. Mitchison, and T. Tada. 326 pages. 1988.

Vol. 24: **Bacteria, Complement and the Phagocytic Cell.**
Edited by F. C. Cabello und C. Pruzzo. 372 pages. 1988.

Vol. 25: **Nicotinic Acetylcholine Receptors in the Nervous System.**
Edited by F. Clementi, C. Gotti, and E. Sher. 424 pages. 1988.

Vol. 26: **Cell to Cell Signals in Mammalian Development.**
Edited by S. W. de Laat, J. G. Bluemink, and C. L. Mummery. 322 pages. 1989.

Vol. 27: **Phytotoxins and Plant Pathogenesis.**
Edited by A. Graniti, R. D. Durbin, and A. Ballio. 508 pages. 1989.

Vol. 28: **Vascular Wilt Diseases of Plants. Basic Studies and Control.**
Edited by E. C. Tjamos and C. H. Beckman. 590 pages. 1989.

Vol. 29: **Receptors, Membrane Transport and Signal Transduction.**
Edited by A. E. Evangelopoulos, J. P. Changeux, L. Packer, T. G. Sotiroudis, and K. W. A. Wirtz. 387 pages. 1989.

Vol. 30: **Effects of Mineral Dusts on Cells.**
Edited by B. T. Mossman and R. O. Bégin. 470 pages. 1989.

Vol. 31: **Neurobiology of the Inner Retina.**
Edited by R. Weiler and N. N. Osborne. 529 pages. 1989.

Vol. 32: **Molecular Biology of Neuroreceptors and Ion Channels.**
Edited by A. Maelicke. 675 pages. 1989.

Vol. 33: **Regulatory Mechanisms of Neuron to Vessel Communication in Brain.**
Edited by F. Battaini, S. Govoni, M. S. Magnoni, and M. Trabucchi. 416 pages. 1989.

Vol. 34: **Vectors as Tools for the Study of Normal and Abnormal Growth and Differentiation.**
Edited by H. Lother, R. Dernick, and W. Ostertag. 477 pages. 1989.

Vol. 35: **Cell Separation in Plants: Physiology, Biochemistry and Molecular Biology.**
Edited by D. J. Osborne and M. B. Jackson. 449 pages. 1989.

Vol. 36: **Signal Molecules in Plants and Plant-Microbe Interactions.**
Edited by B. J. J. Lugtenberg. 425 pages. 1989.

Vol. 37: **Tin-Based Antitumour Drugs.** Edited by M. Gielen. 226 pages. 1990.

Vol. 38: **The Molecular Biology of Autoimmune Disease.**
Edited by A. G. Demaine, J-P. Banga, and A. M. McGregor. 404 pages. 1990.

Vol. 39: **Chemosensory Information Processing.** Edited by D. Schild. 403 pages. 1990.

Vol. 40: **Dynamics and Biogenesis of Membranes.**
Edited by J. A. F. Op den Kamp. 367 pages. 1990.

Vol. 41: **Recognition and Response in Plant-Virus Interactions.**
Edited by R. S. S. Fraser. 467 pages. 1990.

# NATO ASI Series H

Vol. 42: **Biomechanics of Active Movement and Deformation of Cells.**
Edited by N. Akkaş. 524 pages. 1990.

Vol. 43: **Cellular and Molecular Biology of Myelination.**
Edited by G. Jeserich, H. H. Althaus, and T. V. Waehneldt. 565 pages. 1990.

Vol. 44: **Activation and Desensitization of Transducing Pathways.**
Edited by T. M. Konijn, M. D. Houslay, and P. J. M. Van Haastert. 336 pages. 1990.

Vol. 45: **Mechanism of Fertilization: Plants to Humans.**
Edited by B. Dale. 710 pages. 1990.

Vol. 46: **Parallels in Cell to Cell Junctions in Plants and Animals.**
Edited by A. W. Robards, W. J. Lucas, J. D. Pitts, H. J. Jongsma, and D. C. Spray. 296 pages. 1990.

Vol. 47: **Signal Perception and Transduction in Higher Plants.**
Edited by R. Ranjeva and A. M. Boudet. 357 pages. 1990.

Vol. 48: **Calcium Transport and Intracellular Calcium Homeostasis.**
Edited by D. Pansu and F. Bronner. 456 pages. 1990.

Vol. 49: **Post-Transcriptional Control of Gene Expression.**
Edited by J. E. G. McCarthy and M. F. Tuite. 671 pages. 1990.

Vol. 50: **Phytochrome Properties and Biological Action.**
Edited by B. Thomas and C. B. Johnson. 337 pages. 1991.

Vol. 51: **Cell to Cell Signals in Plants and Animals.**
Edited by V. Neuhoff and J. Friend. 404 pages. 1991.

Vol. 52: **Biological Signal Transduction.**
Edited by E. M. Ross and K. W. A. Wirtz. 560 pages. 1991.

Vol. 53: **Fungal Cell Wall and Immune Response.**
Edited by J. P. Latgé and D. Boucias. 472 pages. 1991.

Vol. 54: **The Early Effects of Radiation on DNA.**
Edited by E. M. Fielden and P. O'Neill. 448 pages. 1991.

Vol. 55: **The Translational Apparatus of Photosynthetic Organelles.**
Edited by R. Mache, E. Stutz, and A. R. Subramanian. 260 pages. 1991.

Vol. 56: **Cellular Regulation by Protein Phosphorylation.**
Edited by L. M. G. Heilmeyer, Jr. 520 pages. 1991.

Vol. 57: **Molecular Techniques in Taxonomy.**
Edited by G. M. Hewitt, A. W. B. Johnston, and J. P. W. Young. 420 pages. 1991.

Vol. 58: **Neurocytochemical Methods.**
Edited by A. Calas and D. Eugène. 352 pages. 1991.

www.ingramcontent.com/pod-product-compliance
Lightning Source LLC
Chambersburg PA
CBHW061157240926
51846CB00006B/1040
* 9 7 8 3 6 4 2 8 4 3 0 0 6 *